T0135442

# Set-theoretic Approaches to the Aperiodic Control of Linear Systems

Von der Fakultät Konstruktions-, Produktions- und Fahrzeugtechnik
und dem Stuttgart Research Centre for Simulation Technology
der Universität Stuttgart zur Erlangung der Würde eines
Doktor-Ingenieurs (Dr.-Ing.) genehmigte Abhandlung

Vorgelegt von

## Florian David Brunner

aus Ludwigsburg

Hauptberichter:    Prof. Dr.-Ing. F. Allgöwer
Mitberichter:      prof.dr.ir. W.P.M.H. Heemels
                   Prof. Dr.-Ing. S. Hirche

Tag der mündlichen Prüfung: 15. September 2017

Institut für Systemtheorie und Regelungstechnik
Universität Stuttgart
2017

Bibliografische Information der Deutschen Nationalbibliothek

Die Deutsche Nationalbibliothek verzeichnet diese Publikation in der
Deutschen Nationalbibliografie; detaillierte bibliografische Daten sind
im Internet über http://dnb.d-nb.de abrufbar.


D 93

ISBN 978-3-8325-4622-9


Logos Verlag Berlin GmbH
Comeniushof, Gubener Str. 47,
10243 Berlin
Tel.: +49 (0)30 42 85 10 90
Fax: +49 (0)30 42 85 10 92
INTERNET: http://www.logos-verlag.de

# Acknowledgments

The present thesis is the outcome of my research activities at the Institute for Systems Theory and Automatic Control (IST) of the University of Stuttgart where I was employed from 2012 to 2017. This undertaking would not have been successful without the support from many people to which I hereby extend my deepest gratitude. First and foremost I want to thank Prof. Dr.-Ing Frank Allgöwer for giving me the opportunity to work at his institute, for providing a truly fruitful and unconstrained research environment, and for his continuous encouragement. Second, I want to thank prof.dr.ir. Maurice Heemels, Prof. Dr.-Ing Sandra Hirche, and Prof. Dr.-Ing. Oliver Riedel for their interest in my work and for being part of my examination committee. I am particularly indebted to Maurice for his input and encouragement in our numerous collaborations which shaped my research to a large degree. He was also a great host during my visit to the Control System Technology group of the Eindhoven University of Technology which he made possible. In this respect, many thanks are also in order to Tom Gommans for helping me with the logistics of my stay. I also want to thank Mircea Lazar for his support in the initial phase of my doctoral studies, his collaboration on several projects, and for hosting me at the occasion of a—separate—research visit to Eindhoven. Other collaborators to which I want to express my appreciation are Duarte Antunes and Tom Gommans of the Eindhoven University of Technology and, of the University of Stuttgart, Emre Aydiner, Florian Bayer, Hans-Bernd Dürr, Christian Ebenbauer, Matthias Müller, and Marcus Reble, who, although last in this list, was the first one to collaborate with (actually supervise) me on an academic research project, with the consequence of me pursuing a doctorate at the IST. To all professors, researchers, and staff members of the IST I extend my gratitude for the great working environment. I enjoyed (almost) every single day spent at the institute. I want to thank my office neighbors Daniel Zelazo and Max Montenbruck in particular for bearing with me. Above all Max had to endure years of discussions on technical and, probably to a similar degree, non-technical subject-matters. Other colleagues especially pestered were Emre Aydiner, Florian Bayer, Hans-Bernd Dürr, Christian Feller, Steffen Linsenmayer, Matthias Lorenzen, Matthias Müller, and Shen Zeng, with Wolfgang Halter deserving a special mention (sorry for poisoning your coffee with citric acid!). Special thanks go to Florian Bayer, Steffen Linsenmayer, Matthias Lorenzen, Max Montenbruck, Matthias Müller, and Shen Zeng for nevertheless proof-reading my thesis. Finally, I want to express my eternal gratitude to my parents Ursula and David for their love and support throughout my whole life and my doctoral studies in particular.

Freiberg am Neckar, October 2017
Florian Brunner

# Contents

1. **Introduction**     **1**

2. **Background and Preliminaries**     **7**
   2.1. Dynamical systems and stability . . . . . . . . . . . . . . . . . . . . 7
   2.2. Discrete-time linear systems . . . . . . . . . . . . . . . . . . . . . 9
   2.3. Event-triggered and self-triggered control . . . . . . . . . . . . . . 11
   2.4. Set-valued estimation . . . . . . . . . . . . . . . . . . . . . . . . . 13
       2.4.1. Set-valued estimates from linear estimator dynamics . . . . . . 14
       2.4.2. Set-valued moving-horizon estimation . . . . . . . . . . . . . 15
   2.5. Model predictive control . . . . . . . . . . . . . . . . . . . . . . . 16

3. **Linear Systems Perturbed by Bounded Disturbances**     **23**
   3.1. Preliminaries . . . . . . . . . . . . . . . . . . . . . . . . . . . . 24
   3.2. Lyapunov-based approach . . . . . . . . . . . . . . . . . . . . . . 26
       3.2.1. Theoretical results . . . . . . . . . . . . . . . . . . . . . . 26
       3.2.2. Output feedback . . . . . . . . . . . . . . . . . . . . . . . 28
       3.2.3. Computational aspects . . . . . . . . . . . . . . . . . . . . 33
       3.2.4. Numerical examples . . . . . . . . . . . . . . . . . . . . . 37
   3.3. Set-based approach . . . . . . . . . . . . . . . . . . . . . . . . . . 42
       3.3.1. State feedback . . . . . . . . . . . . . . . . . . . . . . . . 42
       3.3.2. Analysis of given aperiodic schemes . . . . . . . . . . . . . 43
       3.3.3. Output feedback . . . . . . . . . . . . . . . . . . . . . . . 47
       3.3.4. Computational aspects . . . . . . . . . . . . . . . . . . . . 48
       3.3.5. Numerical examples . . . . . . . . . . . . . . . . . . . . . 53
   3.4. Summary . . . . . . . . . . . . . . . . . . . . . . . . . . . . . . . 57

4. **Stochastic Threshold Design in Event-triggered Control**     **59**
   4.1. Threshold design for arbitrarily distributed disturbances . . . . . . . 61
       4.1.1. Probability assigment . . . . . . . . . . . . . . . . . . . . . 63
       4.1.2. Expected value assignment . . . . . . . . . . . . . . . . . . 67
   4.2. Stochastic thresholds for Gaußian noise disturbances . . . . . . . . . 69
       4.2.1. State-Feedback . . . . . . . . . . . . . . . . . . . . . . . . 70
       4.2.2. Output feedback . . . . . . . . . . . . . . . . . . . . . . . 76
   4.3. Summary . . . . . . . . . . . . . . . . . . . . . . . . . . . . . . . 78

**5. Aperiodic Model Predictive Control of Constrained Linear Systems    79**
   5.1. Lyapunov-based approach . . . . . . . . . . . . . . . . . . . . . . 81
      5.1.1. A Lyapunov function for robust MPC . . . . . . . . . . . . 81
      5.1.2. Relaxing the rate of decrease . . . . . . . . . . . . . . . . . 82
      5.1.3. Aperiodic control algorithms . . . . . . . . . . . . . . . . . 84
      5.1.4. Implementation . . . . . . . . . . . . . . . . . . . . . . . . 87
   5.2. Mixed set–Lyapunov approach . . . . . . . . . . . . . . . . . . . 92
      5.2.1. Feasibility by value function decrease, stability by set-membership
            condition . . . . . . . . . . . . . . . . . . . . . . . . . . . . 92
      5.2.2. Aperiodic control algorithms . . . . . . . . . . . . . . . . . 93
      5.2.3. Implementation . . . . . . . . . . . . . . . . . . . . . . . . 97
   5.3. Purely set-based approach . . . . . . . . . . . . . . . . . . . . . . 100
      5.3.1. Feasibility from set-membership conditions . . . . . . . . . . 100
      5.3.2. Aperiodic control algorithms . . . . . . . . . . . . . . . . . 102
      5.3.3. Implementation . . . . . . . . . . . . . . . . . . . . . . . . 105
   5.4. Threshold-based event-triggered MPC: analysis and stochastic design  . . 106
   5.5. Numerical example . . . . . . . . . . . . . . . . . . . . . . . . . . 108
   5.6. Summary . . . . . . . . . . . . . . . . . . . . . . . . . . . . . . . 111

**6. Output-feedback Event-triggered Model Predictive Control    113**
   6.1. Set-valued moving horizon estimation in model predictive control  . . . . 117
      6.1.1. General results . . . . . . . . . . . . . . . . . . . . . . . . 117
      6.1.2. Realization with set-valued moving horizon estimation  . . . . . . 122
      6.1.3. Implementation . . . . . . . . . . . . . . . . . . . . . . . . 126
      6.1.4. Numerical example . . . . . . . . . . . . . . . . . . . . . . . 127
   6.2. Event-triggered output-feedback control . . . . . . . . . . . . . . . . 131
      6.2.1. Closed-loop properties . . . . . . . . . . . . . . . . . . . . . 131
      6.2.2. Implementation . . . . . . . . . . . . . . . . . . . . . . . . 133
      6.2.3. Numerical Examples . . . . . . . . . . . . . . . . . . . . . . 135
      6.2.4. Outlook: extension to self-triggered control . . . . . . . . . . . 137
   6.3. Summary . . . . . . . . . . . . . . . . . . . . . . . . . . . . . . . 138

**7. Conclusions    139**

**A. Auxiliary Results    143**

**B. Proofs of Statements    149**
   B.1. Proof of Theorem 2.1 . . . . . . . . . . . . . . . . . . . . . . . . 149
   B.2. Proof of Theorem 2.3 . . . . . . . . . . . . . . . . . . . . . . . . 149
   B.3. Proof of Proposition 2.1 . . . . . . . . . . . . . . . . . . . . . . . 150
   B.4. Proof of Lemma 3.1 . . . . . . . . . . . . . . . . . . . . . . . . . 151
   B.5. Proof of Lemma 3.2 . . . . . . . . . . . . . . . . . . . . . . . . . 152

B.6. Proof of Theorem 3.5 . . . . . . . . . . . . . . . . . . . . . . . . . . . 153
B.7. Proof of Lemma 4.1 . . . . . . . . . . . . . . . . . . . . . . . . . . . . 154
B.8. Proof of Lemma 4.2 . . . . . . . . . . . . . . . . . . . . . . . . . . . . 154
B.9. Proof of Theorem 4.1 . . . . . . . . . . . . . . . . . . . . . . . . . . . 155
B.10. Proof of Lemma 4.3 . . . . . . . . . . . . . . . . . . . . . . . . . . . . 156
B.11. Proof of Lemma 4.4 . . . . . . . . . . . . . . . . . . . . . . . . . . . . 157
B.12. Proof of Theorem 4.2 . . . . . . . . . . . . . . . . . . . . . . . . . . . 159
B.13. Proof of Corollary 4.1 . . . . . . . . . . . . . . . . . . . . . . . . . . . 162
B.14. Proof of Lemma 5.1 . . . . . . . . . . . . . . . . . . . . . . . . . . . . 163
B.15. Proof of Lemma 5.2 . . . . . . . . . . . . . . . . . . . . . . . . . . . . 164
B.16. Proof of Proposition 5.1 . . . . . . . . . . . . . . . . . . . . . . . . . 165
B.17. Proof of Theorem 5.3 . . . . . . . . . . . . . . . . . . . . . . . . . . . 165
B.18. Proof of Theorem 5.5 . . . . . . . . . . . . . . . . . . . . . . . . . . . 166
B.19. Proof of Lemma 5.3 . . . . . . . . . . . . . . . . . . . . . . . . . . . . 167
B.20. Proof of Lemma 5.4 . . . . . . . . . . . . . . . . . . . . . . . . . . . . 168
B.21. Proof of Theorem 5.6 . . . . . . . . . . . . . . . . . . . . . . . . . . . 169
B.22. Proof of Theorem 5.8 . . . . . . . . . . . . . . . . . . . . . . . . . . . 169
B.23. Proof of Lemma 5.5 . . . . . . . . . . . . . . . . . . . . . . . . . . . . 169
B.24. Proof of Theorem 5.9 . . . . . . . . . . . . . . . . . . . . . . . . . . . 171
B.25. Proof of Lemma 5.6 . . . . . . . . . . . . . . . . . . . . . . . . . . . . 172
B.26. Proof of Lemma 5.7 . . . . . . . . . . . . . . . . . . . . . . . . . . . . 175
B.27. Proof of Theorem 6.1 . . . . . . . . . . . . . . . . . . . . . . . . . . . 175
B.28. Proof of Lemma 6.2 . . . . . . . . . . . . . . . . . . . . . . . . . . . . 176
B.29. Proof of Lemma 6.3 . . . . . . . . . . . . . . . . . . . . . . . . . . . . 177
B.30. Proof of Lemma 6.4 . . . . . . . . . . . . . . . . . . . . . . . . . . . . 179
B.31. Proof of Theorem 6.2 . . . . . . . . . . . . . . . . . . . . . . . . . . . 181

**C. Numerical Data for the Examples in Chapter 5**　　　　　　　　**185**

**Bibliography**　　　　　　　　**201**

# Nomenclature and Conventions

| | |
|---|---|
| $A := B$ | equality of $A$ and $B$ by definition |
| $\mathbb{N}$ | set of natural numbers including zero |
| $\mathbb{Z}$ | set of integers |
| $\mathbb{R}$ | set of real numbers |
| $\mathbb{R}^{(m+n)}$ or $\mathbb{R}^m \times \mathbb{R}^n$ | $(m+n)$-dimensional vector space over the real numbers |
| $(\mathbb{R}^n)^{\mathbb{N}}$ | set of sequences in $\mathbb{R}^n$ |
| $[a,b], (a,b], [a,b), (a,b)$ | closed, left-open and right-closed, left-closed and right-open, and open intervals of the real line |
| $\{m, \ldots\}$ | the set of integers larger than or equal to $m$ |
| $\{m, \ldots, n\}$ | the set of integers larger than or equal to $m$ and less than or equal to $n$ (less than $n$ if $n = \infty$) |
| $0$ | zero, a zero vector, or a zero matrix of appropriate dimensions |
| $I$ | identity matrix of appropriate dimensions |
| $1_n$ | a vector in $\mathbb{R}^n$ where every entry is 1 |
| $\lvert v \rvert$ | absolute value of $v$ if $v \in \mathbb{R}$, maximum norm of $v$ if $v = (v_1, \ldots, v_n) \in \mathbb{R}^n$, that is, $\lvert v \rvert := \max_{i \in \{1, \ldots, n\}} \lvert v_i \rvert$ |
| $\mathscr{B}_n$ | $n$-dimensional unit maximum-norm ball, that is, $\mathscr{B}_n := \{v \in \mathbb{R}^n \mid \lvert v \rvert \leq 1\}$ |
| $A^{\mathsf{T}}$ | transpose of the matrix $A$ |
| $\mathrm{rk}(A)$ | rank of the matrix $A$ |
| $\lvert A \rvert$ | induced matrix norm $\max\{\lvert Ax \rvert \mid x \in \mathbb{R}^m, \lvert x \rvert = 1\}$ of the matrix $A \in \mathbb{R}^{n \times m}$ |
| $A^{\dagger}$ | Moore-Penrose pseudoinverse of the matrix $A \in \mathbb{R}^{n \times m}$ |
| $\det^{\dagger}(A)$ | product of all nonzero eigenvalues of a matrix $A \in \mathbb{R}^{n \times n}$ that is not nilpotent |
| $\lambda_{\max}(P), \lambda_{\min}(P)$ | maximal and minimal eigenvalues of the symmetric matrix $P \in \mathbb{R}^{n \times n}$ |

| | |
|---|---|
| $a \leq b,\ a \geq b$ | element-wise inequalities for the real vectors $a$, $b$ |
| $A \leq B,\ A \geq B$ | element-wise inequalities for the real matrices $A$, $B$ |
| "$A$ is stable" | the eigenvalues of the matrix $A \in \mathbb{R}^{n \times n}$ are contained in the interior of the complex unit disc |
| $A \succ B\ (A \succeq B)$ | for the symmetric matrices $A, B \in \mathbb{R}^{n \times n}$, their difference $A - B$ is positive definite (semi-definite) |
| $A \prec B\ (A \preceq B)$ | for the symmetric matrices $A, B \in \mathbb{R}^{n \times n}$, their difference $B - A$ is positive definite (semi-definite) |
| $\lvert v \rvert_{\mathscr{X}}$ | the distance $\lvert v \rvert_{\mathscr{X}} := \inf_{x \in \mathscr{X}} \lvert v - x \rvert$ of the vector $v \in \mathbb{R}^n$ from the set $\mathscr{X} \subseteq \mathbb{R}^n$ |
| $W^{\perp}$ | orthogonal complement of $W$ for $W$ a subspace of some inner product space |
| $\mathrm{int}(X)$ | interior of the set $\mathscr{X} \subseteq \mathbb{R}^n$ |
| $\mathrm{convh}(\mathscr{X})$ | convex hull of the set $\mathscr{X} \subseteq \mathbb{R}^n$ |
| $\mathrm{vol}(\mathscr{X})$ | Lebesgue measure of the set $\mathscr{X} \subseteq \mathbb{R}^n$ |
| polyhedron | a subset $\mathscr{P}$ of $\mathbb{R}^n$ that can be defined by $\mathscr{P} = \{x \in \mathbb{R}^n \mid Hx \leq h\}$ for a matrix $H \in \mathbb{R}^{m \times n}$ and a vector $h \in \mathbb{R}^m$ |
| polytope | a polyhedron $\mathscr{P}$ that satisfies $\mathscr{P} = \mathrm{convh}\left(\bigcup_{i=1}^{p}\{v_i\}\right)$ for some $p \in \mathbb{N}$ and $v_i \in \mathbb{R}^n$, $i \in \{1, \dots, p\}$ |
| $\mathscr{X} \oplus \mathscr{Y}$ | Minkowski sum $\{x + y \mid x \in \mathscr{X}, y \in \mathscr{Y}\}$ of the sets $\mathscr{X}, \mathscr{Y} \subseteq \mathbb{R}^n$ |
| $\bigoplus_{i=a}^{b} \mathscr{X}_i$ | the set $\mathscr{X}_a \oplus \cdots \oplus \mathscr{X}_b = \left\{\sum_{i=a}^{b} x_i \;\middle|\; x_i \in \mathscr{X}_i, i \in \{a, \dots, b\}\right\}$ for a finite collection $\{\mathscr{X}_i \mid i \in \{a, \dots, b\}\}$ with $\mathscr{X}_i \subseteq \mathbb{R}^n$ and $a \leq b$; equal to $\{0\}$ if $a > b$ |
| $\bigoplus_{i=0}^{\infty} \mathscr{X}_i$ | the set $\{\sum_{i=0}^{\infty} x_i \mid x_i \in \mathscr{X}_i, i \in \mathbb{N}\}$ for a sequence $(\mathscr{X}_i)_{i \in \mathbb{N}}$ with $\mathscr{X}_i \subseteq \mathbb{R}^n$, $i \in \mathbb{N}$, if all sequences $\sum_{i=0}^{\infty} x_i$ with $x_i \in \mathscr{X}_i$, $i \in \mathbb{N}$, converge |
| $\mathscr{X} \ominus \mathscr{Y}$ | Pontryagin difference $\{z \in \mathbb{R}^n \mid \{z\} \oplus \mathscr{Y} \subseteq \mathscr{X}\}$ of the sets $\mathscr{X}, \mathscr{Y} \subseteq \mathbb{R}^n$, compare [Kolmanovsky and Gilbert, 1998] |
| $A\mathscr{X}$ | the set $\{Ax \mid x \in \mathscr{X}\}$ for $\mathscr{X} \subseteq \mathbb{R}^n$ and $A$ either a scalar or a matrix in $\mathbb{R}^{m \times n}$ |
| $A^{-1}\mathscr{X}$ | the set $\{x \in \mathbb{R}^m \mid Ax \in \mathscr{X}\}$ for $\mathscr{X} \subseteq \mathbb{R}^n$ and $A \in \mathbb{R}^{n \times m}$ |

| | |
|---|---|
| class $\mathcal{K}$ | sets of functions $\alpha : [0, \infty) \to [0, \infty)$ that are continuous, strictly increasing and satisfy $\alpha(0) = 0$, compare [Hahn, 1967] (also for the next two items) |
| class $\mathcal{K}_\infty$ | sets of surjective $\mathcal{K}$-functions |
| class $\mathcal{K}\mathcal{L}$ | set of functions $\beta : [0, \infty) \times \mathbb{N} \to [0, \infty)$ that are of class $\mathcal{K}$ in their first argument for any fixed value of their second argument, are decreasing functions in their second argument for any fixed value of their first argument, and satisfy $\beta(r, s) \to 0$ as $s \to \infty$ for any $r \in [0, \infty)$ |
| $\mathbb{P}(e)$ | Probability of the event $e$ |
| $\mathbb{E}[x]$ | Expected value of the real-valued random variable $x$ |
| $\mathcal{N}(\mu, \Sigma)$ | (multivariate) normal distribution with mean $\mu \in \mathbb{R}^n$ and covariance matrix $\Sigma \in \mathbb{R}^{n \times n}$ |

# Abstract

Feedback is the key concept in control theory: based on the measurements of the outputs of a plant, a controller computes appropriate inputs which influence the dynamics of the plant in a way such that the overall system exhibits a desirable behavior. In modern control systems, the information exchange between plant and controller often takes place over a communication network. We consider the case where the usage of this network comes with a certain cost—defined by energy consumption or the need for infrequent transmissions ensuing, for example, from bandwidth limitations or stealth considerations. Consequently, we take into account the amount of communication required by the controller when evaluating its performance. It has been found that aperiodic communication schemes where the transmission of information over the network depends on the current state of the system often outperform schemes with periodic or continuous communication. Two aperiodic scheduling paradigms, known as event-triggered control and self-triggered control, have received particular attention in this respect. The idea in event-triggered control is to send information at a given point in time only when certain conditions—usually certain quantities in the closed-loop system deviating from their setpoints—are met. In self-triggered control, at each transmission instant, the next transmission instant is computed as a function of the current system state, scheduling longer inter-event times if permissible under the current circumstances.

In this thesis, we employ set-theoretic properties of additively disturbed linear discrete-time systems to develop stabilizing event-triggered and self-triggered controllers with *a priori* guarantees on closed-loop characteristics such as stability, asymptotic bound, and average communication rate. Different models for the disturbances are taken into account, namely arbitrary disturbances of which only a bound in the form of a compact set is known and stochastic disturbances with known probability distribution. For setups with hard constraints on the states and inputs, we propose aperiodic schemes based on robust model predictive control methods. Both the full information (state-feedback) case, as well as the limited information (output-feedback) case are investigated. It is demonstrated that the proposed controllers achieve a considerable reduction in the required network usage with only moderate or non-existing deterioration of the closed-loop properties guaranteed by comparable controllers that transmit information at every point in time.

# Deutsche Kurzfassung

Das Hauptziel der Regelungstechnik ist der Entwurf dynamischer Systeme mit gewünschten Eigenschaften. Tatsächlich ist gewöhnlicherweise ein Großteil des Systems in Form einer Regelstrecke mit festgelegten Schnittstellen – Ein- und Ausgängen – gegeben und der vorhandene Freiheitsgrad ist die Wahl eines Reglers. Dieser ist im Allgemeinen selbst ein dynamisches System, dessen Aufgabe darin besteht, die Eingänge der Regelstrecke aufgrund von aus den Ausgängen gewonnenen Informationen zu bestimmen. In Abbildung 1 ist diese Struktur, Rückkopplung genannt, dargestellt. In einem etwas konkreteren Regelkreis werden die Eingänge von realen Vorrichtungen, Aktoren genannt, erzeugt, während die Ausgänge von Sensoren gemessen werden. In dieser Abhandlung liegt das Interesse auf dem Fall, in dem sich die Aktoren und Sensoren an verschiedenen Orten befinden und der Informationsaustausch im Regelkreis über ein physisches Netzwerk stattfindet, wie in Abbildung 2 dargestellt. Bedenken um Energieverbrauch, Bandbreitengrenzen und Geheimhaltung haben zu einem Interesse an Reglerentwürfen geführt, die die Kommunikation im Regelkreis minimieren [Hespanha et al., 2007]. Falls die Regelstrecke instabil oder von beträchtlicher Unsicherheit betroffen ist, steht das Ziel, Übertragung über das Netzwerk zu reduzieren, der Notwendigkeit stabilisierender Rückkopplung entgegen, was unausweichlich zu einer gewissen Abwägung zwischen den Leistungsspezifikationen des geschlossenen Regelkreises und dem Maß an Kommunikation über das Netzwerk führt. Es hat sich ergeben, dass bezüglich solcher Abwägungen Regelkonzepte, bei denen Kommunikation an periodisch festgelegten Zeitpunkten stattfindet, oft von *aperiodischen* Regelkonzepten[1] übertroffen werden, bei welchen die Zeitpunkte, an denen Informationen über das Netzwerk ausgetauscht werden, vom Zustand des Regelkreises abhängen. Quantitativ werden die Vorteile aperiodischer Abtastung zum Beispiel in [Åström and Bernhardsson, 2002, Antunes and Heemels, 2014] gezeigt,[2] (siehe [Brunner et al., 2016b]). In – zweckmäßig entworfenen – aperiodischen Regelsystemen wird die Kommunikation über das Netzwerk in einer Weise geplant, so dass nur zu Zeitpunkten Informationen von der Sensorseite zur Aktorseite übertragen werden, an denen ansonsten der gegenwärtig anliegende Eingangswert zu unbefriedigendem Verhalten führen könnte.

---

[1] Zeitkontinuierliche Kommunikation spielt in modernen – digitalen – Regelsystemen und Kommunikationsnetzwerken keine Rolle.

[2] Dagegen werden in [Blind and Allgöwer, 2011, Blind and Allgöwer, 2013] Anordnungen beschrieben, bei denen periodische Kommunikation, aufgrund der – gewollten – Eigenschaft aperiodischer Regelung die Kommunikationszeitpunkte nicht *a priori* festzulegen, tatsächlich die bessere Wahl darstellen könnte

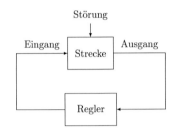

Abbildung 1: Abstrakte Rückkopplung.

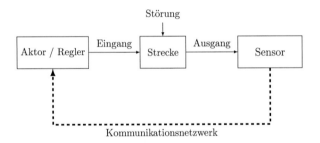

Abbildung 2: Rückkopplung mit Sensor, Aktor und Kommunikationsnetzwerk.

In dieser Abhandlung liegt das Hauptaugenmerk auf den zwei zur Zeit populärsten Ansätzen aperiodisch in einem Regelkreis zu kommunizieren, und zwar auf der ereignisbasierten Regelung[3] und der selbstgetriggerten Regelung[4]. Die Werke [Heemels et al., 2012, Cassandras, 2014, Mahmoud and Memon, 2015, Hetel et al., 2017] bieten einen aktuellen und geschichtlichen Überblick über die zu diesen Regelverfahren im Zusammenhang stehenden Ergebnisse. Die wesentliche Idee bei der ereignisbasierten Regelung ist es, die Regelstrecke auf der Sensorseite kontinuierlich oder periodisch zu überwachen und Information nur dann in Richtung des Aktors zu senden, wenn bestimmte Bedingungen erfüllt sind, etwa dass der Ausgang der Regelstrecke sich über einen gegeben Schwellwert hinaus vom gewünschten Arbeitspunkt entfernt. Bei der selbstgetriggerten Regelung wird zu jedem Kommunikationszeitpunkt der nächste Kommunikationszeitpunkt als eine Funktion des gegenwärtigen Zustands des Regelkreises bestimmt.

Die in dieser Arbeit untersuchten Regelstrecken sind zeitdiskrete lineare Systeme mit Unsicherheiten in Form von additiven Störungen. Um Regler mit garantierten Eigenschaften zu entwerfen, ist es von Nutzen zusätzliches Wissen über diese Störungen

---

[3]Englisch: "event-triggered control" oder "event-based control".
[4]Englisch: "self-triggered control".

miteinzubeziehen. Im größten Teil der Abhandlung ist dieses Wissen auf konkrete kompakte Mengen, zu denen die Störungen gehören, beschränkt; in einem Kapitel wird angenommen, dass die Störung durch zu jedem Zeitpunkt unabhängige Zufallsvariablen mit bekannten Verteilungen bestimmt ist. Das Verhalten linearer Systeme unter beschränkten additiven Störungen wird durch die Dynamik von Mengen beschrieben. Die Regelung solcher Systeme erfordert den Einsatz angepasster mengentheoretischer Methoden, welche bereits in [Schweppe, 1968, Glover and Schweppe, 1971] (mit Verweis auf [Witsenhausen, 1968, Delfour and Mitter, 1969]) untersucht wurden aber auch gegenwärtig Beachtung finden [Blanchini and Miani, 2008]. Die Anwendung dieser Methoden auf die Untersuchung und den Entwurf aperiodischer Regelsysteme wurde bisher jedoch nur spärlich betrieben; erst in jüngster Vergangenheit finden sich vermehrt entsprechende Ansätze, welche ihre Erwähnung an den entsprechenden Stellen der Arbeit finden.

# Beiträge und Gliederung der Arbeit

Der Hauptbeitrag der Arbeit ist der Entwurf von ereignisbasierten und selbstgetriggerten Reglern auf Basis mengentheoretischer Eigenschaften linearer Systeme mit additiven Störungen. Insbesondere werden in der Abhandlung strukturierte Ansätze vorgestellt, basierend auf den von gegebenen stabilisierenden periodischen Reglern hervorgerufenen Mengendynamiken, die es ermöglichen, Bedingungen zu definieren, unter denen zugeordnete aperiodische Regelkreise stabil bleiben. Als bestimmendes Merkmal erscheinen die Gütekriterien der jeweiligen geschlossenen Kreise als explizite Einstellparameter in den Regelkonzepten.

Die genauen Inhalte und Beiträge der Abhandlung setzen sich wie folgt zusammen.

## Kapitel 2: Background and Preliminaries

Dieses Kapitel enthält den zum Verständnis der Arbeit notwendigen theoretischen Rahmen. Insbesondere wird das Stabilitätskonzept, das in der Arbeit verwendet wird, vorgestellt und einfache Werkzeuge präsentiert, um diese Eigenschaft in gegebenen dynamischen Systemen nachzuweisen. Außerdem ist eine kurze Einführung in mehrere relevante Konzepte der Regelungstheorie enthalten, und zwar in die Dynamik linearer Systeme mit additiven beschränkten Störungen, Mengenwertige Zustandsschätzung und modellbasierte prädiktive Regelung.

Die Beiträge in diesem Kapitel, speziell das verwendete Stabilitätskonzept und die zugehörigen Ergebnisse, sind zum Teil in [Brunner and Allgöwer, 2016] und [Brunner et al., 2017b] enthalten.

## Kapitel 3: Linear Systems Perturbed by Bounded Disturbances

Basierend auf dem Stabilitätskonzept und den Methoden für seinen Nachweis, die in Kapitel 2 eingeführt wurden, werden hier stabilisierende ereignisbasierte und selbstgetriggerte Regler für Systeme mit beschränkten Störungen entwickelt. Im ersten Teil des Kapitels sind die Bedingungen für Kommunikation explizit durch eine Lyapunov-Funktion mit vorgegebener Abnahme definiert. Im zweiten Teil wird stattdessen die Zugehörigkeit gewisser Punkte im Zustandsraum zu bestimmten Mengen verwendet, welche viel einfacher überprüft werden kann, wobei eine gewisse Konservativität in Kauf genommen wird. Beide Ansätze werden zuerst als Zustandsrückführung vorgestellt. Die Erweiterung auf Ausgangsrückführung erfolgt durch die Einführung eines stabilen Zustandsschätzers und indem der gemeinsame Zustandsraum von Schätzer und Schätzfehlerdynamik betrachtet wird.

Die Inhalte dieses Kapitels sind zum größten Teil zugleich in [Brunner and Allgöwer, 2016, Brunner et al., 2016c, Brunner et al., 2017b] enthalten.

## Kapitel 4: Stochastic Threshold Design in Event-triggered Control

In diesem Kapitel wird gezeigt, wie zusätzliches Wissen über die Störung, nämlich ihre Wahrscheinlichkeitsverteilung, eingesetzt werden kann, um ereignisbasierte Regler mit gewünschten stochastischen Eigenschaften der Zeiten zwischen Ereignissen zu entwerfen. Für durch beliebige beschränkte Wahrscheinlichkeitsdichten beschriebene Störungen werden numerische Methoden für die Berechnung geeigneter Ereignisschwellwerte vorgestellt. Für den Spezialfall normalverteilter Störungen werden stochastische Ereignisbedingungen eingesetzt, die zuerst in anderem Kontext – ereignisbasierter Zustandsschätzung – vorgeschlagen worden waren, und welche es erlauben, die Ereignisbedingungen als expliziten Ausdruck der gewünschten Verteilung der Zeiten zwischen den Ereignissen zu definieren.

Der Inhalt dieses Kapitels überschneidet sich teilweise mit [Brunner et al., 2015b, Brunner et al., 2017c, Brunner et al., 2017a].

## Kapitel 5: Aperiodic Model Predictive Control of Constrained Linear Systems

Über die Anforderung, Stabilität im geschlossenen Regelkreis zu garantieren, hinaus werden in diesem Kapitel Beschränkungen des Zustandes und der Eingänge des Systems betrachtet. Auf Basis einer erfolgreichen Methode zur Lösung solcher Probleme – robuster modellbasierter prädiktiver Regelung – und unter Einsatz der in Kapitel 2 und Kapitel 3 eingeführten Konzepte, werden mehrere ereignisbasierte und selbstgetriggerte Zustandsrückführungen entwickelt. Einer der vorgestellten ereignisbasierten Regler ist in vollem Umfang mit den Ergebnissen (für beliebige Wahrscheinlichkeitsdichten) aus Kapitel 4 kompatibel, was es erlaubt, zusätzliche stochastische Informationen in den Entwurfsprozess miteinzubeziehen. Die vorgeschlagenen Regler unterscheiden sich in ihrer

Komplexität, Konservativität und Effektivität in der Reduzierung der benötigten Kommunikation, was Möglichkeiten zur Abwägung zwischen den zugehörigen Anforderungen an den geschlossenen Regelkreis bietet.

Dieses Kapitel basiert zum Teil auf den Ergebnissen in [Brunner et al., 2014, Brunner et al., 2015b, Brunner et al., 2016a, Brunner et al., 2016d, Brunner et al., 2017c].

**Kapitel 6: Output-feedback Event-triggered Model Predictive Control**

Der erste Teil dieses Kapitels ist der Entwicklung eines neuartigen modellbasierten prädiktiven Reglers mit Ausgangsrückführung gewidmet, der eine Lücke in der vorhandenen Literatur schließt. Hierbei existieren zum einen nicht konservative aber hochkomplexe bis nicht implementierbare Entwürfe und zum anderen einfache aber konservative Entwürfe. Speziell wird ein Reglerkonzept vorgeschlagen, bei dem, zusätzlich zu dem Ausgang eines linearen Schätzers, eine gewisse Anzahl vergangener Messungen eingesetzt wird, um eine mengenwertige Schätzung des gegenwärtigen Zustands zu berechnen. Diese Anzahl ist ein Einstellparameter für die Abwägung zwischen Komplexität und Konservativität, wobei die Parametrierung mit geringster Komplexität mit einem bestehenden (einfachen aber konservativen) Reglerkonzept aus der Literatur zusammenfällt.

Im zweiten Teil des Kapitels wird das Konzept auf eine aperiodische, genauer, ereignisbasierte Implementierung erweitert, mithilfe der in Kapitel 5 unter Annahme einer Zustandsrückführung erarbeiteten Methoden.

Der erste Teil dieses Kapitels überdeckt sich größtenteils mit [Brunner et al., 2016e] und [Brunner et al., 2017d].

**Kapitel 7: Conclusions**

Dieses Kapitel besteht aus abschließenden Bemerkungen zu den in der Abhandlung vorgestellten Ergebnissen und Anknüpfungspunkten für weitere Untersuchungen.

# 1. Introduction

The focus of control theory is the design of dynamical systems with desired properties. Usually, the largest parts of the dynamical systems to be designed are in fact given in the form of a plant with fixed interfaces—inputs and outputs—and the degree of freedom available is the choice of a controller, in general a dynamical system itself, whose task is to determine the inputs to the plant based on information gained from its outputs. This type of structure, illustrated in Figure 1.1, is known as feedback. In a slightly more concrete feedback loop the inputs to the plant are generated by physical devices, called actuators, while the outputs are measured by sensors. In this thesis, we are interested in the case where the actuators and sensors are not located in the same place and information exchange in the loop is carried out over a physical network, as depicted in Figure 1.2. Concerns for energy consumption, limited network bandwidth, and concealment have led to an interest in controller designs that minimize communication in the feedback loop [Hespanha et al., 2007]. If the plant is unstable or subject to significant uncertainty, the goal of reducing transmissions over the network is opposed to the necessity of stabilizing feedback, inevitably leading to a certain trade-off between the performance specifications of the closed-loop system and the overall amount of communication over the network. Here, it has been found that control schemes where communication takes place at periodically scheduled points in time are often outperformed by *aperiodic* control schemes[1], where the time instants at which information is exchanged over the network depend on the state of the closed-loop system. See, for example, [Åström and Bernhardsson, 2002, Antunes and Heemels, 2014], where the advantages of aperiodic sampling are demonstrated quantitatively[2] (compare [Brunner et al., 2016b]). In—sensibly designed—aperiodic control systems, communication over the network is scheduled in a way such that information from the sensor side of the system is transmitted to the actuator side only at times when the currently applied input to the system could lead to unsatisfactory behavior otherwise.

In the present thesis, the focus is on the two currently most popular approaches to communicating aperiodically in a feedback loop, being *event-triggered control* and *self-triggered control*. Recent overviews related to these approaches, including the historical perspective, are available in [Heemels et al., 2012, Cassandras, 2014, Mahmoud and

---

[1]Continuous communication usually plays no role in modern—digital—control systems and communication networks.

[2]However, in [Blind and Allgöwer, 2011, Blind and Allgöwer, 2013], setups are considered where periodic control in fact may be the better choice due to the—intentional—feature of the communication instants not being known *a priori* in aperiodic control.

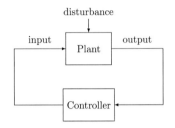

Figure 1.1.: Abstract feedback loop.

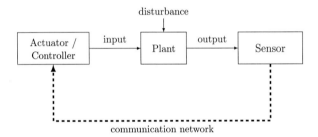

Figure 1.2.: Feedback loop with sensor, actuator, and communication network.

Memon, 2015, Hetel et al., 2017], where the latter of these works offers an overview of aperiodic sampling in general. In event-triggered control, the main idea is to monitor the plant at the sensor side, continuously or periodically, and to transmit information in the direction of the actuator only if certain conditions are met, such as the plant output deviating from a desired setpoint beyond a given threshold. In self-triggered control, at each communication instant, the next one is determined as a function of the current state of the closed-loop system. Due to the structure of the control loop in Figure 1.2, the input applied to the plant between two communication instants in self-triggered control is necessarily also a function of the state of the closed-loop system at the most recent communication instant; the states at times between communication instants have no influence on the current input. From this fact follows the main advantage of event-triggered control over self-triggered control, that is, its ability to react instantaneously to unforeseen occurrences in the closed-loop system, for example due to external disturbances, as the plant state is monitored at every time instant. On the other hand, self-triggered control allows a shutdown of all sensors and the whole communication system in the time span between communications, potentially leading to benefits over event-triggered control if operating the sensors or the communication network (in idle mode) entail comparatively high costs (the actuator side needs to be able to receive transmissions at all times if the next communication instant is not known). See also [Heemels et al., 2012, Gommans, 2016] and [Araújo et al., 2012] with reference to [Anta and Tabuada, 2010] for similar comparisons between the two control paradigms.

A structured approach to designing event-triggered and self-triggered controllers, favored in the present thesis, is to "emulate" periodically or continuously updated controllers in an aperiodic fashion. The idea is to obtain, in a first step, a periodically or continuously updated controller such that the closed-loop system has certain desired properties. In a second step, the rules for input generation and communication scheduling are devised such that these properties are retained while communication is reduced as much as possible, possibly relaxing the properties guaranteed in the first step. See, for example, [Laila et al., 2002, Lemmon, 2010, Donkers et al., 2014, Hetel et al., 2017] for a definition of emulation and [Heemels et al., 2012] for an overview of several emulation-based event-triggered and self-triggered control schemes. In the literature, the term "emulation" is sometimes meant to imply that the input generated by the aperiodic controller is equal to the input supplied by the periodic or continuous controller at the times of communication, but is held constant in the time spans between communication instants [Hetel et al., 2017]. We only consider such sample-and-hold implementations as a special case and adopt the looser definition above, being close to the notion of emulation in [Donkers et al., 2014] in that we first design a controller without considering the communication network and then attempt to mimic the properties of the associated closed-loop system under reduced transmission rates.

The plants under consideration in the thesis are discrete-time linear systems subject to uncertainties in the form of additive disturbances. In order to design controllers with guaranteed characteristics, it is beneficial to take into account additional knowledge about these disturbances acting on the system: in most parts in the thesis, this knowledge will only consist of a compact set to which the disturbances are confined; in one chapter we assume the disturbance at each point in time to be an independently sampled random variable with a known probability distribution. Having only set-valued information available naturally leads to the consideration of set-valued dynamics—and set-theoretic control concepts in general—which have already been studied in the context of linear systems in [Schweppe, 1968, Glover and Schweppe, 1971] with reference to [Witsenhausen, 1968, Delfour and Mitter, 1969], and have been receiving continued interest [Blanchini and Miani, 2008]. The application of these methods to event- and self-triggered control has received relatively little attention until recently; see [Heemels et al., 2008, Iino et al., 2009, Teixeira et al., 2010, Lunze and Lehmann, 2010, Grüne et al., 2010, Lehmann and Lunze, 2011, Nghiem, 2012, Kögel and Findeisen, 2014, Nowzari and Cortés, 2014, Silvestre et al., 2015, Zhe et al., 2015, Kolarijani et al., 2015, Ge et al., 2016, Boisseau et al., 2017, Hashimoto et al., 2017] for examples where such methods are employed. Most works that assume arbitrary disturbances acting on the system use concepts such as input-to-state stability, (ultimate) boundedness, and $\mathscr{L}_p$ stability; see, for example, [Wang and Lemmon, 2009, Mazo et al., 2010, Sijs et al., 2010, Eqtami et al., 2011a, Heemels et al., 2012, Donkers and Heemels, 2012, Di Benedetto et al.,

2013, Stöcker and Lunze, 2013, Lehmann et al., 2013, Aggoune et al., 2014, Mishra et al., 2014, Aggoune et al., 2014, Kiener et al., 2014, Almeida et al., 2015, Wu et al., 2016b, Selivanov and Fridman, 2016, Heemels et al., 2014, Yajie and Wei, 2015, Senejohnny et al., 2016, Kishida et al., 2016, Liu et al., 2017, Poveda and Teel, 2017, Abdelrahim et al., 2017, Dolk et al., 2017]. Approaches that take stochastic properties of the disturbances into account can be found, for example, in [Åström and Bernhardsson, 2002, Imer and Başar, 2006, Cervin and Henningsson, 2008, Henningsson et al., 2008, Cogill, 2009, Rabi and Johansson, 2009, Sijs et al., 2010, Meng and Chen, 2012, Battistelli et al., 2012, Molin and Hirche, 2013, Wu et al., 2013, Lipsa and Martins, 2013, Trimpe and Andrea, 2014, Antunes and Heemels, 2014, Han et al., 2015, Ebner and Trimpe, 2016, Khashooei et al., 2017].

# Contribution and outline of the thesis

The main contribution of this thesis is the design of event-triggered and self-triggered controllers based on set-theoretic properties of additively perturbed linear systems. In particular, we provide frameworks based on the set dynamics induced by given stabilizing and periodically updated controllers, which allow us to define—following the emulation paradigm—conditions under which corresponding aperiodic feedback loops remain stable. As a defining feature, the performance guarantees of the respective closed-loop systems appear as explicit tuning parameters, leading to a structured approach to designing aperiodic control schemes. The specific contents and contributions of the individual chapters in the thesis are as follows.

### Chapter 2: Background and Preliminaries

In this chapter, we provide the theoretic framework necessary for understanding the thesis; in particular, we present the specific notion of stability used throughout the work and basic tools for establishing this property in given dynamical systems. We also offer a brief introduction to several control theoretic concepts, namely to the dynamics of linear systems under bounded disturbances, set-valued estimation, and model predictive control.

The contributions in this chapter, namely the employed stability concept and the related results, are, in parts, contained in [Brunner and Allgöwer, 2016] and [Brunner et al., 2017b].

### Chapter 3: Linear Systems Perturbed by Bounded Disturbances

Based on the stability properties and methods for verifying them introduced in Chapter 2, we develop stabilizing event-triggered and self-triggered controllers for systems subject to bounded disturbances. In the first part of the chapter, the trigger conditions are defined explicitly in terms of a Lyapunov function for which we enforce certain decrease conditions. In the second part, we employ set-membership conditions instead, which can be evaluated more easily, while introducing a certain level of conservatism. Both methods are first shown for the state-feedback case; the extension to output feedback is achieved by introducing a stable estimator and considering the joint state space of the estimator and the estimation error dynamics.

Most of the content of this chapter is likewise contained in [Brunner and Allgöwer, 2016, Brunner et al., 2016c, Brunner et al., 2017b].

## Chapter 4: Stochastic Threshold Design in Event-triggered Control

In this chapter, we show how additional knowledge about the disturbances, namely their probability distributions, can be employed to design event-triggered controllers with desired stochastic properties of the inter-event times. For arbitrary bounded probability density functions governing the disturbances, we suggest numerical methods for the computation of appropriate trigger thresholds. For the special case of normal distributions, we employ stochastic trigger rules that were previously proposed in the context of event-triggered estimation and that allow the trigger conditions to be defined explicitly in terms of the desired distribution of inter-event times.

The contents of this chapter are partly also contained in [Brunner et al., 2015b, Brunner et al., 2017c, Brunner et al., 2017a].

## Chapter 5: Aperiodic Model Predictive Control of Constrained Linear Systems

In addition to guaranteeing stability in the closed-loop system, we consider hard constraints on the state and input of the system in this chapter. Several state-feedback event-triggered and self-triggered control schemes are developed based on a successful method for dealing with this kind of problem, namely robust model predictive control. We follow the concepts introduced in Chapter 2 and Chapter 3 in the construction of the specific control laws. One type of the presented event-triggered controllers is entirely compatible with the results (for arbitrary distributions) in Chapter 4, allowing additional stochastic information to be used in the design process. The proposed controllers differ in complexity, conservatism, and effectiveness in reducing the required communication, providing trade-off options between the associated demands on the closed-loop system.

This chapter is based, in parts, on the results in [Brunner et al., 2014, Brunner et al., 2015b, Brunner et al., 2016a, Brunner et al., 2016d, Brunner et al., 2017c].

## Chapter 6: Output-feedback Event-triggered Model Predictive Control

The first part of this chapter is devoted to the development of a novel output-feedback predictive control scheme, filling a gap in the existing literature between non-conservative but highly complex (or even non-implementable) schemes, and simple but conservative schemes. In particular, we propose a scheme where, in addition to the output of a linear estimator, a certain number of recent measurements is employed to compute a set-valued estimate of the current system state. This number is a tuning parameter for the trade-off between complexity and conservatism, where the parameterization with the lowest complexity coincides with one of the established (simple but conservative) approaches from the literature. In the second part of the chapter, we extend the control scheme to an aperiodic—namely event-triggered—implementation, based on the methods developed in Chapter 5 for the state-feedback case.

The first part of this chapter is mostly also contained in [Brunner et al., 2016e] and [Brunner et al., 2017d].

**Chapter 7: Conclusions**

This chapter contains concluding remarks on the results presented in the thesis and potential starting points for further investigation.

**Appendices**

In Appendix A we summarize several auxiliary results used at various points in the thesis. Proofs of mathematical statements are—if not indicated otherwise—provided in Appendix B.

The numerical experiments were performed in Matlab R2015a, with the help of the Multi-Parametric Toolbox 3.0 ([Herceg et al., 2013]), the IBM ILOG CPLEX Optimization Studio ([IBM, 2014]), and YALMIP ([Löfberg, 2004]).

# 2. Background and Preliminaries

In this chapter, we provide the control theoretic background for the approaches presented in the thesis and provide several preliminary results that are used at various points throughout.

The definition of $\theta$-uniform asymptotic stability and the related sufficient conditions in terms of Lyapunov functions and set-membership constraints were introduced in [Brunner and Allgöwer, 2016], [Brunner et al., 2016c], and [Brunner et al., 2017b].

## 2.1. Dynamical systems and stability

In an abstract sense, a dynamical system is a mapping to a set of sequences. We consider the following general definition in the thesis, compare [Hahn, 1967, Section I].

**Definition 2.1.** *A* dynamical system *is a mapping* $\mathfrak{f} : \mathcal{S} \to (\mathbb{R}^n)^{\mathbb{N}}$ *for some set* $\mathcal{S}$*.*

The set $\mathcal{S}$ may include, for example, initial conditions, controlled input sequences, and external disturbances. Note that we do not assume that the system can necessarily be described by a state-space model of dimension $n$; it may, for example, be described by a state-space model of higher dimension where only a subset of the system states are relevant for the performance evaluation. In the thesis, a dynamical system "generating a sequence $(x_t)_{t \in \mathbb{N}}$ for a (closed-loop) system", refers to any pair $(\mathcal{S}, \mathfrak{f})$ such that $\mathfrak{f}(S)$ is the set of all possible realizations of $(x_t)_{t \in \mathbb{N}}$ consistent with the description of the (closed-loop) system under consideration. In particular, we will investigate closed-loop systems where the internal state of the controller has an influence on the overall system dynamics, but we are only concerned with the state of the system being controlled. Here, we wish to be able to compare, in a meaningful sense, different controllers—with possibly different dimensions of their internal state—for the same system being controlled: we will even consider systems that can be stabilized by static state feedback, but which we interconnect with dynamic controllers nonetheless in order to achieve different objectives. Therefore, we characterize the stability properties of the closed loop solely in terms of the sequences the dynamical system being controlled maps to by introducing the following definition.

**Definition 2.2.** *Let a dynamical system* $\mathfrak{f} : \mathcal{S} \to (\mathbb{R}^n)^{\mathbb{N}}$ *be given. Let further* $\mathcal{X} \subseteq \mathbb{R}^n$ *be compact and let* $\mathcal{X} \subseteq \mathcal{X}_\mathfrak{f}$ *for some set* $\mathcal{X}_\mathfrak{f} \subseteq \mathbb{R}^n$*. Finally, let* $\theta \in \{1, \dots\} \cup \{\infty\}$*. The set* $\mathcal{X}$ *is* $\theta$*-uniformly asymptotically stable for the dynamical system with* $\mathcal{X}_\mathfrak{f}$ *belonging to its* region of attraction *if there exists a $\mathcal{KL}$-function $\beta$ such that for every $(x_t)_{t\in\mathbb{N}} = \mathfrak{f}(s)$ with $s \in \mathcal{S}$ and $x_0 \in \mathcal{X}_f$, any $t_0 \in \mathbb{N}$, and any $t \in \{t_0,\dots\}$ it holds that $|x_t|_{\mathcal{X}} \leq \max_{\tau\in\{0,\dots,\min\{t_0,\theta-1\}\}} \beta(|x_{t_0-\tau}|_{\mathcal{X}}, t - t_0 + \tau)$.*

**Remark 2.1.**

(i) *For $\theta = 1$, one obtains the standard characterization of uniform asymptotic stability. See for example [Hahn, 1967, Chapter V], where the function $\beta$ is chosen as a product of a $\mathcal{K}$-function and a monotonically decreasing function. For $\theta > 1$, the condition is relaxed and potentially allows $|x_{t_0}|_{\mathcal{X}}$ to become very small or even zero for some $t_0 \in \mathbb{N}$ without enforcing the same for subsequent states; in contrast, the case $\theta = 1$ guarantees just this relationship between states at different points in time. In other words, $\theta$-uniform asymptotic stability (only) ensures that if the state of the system is close to the set $\mathcal{X}$ for $\theta$ subsequent time steps, then the state will remain close to the set $\mathcal{X}$ for all further time steps. It seems to be desirable to design controllers which achieve a low $\theta$ in the closed-loop system in order to minimize the influence of past system states on the future evolution. With Definition 2.2, a trade-off between this property and other meaningful criteria, such as the amount of communication in the feedback loop, becomes accessible.*

(ii) *If the dynamical system describes the evolution of the state of an autonomous difference equation, say, $x_{t+1} = f(x_t)$ for $t \in \mathbb{N}$, then the stability concept in Definition 2.2 is equivalent for all $\theta \in \{1,\dots\} \cup \{\infty\}$.*

(iii) *Definition 2.2 resembles stability definitions for time-delay systems, where $\theta$ is the time-delay and the condition for asymptotic stability, in terms of a $\mathcal{KL}$-function, reads $|x_t|_{\mathcal{X}} \leq \beta(\max_{\tau\in\{0,\dots,\theta\}} |x_{t_0-\tau}|_{\mathcal{X}}, t - t_0)$, compare [Liz and Ferreiro, 2002] and also [Hahn, 1967, Section 44] for a continuous-time counterpart. The precise relation between Definition 2.2 and the stability definition for time-delay systems is given in [Brunner et al., 2017b] (where this contribution is largely due to the second author of the work). In the literature, there exists a range of techniques for the analysis of aperiodically sampled systems (of which event-triggered and self-triggered control loops are a special case) based on modeling the closed loop as a time-delay system, see [Hetel et al., 2017, Section 4.1].*

It is possible to use Lyapunov-type functions in order to prove $\theta$-uniform asymptotic stability for a system. For simplicity, and because we only consider the control of linear systems in the thesis, we restrict the discussion to *exponential stability* (see, for example, [Hahn, 1967]).

**Theorem 2.1.** *Let a dynamical system $\mathfrak{f} : \mathcal{S} \to (\mathbb{R}^n)^{\mathbb{N}}$ be given. Let further $\mathcal{X} \subseteq \mathbb{R}^n$ be compact and let $\mathcal{X} \subseteq \mathcal{X}_f$ for a set $\mathcal{X}_f \subseteq \mathbb{R}^n$. Assume that for all $(x_t)_{t\in\mathbb{N}} = \mathfrak{f}(s)$ with $s \in \mathcal{S}$ and $x_0 \in \mathcal{X}_f$, it holds that $x_t \in \mathcal{X}_f$ for all $t \in \mathbb{N}$. Let $V : \mathcal{X}_f \to \mathbb{R}$ and let there exist scalars $c_1, c_2, a \in (0,\infty)$ and $\eta \in (0,1)$ such that*

- $\forall x \in \mathscr{X}_{\mathrm{f}}, \ c_1 |x|_{\mathscr{X}}^a \leq V(x) \leq c_2 |x|_{\mathscr{X}}^a$ and

- $\forall (x_t)_{t \in \mathbb{N}} = \mathfrak{f}(s)$ with $s \in \mathscr{S}$ and $x_0 \in \mathscr{X}_{\mathrm{f}}$,

  $\forall t \in \mathbb{N}, \ V(x_{t+1}) \leq \max_{\tau \in \{1,\ldots,\min\{t+1,\theta\}\}} \eta^\tau V(x_{t+1-\tau})$.

*Then, the set $\mathscr{X}$ is $\theta$-uniformly asymptotically stable for the dynamical system with $\mathscr{X}_{\mathrm{f}}$ belonging to its region of attraction. Further, one may choose $\beta \colon (r,t) \mapsto (\eta^{1/a})^t \, (c_2/c_1)^{1/a} \, r$.*

**Remark 2.2.** *The function $V$ employed in Theorem 2.1 is defined similarly to the finite-step Lyapunov functions studied in [Geiselhart et al., 2014]. However, we do not assume $(x_t)_{t \in \mathbb{N}}$ to be generated by a state-space model $x_{t+1} = f(x_t)$, and, in particular, cannot guarantee a bound of the form $|x_{t+1}|_{\mathscr{X}} \leq \alpha(|x_t|_{\mathscr{X}})$, $t \in \mathbb{N}$, for a $\mathscr{K}$-function $\alpha$, as was assumed in [Geiselhart et al., 2014]. If one could establish such a bound, say $|x_{t+1}|_{\mathscr{X}} \leq c_3 |x_t|_{\mathscr{X}}$ for some $c_3 \in \mathbb{R}$ and all $t \in \mathbb{N}$, then for all finite $\theta$, the prerequisites of Theorem 2.1 would imply 1-uniform asymptotic stability of $\mathscr{X}$, which follows along the same lines as in that work.*

*Further, Theorem 2.1 is similar to the concept of Razumikhin functions used for the stability analysis of time-delay systems, see, for example, [Liu and Marquez, 2007, Gielen, 2013].*

In the next section, we discuss disturbed linear systems and a sufficient condition for $\theta$-uniform asymptotic stability based on set membership conditions.

## 2.2. Discrete-time linear systems

In the thesis, we consider dynamical systems described by difference equations of the form

$$x_{t+1} = Ax_t + Bu_t + w_t \tag{2.1a}$$
$$y_t = Cx_t + v_t. \tag{2.1b}$$

Here, $x_t \in \mathbb{R}^{n_{\mathrm{x}}}$ is the state of the system, $u_t \in \mathbb{R}^{n_{\mathrm{u}}}$ is the controlled input of the system, $y_t \in \mathbb{R}^{n_{\mathrm{y}}}$ is the output of the system, $w_t \in \mathbb{R}^{n_{\mathrm{x}}}$ is a disturbance acting on the state, and $v_t \in \mathbb{R}^{n_{\mathrm{y}}}$ is noise acting on the output, each at time $t \in \mathbb{N}$. The dimensions of the system are given by the integers $n_{\mathrm{x}}$, $n_{\mathrm{u}}$, and $n_{\mathrm{y}}$. The system matrices are $A \in \mathbb{R}^{n_{\mathrm{x}} \times n_{\mathrm{x}}}$, $B \in \mathbb{R}^{n_{\mathrm{x}} \times n_{\mathrm{u}}}$ and $C \in \mathbb{R}^{n_{\mathrm{y}} \times n_{\mathrm{x}}}$. In the framework of Definition 2.1, equation (2.1a) determines a mapping $(x_0, (w_t)_{t \in \mathbb{N}}, (u_t)_{t \in \mathbb{N}}) \mapsto (x_t)_{t \in \mathbb{N}}$ describing the evolution of $x_t$. Equation (2.1b) defines the information available about $(x_t)_{t \in \mathbb{N}}$ at a given time $t$.

We derive most results in the thesis first under the simplifying assumption that the state $x_t$ is available as a measurement at any time point $t \in \mathbb{N}$, which we refer to as the *state-feedback* setup. In the more general *output-feedback* setup, only the output $y_t$ is available as a measurement at time point $t$.

**Bounded disturbances**

We summarize some results for linear systems subject to bounded disturbances in the following theorem. Relevant properties of the Minkowski sum, denoted by $\oplus$, are given in Lemma A.1 in the appendix.

**Theorem 2.2.** *Consider system (2.1) in the state-feedback case with $u_t = Kx_t$, $t \in \mathbb{N}$, for a matrix $K \in \mathbb{R}^{n_u \times n_x}$, that is, a closed-loop system of the form*

$$x_{t+1} = (A + BK)x_t + w_t. \tag{2.2}$$

*Assume that $(A + BK)$ is stable. Further, assume that $w_t \in \mathscr{W}$, $t \in \mathbb{N}$, for a compact set $\mathscr{W} \subseteq \mathbb{R}^{n_x}$. Then it holds that*

$$x_t \in \{(A + BK)^{t-t_0} x_{t_0}\} \oplus \mathscr{F}_{t-t^0} \tag{2.3}$$

*for all $t_0 \in \mathbb{N}$ and $t \in \{t_0, \dots\}$, where $\mathscr{F}_i := \bigoplus_{j=0}^{i-1}(A + BK)^j \mathscr{W}$ for $i \in \mathbb{N} \cup \{\infty\}$. It holds that*

$$(A + BK)^i \mathscr{F}_j \oplus \mathscr{F}_i = \mathscr{F}_{i+j} \tag{2.4}$$

*for all $i \in \mathbb{N}$ and $j \in \mathbb{N} \cup \{\infty\}$ with the convention $i + \infty = \infty$, $i \in \mathbb{N}$. Furthermore, the set $\mathscr{F}_\infty$, which is compact, is 1-uniformly asymptotically stable for any dynamical system generating $(x_t)_{t \in \mathbb{N}}$ in (2.2), with region of attraction $\mathbb{R}^{n_x}$.*

These statements can be found in the literature, or follow from straight-forward manipulations, see, for example, [Glover and Schweppe, 1971, Kuntsevich and Pshenichnyi, 1996, Kolmanovsky and Gilbert, 1998, Blanchini and Miani, 2008, Raković and Mayne, 2005, Raković and Kouramas, 2006a].

**Remark 2.3.**

(i) *A dynamical system generating $(x_t)_{t \in \mathbb{N}}$ in (2.2) is, for example, given by $\mathfrak{f} : (\mathbb{R}^{n_x} \times (\mathbb{R}^{n_x})^{\mathbb{N}}) \to (\mathbb{R}^{n_x})^{\mathbb{N}}$, $(x_0, (w_t)_{t \in \mathbb{N}}) \mapsto ((A + BK)^t x_0 + \sum_{i=0}^{t-1}(A + BK)^{t-1-i} w_i)_{t \in \mathbb{N}}$.*

(ii) *It holds that the set $\mathscr{F}_\infty$ is included in every compact set $\Omega \subseteq \mathbb{R}^{n_x}$ that satisfies $(A + BK)\Omega \oplus \mathscr{W} \subseteq \Omega$ and is hence also called the minimal robustly positive invariant set for the dynamics $x_{t+1} = (A + BK)x_t + w_t$, where $w_t \in \mathscr{W}$, $t \in \mathbb{N}$, compare [Kolmanovsky and Gilbert, 1998].*

In addition to decrease conditions on Lyapunov-like functions in Theorem 2.1, we employ set-membership conditions to guarantee asymptotic stability in this thesis, using the following result.

**Theorem 2.3.** *Let a dynamical system $\mathfrak{f} : \mathcal{S} \to (\mathbb{R}^{n_x})^{\mathbb{N}}$ be given. Let further $\gamma \in [1, \infty)$ and let $\gamma \mathscr{F}_\infty \subseteq \mathcal{X}_{\mathfrak{f}}$ for some set $\mathcal{X}_{\mathfrak{f}} \subseteq \mathbb{R}^{n_x}$. Assume that for all $(x_t)_{t \in \mathbb{N}} = \mathfrak{f}(s)$ with $s \in \mathcal{S}$ and $x_0 \in \mathcal{X}_{\mathfrak{f}}$, it holds that $x_t \in \mathcal{X}_{\mathfrak{f}}$ for $t \in \mathbb{N}$. Finally, assume that for all $t \in \mathbb{N}$ there exists a $\tau \in \{1, \dots, \min\{t+1, \theta\}\}$ with $x_{t+1} \in \{(A + BK)^\tau x_{t+1-\tau}\} \oplus \gamma \mathscr{F}_\tau$. Then, the set $\gamma \mathscr{F}_\infty$ is $\theta$-uniformly asymptotically stable for the dynamical system.*

**Remark 2.4.**

*(i) Theorem 2.3 holds for general dynamical systems, not only for those generating sequences for (2.2). In fact, we will design nonlinear dynamic controllers for system (2.1) in the subsequent chapters, where we will establish stability using this result.*

*(ii) We prove Theorem 2.3 by establishing that the prerequisites of Theorem 2.1 hold. Hence, employing the set-membership conditions in Theorem 2.3 is* more conservative *than employing the conditions in Theorem 2.1 directly.*

## 2.3. Event-triggered and self-triggered control

In the thesis, we are interested in designing controllers for systems as in (2.1) which do not require communication from the sensor to the actuator at every point in time $t \in \mathbb{N}$, compare Figure 1.2 of the introduction. In the following overview, we focus on the state-feedback setup.

Let $t_i \in \mathbb{N}$, $i \in \{0, \ldots, i_{\max}\}$ with $i_{\max} \in \mathbb{N} \cup \{\infty\}$ be the times at which information is transmitted from the sensor to the actuator, referred to as *transmission instants* henceforth. Here, $t_0 = 0$ and $t_{i+1} > t_i$ for all $i \in \{0, \ldots, i_{\max} - 1\}$ if there exists a time $t_{i_{\max}}$ with $i_{\max} \in \mathbb{N}$ after which no further transmissions occur (otherwise, it holds that $i \in \mathbb{N}$). For convenience, if $i = i_{\max}$, we associate expressions such as $\{t_i, \ldots, t_{i+1} - 1\}$ with related infinite sets, ($\{t_i, \ldots\}$ in this case). At any transmission time $t_i$, the information possibly available at the sensor consist of the system state at all time points up until $t_i$, that is $(x_0, x_1, \ldots, x_{t_i-1}, x_{t_i})$. Assuming this information is transmitted at time $t_i$ to the actuator, the information available there for all time points $t \in \{t_i, \ldots, t_{i+1} - 1\}$ is also $(x_0, x_1, x_2, \ldots, x_{t_i})$, as no new information about the system state arrives until the time $t_{i+1}$. Hence, disregarding for the moment which information actually is sent over the network at the transmission times $t_i$, we restrict the inputs generated by the actuator in the time span $t \in \{t_i, \ldots, t_{i+1} - 1\}$ to be of the form

$$u_t = \kappa_{t_i}(x_0, x_1, \ldots, x_{t_i-1}, x_{t_i}, t), \qquad (2.5)$$

for some function $\kappa_{t_i} : \mathbb{R}^{(t_i+1)n_x+1} \to \mathbb{R}^{n_u}$. Control laws with periodically scheduled updates are a special case of this framework, where $t_{i+1} = t_i + M$, $i \in \mathbb{N}$, for a fixed $M \in \{1, \ldots\}$. Here, we are interested in *aperiodic* control laws, particularly in control laws where the time between transmission instants depends on the evolution of the system state, and, hence, on the realization of the disturbances $w_t$. The structure of the closed-loop system under such aperiodic control laws is depicted in Figure 2.1. In the remainder of the thesis, we refer to control laws that are updated at every point in time ($M = 1$), such as the state feedback $u_t = Kx_t$, $t \in \mathbb{N}$ as *periodic*.

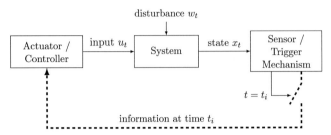

Figure 2.1.: Aperiodic control structure for the state-feedback case. Both the "Sensor /
Trigger Mechanism" -block and the "Actuator / Controller"-block may be
dynamical systems. Communication between these blocks is allowed only at
the transmission instants $t_i$, which are decided at the sensor side.

Two methods for defining aperiodic control laws are event-triggered and self-triggered
control. In event-triggered control, at every time point $t \in \{1, \ldots\}$ a decision is made
at the sensor, based on the currently available information, whether a transmission is
scheduled or not, leading to a recursion of the form

$$t_{i+1} = \inf\{t \in \{t_i + 1, \ldots\} \mid \delta_t(x_0, x_1, \ldots, x_{t_i-1}, x_t, t_i) = 1\}, \qquad (2.6)$$

where $\delta_t : \mathbb{R}^{(t+1)n_x} \times \mathbb{N} \to \{0, 1\}$, $t \in \{1, \ldots\}$, are the *event-generating functions*, compare
[Molin and Hirche, 2009]. As $x_t$ is an argument of the function, event-triggered control
generally requires measurement of the state at every point in time. In self-triggered control, at every transmission instant, the next transmission instant is computed explicitly
as a function of available information at that time, leading to a recursion of the form
(compare [Heemels et al., 2012])

$$t_{i+1} = \mu_i(x_{t_0}, x_{t_1}, \ldots, x_{t_{i-1}}, x_{t_i}, t_0, t_1, \ldots, t_{i-1}, t_i) \qquad (2.7)$$

where $\mu_i : \mathbb{R}^{(i+1)n_x} \times \mathbb{N}^{i+1} \to \mathbb{N}$, $i \in \mathbb{N}$, are the *scheduling functions*. Furthermore, we
restrict the control laws in the self-triggered case to be of the form

$$u_t = \kappa_{t_i}(x_{t_0}, x_{t_1}, \ldots, x_{t_{i-1}}, x_{t_i}, t), \qquad t \in \{t_i, \ldots, t_{i+1} - 1\}, \ i \in \{0, \ldots, i_{\max}\}, \qquad (2.8)$$

that is, to be only explicitly dependent on the state at transmission times. This implies
that the system state $x_t$ for $t \in \{t_j + 1, \ldots, t_{j+1} - 1\}$, $j \in \{0, \ldots, i-1\}$ neither has an
influence on the value of $t_{i+1}$, nor on $u_t$, and, hence, no measurements are required at
these points in time.

**Remark 2.5.**

*(i) In this general form, both the event-triggered and the self-triggered setup involve functions with arguments whose dimension grows unbounded with time. It is generally impossible to implement such functions in practice and in the respective sections in the thesis we will provide particular event-generating functions and scheduling functions where the dimension of the arguments remain bounded.*

*(ii) It is possible to define self-triggered control in a way that information about the states at every point in time is available. Here, we stick to the more restrictive (and more challenging) setting presented above in order to emphasize the advantage of self-triggered control over its event-triggered counterpart, that is, that no measurements need to be taken between transmissions instants.*

The aperiodic control schemes developed in the thesis generally rely on the assumption that information about the input $u_t$ applied to the system is available to the trigger mechanism at each time point $t$. Especially for the control laws in Chapter 5 and Chapter 6, computing $u_t$ from $x_t$ or $x_{t_i}$ may be a nontrivial task involving the solution of optimal control problems. Hence, in order to save hardware or energy, one might consider computing a sequence $(u_{t_i}[0], u_{t_i}[1], \ldots, u_{t_i}[M])$ at time $t_i$ at the sensor side and transmitting it as a whole to the actuator side at time $t_i$. The input applied to the plant would then be $u_t = u_{t_i}[t - t_i]$ for $t \in \{t_i, \ldots, t_{i+1} - 1\}$. In certain communication networks, information is sent in *packets*, with a relatively high overhead in terms of energy consumption; hence, the same overall amount of information may be transmitted with a reduced energy cost if packets are only sent infrequently (but containing more data each), justifying a control system with this structure, compare [Feeney and Nilsson, 2001, Georgiev and Tilbury, 2004, Bernardini and Bemporad, 2012]. Alternatively, one might compute the sequence of inputs at the actuator side and transmit it *back* to the trigger mechanism (together with possible additional information, as required for the control laws in Chapter 5 and Chapter 6); a communication structure along these lines is considered in [Bernardini and Bemporad, 2012]. In the thesis, we stick to the structure depicted in Figure 2.1.

## 2.4. Set-valued estimation

Consider system (2.1) where only $y_t$, but not $x_t$, is available as a measurement at time $t$ and $w_t$ and $v_t$ are known to be contained in compact sets $\mathcal{W} \subseteq \mathbb{R}^{n_x}$ and $\mathcal{V} \subseteq \mathbb{R}^{n_y}$, respectively. In this setting, if no additional characteristics of the sequences $(w_t)_{t \in \mathbb{N}}$ and $(v_t)_{t \in \mathbb{N}}$ are given, the best estimate of the state $x_t$ one can obtain from the measurements $(y_0, \ldots, y_t)$ and the inputs $(u_0, \ldots, u_{t-1})$ is that of a set $\mathcal{X}_t \subseteq \mathbb{R}^{n_x}$ known to contain $x_t$ and which is as small as possible, see [Schweppe, 1968]. A sketch of this estimation structure

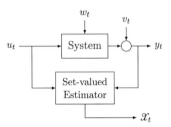

Figure 2.2.: Set-valued estimation of the state of a dynamical system based on the measurement of its input and output. Note that the estimator is a dynamical system itself.

is depicted in Figure 2.2. A recursion defining the optimal $\mathcal{X}_t$ is readily obtained, namely

$$\mathcal{X}_{t+1} := (A\mathcal{X}_t \oplus \{Bu_t\} \oplus \mathcal{W}) \cap C^{-1}(\{y_{t+1}\} \oplus (-\mathcal{V})), \ t \in \mathbb{N}, \tag{2.9}$$

with $\mathcal{X}_0$ initialized with prior knowledge of $x_0$, compare [Schweppe, 1968]. With this recursion, however, the complexity of $\mathcal{X}_t$, determined by the number of inequalities needed to describe the set, may grow unbounded with time, such that one has to be content with outer approximations of $\mathcal{X}_t$, for example outer bounding ellipsoids as proposed for this reason in [Schweppe, 1968].

In this thesis, we mainly employ two means of overapproximating the optimal $\mathcal{X}_t$: the first is based on the error dynamics of a linear estimator, the second on a moving-horizon implementation of (2.9). For an overview of set-valued estimation in general, we refer to [Blanchini and Miani, 2008, Chapter 10].

## 2.4.1. Set-valued estimates from linear estimator dynamics

Let a linear estimator for the state $x_t$ of system (2.1) in the form

$$\hat{x}_{t+1} = A\hat{x}_t + Bu_t + L(C(A\hat{x}_t + Bu_t) - y_{t+1}) \tag{2.10}$$

be given, where $L \in \mathbb{R}^{n_x \times n_y}$ and the matrix $A + LCA$ is stable. Define the estimation error $\tilde{x}_t := x_t - \hat{x}_t, \ t \in \mathbb{N}$, and let sets $\mathcal{E}_t, \ t \in \mathbb{N}$, be given such that $\tilde{x}_t \in \mathcal{E}_t$ for $t \in \mathbb{N}$. Then it holds that $x_t \in \{\hat{x}_t\} \oplus \mathcal{E}_t$ for all $t \in \mathbb{N}$. Set-valued estimators of this form were proposed in [Chisci and Zappa, 2002] and [Mayne et al., 2009] in the context of output-feedback predictive control. A particular advantage of this approach is that the sets $\mathcal{E}_t$ can be determined *a priori*: it holds that

$$\tilde{x}_{t+1} = (A + LCA)\tilde{x}_t + (I + LC)w_t + Lv_{t+1}, \tag{2.11}$$

such that any recursion satisfying

$$\mathscr{E}_{t+1} \supseteq (A + LCA)\mathscr{E}_t \oplus (I + LC)\mathscr{W} \oplus L\mathscr{V} \tag{2.12}$$

yields valid bounds on the estimation error, compare [Mayne et al., 2009]. In particular, $\mathscr{E}_{t+1}$ may be defined as any low-complexity overapproximation of the right-hand side of (2.12). It is important to note that the dynamics defined by the right-hand side of (2.12) are convergent (see also [Mayne et al., 2009]) such that one may define $\mathscr{E}_t = \mathscr{E}$ for all $t \in \{\bar{t}, \ldots\}$ and a fixed $\bar{t} \in \mathbb{N}$, for an appropriately chosen (that is, large enough) set $\mathscr{E}$, compare [Kögel and Findeisen, 2016, Subramanian et al., 2016]. One may even choose $\bar{t} = 0$ as in [Chisci and Zappa, 2002].

## 2.4.2. Set-valued moving-horizon estimation

A second way of outer-bounding $\mathscr{X}_t$ with a set of finite complexity is to perform the iteration in (2.9) only over a finite number of steps $M$. For $t \geq M$ we may thus define

$$\mathscr{X}_t' := \mathscr{X}_{M|t-M}' \tag{2.13}$$

where

$$\mathscr{X}_{i+1|t-M}' := (A\mathscr{X}_{i|t-M}' \oplus \{Bu_{t-M+i}\} \oplus \mathscr{W}) \cap C^{-1}\left(\{y_{t-M+i+1}\} \oplus (-\mathscr{V})\right),$$
$$i \in \{0, \ldots, M-1\} \tag{2.14}$$

and $x_{t-M} \in \mathscr{X}_{0|t-M}'$ for all $t \in \{M, \ldots\}$ and given sets $\mathscr{X}_{0|t-M}'$ of bounded complexity. For example, one may define $\mathscr{X}_{0|t-M}' := \{\hat{x}_{t-M}\} \oplus \mathscr{E}_{t-M}$, as described in the previous subsection, or simply $\mathscr{X}_{0|t-M}' := \mathbb{R}^{n_x}$, which implies that no *a priori* information about the state at time $t - M$ is employed. It readily follows that $x_t \in \mathscr{X}_t'$ for all $t \in \mathbb{N}$. With these definitions, only a bounded number of measurements, namely $(y_{t-M+1}, y_{t-M+2}, \ldots, y_t)$ are explicitly employed in the set iterations, whence the name "moving-horizon set-valued estimation". A set-valued estimation scheme of this type is also described in [Blanchini and Miani, 2008, Section 10.1.2].

**Remark 2.6.** *"Moving-horizon estimation" generally refers to a technique where a state estimate is obtained at each point in time from such a finite set of measurements by means of a model of the system and the minimization of some cost function, see, for example [Rawlings and Mayne, 2009, Chapter 4.3].*

We have the following characterization of $\mathscr{X}_{M|t-M}'$.

**Proposition 2.1.** *Let*

$$
\mathscr{E}_t^j := \left\{ \tilde{x}_t' \in \mathbb{R}^{n_x} \,\middle|\, 
\begin{array}{r}
\tilde{x}_{t-j}' \in \mathscr{E}_{t-j}, \\
(w_k', v_{k+1}') \in \mathscr{W} \times \mathscr{V}, \\
\tilde{x}_{k+1}' = (A + LCA)\tilde{x}_k' + (I + LC)w_k' + Lv_{k+1}', \\
C(A\hat{x}_k + Bu_k) - y_{k+1} = -CA\tilde{x}_k' - Cw_k' - v_{k+1}', \\
k \in \{t-j, \ldots, t-1\}
\end{array}
\right\}
$$

(2.15)

*for all* $j \in \mathbb{N}$ *and all* $t \in \{j, \ldots\}$. *It holds that* $\{\hat{x}_t\} \oplus \mathscr{E}_t^M = \mathcal{X}_{M|t-M}'$ *if* $\mathcal{X}_{0|t-M}' = \{\hat{x}_{t-M}\} \oplus \mathscr{E}_{t-M}$.

If $\mathscr{W}$, $\mathscr{V}$, and $\mathscr{E}_{t-j}$ are polyhedral, then so is $\mathscr{E}_t^j$: the representation in (2.15) is that of a polyhedral subset of $\mathscr{W}^j \times \mathscr{V}^j \times \mathscr{E}_{t-j} \times \mathbb{R}^{jn_x}$ projected onto $\mathbb{R}^{n_x}$.

## 2.5. Model predictive control

Control problems that include the satisfaction of constraints of the form $(x_t, u_t) \in \mathcal{X} \times \mathcal{U}$, $t \in \mathbb{N}$, for compact sets $\mathcal{X} \subseteq \mathbb{R}^{n_x}$ and $\mathcal{U} \subseteq \mathbb{R}^{n_u}$ are, in general, hard to solve if *a priori* guarantees on the satisfaction of the constraints are to be given. One method that has proven effective in this respect is model predictive control (MPC). The general idea is to solve, at each point in time, a finite horizon optimal control problem based on a model of the system being controlled for which the current state provides the initial condition and any desired constraints on state and input are explicitly included in the optimal control problem. After the solution to the problem has been obtained at a given point in time, the first element of the optimal input sequence is applied to the system. Hence, the controller—by solving the optimal control problem—periodically performs a prediction of the future evolution of the system, whence the name of the method. This concept is illustrated in Figure 2.3. If the optimal control problem has a solution at each point in time, then the constraints on the actual system state and input must necessarily be satisfied. However, *a priori* guaranteeing that a certain constrained optimization problem is feasible for a whole set of parameters (which is infinite, as the system state is a parameter) is not an easy task. In most cases, one is satisfied with showing that the optimal control problem underlying the MPC scheme is *recursively feasible*, meaning that if the problem is feasible at a point in the state space, then any optimal input will drive the system only to points where the problem is again feasible. Hence, recursive feasibility allows to conclude that the MPC scheme is well-defined and the constraints on the state and input are satisfied for all time if the optimal control problem admits a feasible solution at initialization. One way of ensuring that the problem is recursively feasible is to include terminal constraints which are tailored to the dynamics of the system being controlled and the constraints imposed on the input and state. Stability

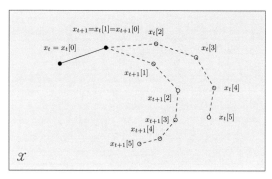

Figure 2.3.: Optimal predicted state trajectories at times $t$ and $t+1$ for a closed-loop system under predictive control where the dynamics are known exactly. As the input applied at time $t$ is equal to the optimal predicted input for the initial condition $x_t$, the value of $x_{t+1}$, the actual state at time $t+1$, is equal to its predicted value, that is $x_t[1]$. Note that the predicted trajectories at subsequent points in time do not necessarily coincide elsewhere.

properties of the closed-loop system can be assigned by a careful choice of the cost function of the optimal control problem. For a more detailed discussion of these issues and an overview of the subject of MPC, we refer to [Mayne et al., 2000], [Rawlings and Mayne, 2009], and [Mayne, 2014].

If the system being controlled is subject to uncertainty, for example due to plant-model mismatch, external disturbances, or measurement noise, then the controller will not be able to predict the future evolution of the system exactly. If bounds on the uncertainty are known, then it is—at least in principle—possible to predict the worst-case behavior of the system regarding constraint satisfaction and to adapt the predicted input sequence accordingly, compare Figure 2.4. Here, it has been found that optimizing over future *feedback policies* instead of inputs is essential in order to prevent the approach from becoming overly conservative: for open-loop unstable plants under uncertainty, the size of the sets containing all possible future states under an assumed input sequence grows exponentially (for linear systems), see [Rawlings and Mayne, 2009, Section 3.1.2]. However, the optimization over general feedback policies has been found to be computationally intractable; one has to parameterize the future feedback applied to the system and perform a trade-off between computational complexity and conservatism, compare [Raković, 2012], which also contains a recent overview of existing robust MPC approaches for linear systems. For such systems in particular, if the constraints are polytopic and the uncertainty enters in the form of a bounded additive disturbance, a family of methods often referred to as "tube MPC" has been shown to offer a particular good trade-off between computational complexity and conservatism. The main idea is to

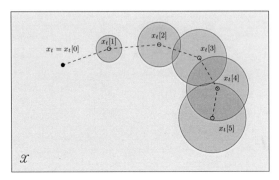

Figure 2.4.: Nominal predicted state trajectories at times $t$ and uncertainty bounds containing all possible future system states under the assumed input.

separate, in the predictions, the dynamics of the nominal system (without disturbances) from the dynamics of the prediction error and, at the same time, separate the (assumed) input applied to the system into a part acting on the nominal system and a part acting, as a feedback, only on the prediction error, compare [Raković, 2012]. This way, the dynamics of the prediction error can be stabilized without influencing the nominal system dynamics. By the superposition principle of linear systems, the reachable sets of these error dynamics are independent of the state of the nominal system and only depend on the assumed feedback law. In the MPC scheme proper, the optimal control problem is based on the nominal system while the constraints are restricted based on the reachable sets of the error system, thereby guaranteeing robust constraint satisfaction. In the simplest schemes, for example [Chisci et al., 2001] and [Mayne et al., 2005], the feedback takes the form of a fixed linear gain such that no additional optimization is required. In fact, the computational complexity of these schemes was shown to be comparable to MPC schemes for undisturbed systems. More advanced schemes such as [Löfberg, 2003, Raković et al., 2012a] involve a (parameterized) optimization over the error feedback, trading a higher computational effort for reduced conservatism.

The method proposed in [Chisci et al., 2001], although being one of the very first tractable solutions to the robust predictive control problem, offers an overall attractive trade-off between computational complexity and performance, compare the discussion in [Raković, 2012]. Therein, various robust MPC schemes are compared in terms of conservatism and computational complexity and the scheme in [Chisci et al., 2001] was ranked best among the schemes whose complexity scales linearly in the prediction horizon. The MPC approaches in this thesis are extensions of that control scheme; in the remainder of this section we present this method in the notation used throughout the thesis, which partly follows [Rawlings and Mayne, 2009].

**The robust MPC scheme in [Chisci et al., 2001]**

The system being controlled is described by (2.1a) and it is assumed that there exists a matrix $K \in \mathbb{R}^{n_u \times n_x}$ such that $A + BK$ is stable. The goal is to construct a set-valued function $\kappa_{\mathrm{MPC}} : \mathcal{X}_\mathrm{f} \to 2^{(\mathbb{R}^{n_u})}$ with $\mathcal{X}_\mathrm{f} \subseteq \mathbb{R}^{n_x}$ such that if $x_0 \in \mathcal{X}_\mathrm{f}$ and $u_t \in \kappa_{\mathrm{MPC}}(x_t)$ for all $t \in \mathbb{N}$, then it holds that $(x_t, u_t) \in \mathcal{X} \times \mathcal{U}$ for all $t \in \mathbb{N}$ and $x_t$ converges to the set $\mathcal{F}_\infty$. Here, $\mathcal{X} \subseteq \mathbb{R}^{n_x}$ and $\mathcal{U} \subseteq \mathbb{R}^{n_u}$ are given compact sets. Define first the set of admissible nominal input sequences for a given point $x \in \mathbb{R}^{n_x}$,

$$
\mathscr{D}(x) := \left\{ (u[i])_{i \in \{0,\dots,N-1\}} \in \mathbb{R}^{Nn_u} \,\Bigg|\, \right.
$$

$$
x[0] = x, \tag{2.16a}
$$
$$
x[i+1] = Ax[i] + Bu[i], \tag{2.16b}
$$
$$
u[i] \in \mathcal{U} \ominus K\mathcal{F}_i, \tag{2.16c}
$$
$$
x[i] \in \mathcal{X} \ominus \mathcal{F}_i, \tag{2.16d}
$$
$$
i \in \{0,\dots,N-1\},
$$
$$
\left. x[N] \in \mathcal{X}_\mathrm{T} \right\} \tag{2.16e}
$$

and the subset

$$
\mathcal{X}_\mathrm{f} := \{ x \in \mathbb{R}^{n_x} \mid \mathscr{D}(x) \neq \emptyset \} \tag{2.17}
$$

of the state space where the constraints are feasible. Here, the *terminal set* $\mathcal{X}_\mathrm{T}$ is assumed[1] to be compact and to satisfy $\mathcal{X}_\mathrm{T} \subseteq \mathcal{X} \ominus \mathcal{F}_N$, $K\mathcal{X}_\mathrm{T} \subseteq \mathcal{U} \ominus K\mathcal{F}_N$, and $(A + BK)\mathcal{X}_\mathrm{T} \subseteq \mathcal{X}_\mathrm{T} \ominus (A + BK)^N \mathcal{W}$. A summary of relevant properties of the Pontryagin difference, denoted by $\ominus$, is given in Lemma A.2.

Define further the cost function $\bar{J} : \mathbb{R}^{n_x} \times \mathbb{R}^{Nn_u}$ by

$$
\bar{J}(x, (u[0],\dots,u[N-1])) = \sum_{k=0}^{N-1} \ell(u[i] - Kx[i]), \tag{2.18}
$$

where $x[0] = x$ and $x[i+1] = Ax[i] + Bu[i]$ for $i \in \{0,\dots,N-2\}$ and $\ell$ is a continuous positive definite[2] function. The MPC value function $J : \mathbb{R}^{n_x} \to \mathbb{R}$ is defined by the optimization problem

$$
J(x) = \inf \left\{ \bar{J}(x, \mathbf{u}) \mid \mathbf{u} \in \mathscr{D}(x) \right\}. \tag{2.19}
$$

---

[1]The terminal set in [Chisci et al., 2001], while defined differently, satisfies these assumptions. In fact, we define the terminal set in the same way as [Ghaemi et al., 2008].

[2]In [Chisci et al., 2001], $\ell$ was chosen to be a particular quadratic function.

Associated with this problem are the set of optimal trajectories of infinite length

$$
\mathscr{T}(x) = \left\{ (\mathbf{x}, \mathbf{u}) \in (\mathbb{R}^{n_{\mathrm{x}}})^{\mathbb{N}} \times (\mathbb{R}^{n_{\mathrm{u}}})^{\mathbb{N}} \;\middle|\; \right.
$$

$$
\begin{aligned}
\mathbf{x} &= (x[i])_{i \in \mathbb{N}}, \\
\mathbf{u} &= (u[i])_{i \in \mathbb{N}}, \\
x[0] &= x, \\
x[i+1] &= Ax[i] + Bu[i], \ i \in \mathbb{N}, \\
(u[0], \dots, u[N-1]) &\in \mathscr{D}(x), \\
u[i] &= Kx[i], \ i \in \{N, \dots\}, \\
\left. \bar{J}(x, (u[0], \dots, u[N-1])) = J(x) \right\} ,
\end{aligned}
\tag{2.20}
$$

where the input for $i \in \{N, \dots\}$ is fixed to $u[i] = Kx[i]$. Finally, we define the set-valued controller

$$
\kappa_{\mathrm{MPC}}(x) = \left\{ u \in \mathbb{R}^{n_{\mathrm{u}}} \;\middle|\; \exists \mathbf{x} \in (\mathbb{R}^{n_{\mathrm{x}}})^{\mathbb{N}}, \ \exists \tilde{\mathbf{u}} \in (\mathbb{R}^{n_{\mathrm{u}}})^{\mathbb{N}}, \ (\mathbf{x}, (u, \tilde{\mathbf{u}})) \in \mathscr{T}(x) \right\}, \tag{2.21}
$$

where $(u, \tilde{\mathbf{u}}))$ denotes the sequence $(u[i])_{i \in \mathbb{N}}$ with $u[0] = u$, $u[i+1] = \tilde{u}[i]$, $i \in \mathbb{N}$, and $(\tilde{u}[i])_{i \in \mathbb{N}} = \tilde{\mathbf{u}}$. That is, the set of admissible controller outputs consists of all $u \in \mathscr{U}$ that can be extended to an optimal input trajectory.

The following results can be found (with slight modifications) in [Chisci et al., 2001].

**Theorem 2.4.** *Let $x_t \in \mathscr{X}_{\mathrm{f}}$ and $u_t \in \kappa_{\mathrm{MPC}}(x_t)$, then $(x_t, u_t) \in \mathscr{X} \times \mathscr{U}$ and $x_{t+1} \in \mathscr{X}_{\mathrm{f}}$ for any $w_t \in \mathscr{W}$, if $x_{t+1} = Ax_t + Bu_t + w_t$. Further, $J(x_{t+1}) \leq J(x_t) - \ell(u_t - Kx_t)$. Finally, if $x_0 \in \mathscr{X}_{\mathrm{f}}$, $u_t \in \kappa_{\mathrm{MPC}}(x_t)$, $x_{t+1} = Ax_t + Bu_t + w_t$, and $w_t \in \mathscr{W}$ for all $t \in \mathbb{N}$, then $\lim_{t \to \infty} |x_t|_{\mathscr{F}_\infty} = 0$.*

**Remark 2.7.** *As the set $\mathscr{D}(x)$ is compact for all $x$ and $\bar{J}$ is continuous, the minimum in (2.19) is attained for all $x \in \mathscr{X}_{\mathrm{f}}$. Hence, $\kappa_{\mathrm{MPC}}(x) \neq \emptyset$ for $x \in \mathscr{X}_{\mathrm{f}}$, compare [Rawlings and Mayne, 2009, Proposition 2.4].*

Note that in [Chisci et al., 2001] the *stability* of the set $\mathscr{F}_\infty$ was neither claimed nor proven. However, under mild technical assumptions stability can, in fact, be established; in Chapter 5 we employ techniques that yield, as a by-product, the stability of $\mathscr{F}_\infty$ under the control scheme.

The implementation of model predictive controllers as described in this section requires the computation of $\kappa_{\mathrm{MPC}}$ or subsets thereof, in turn necessitating the solution of optimization problems as defined by (2.19). If the sets $\mathscr{U}$, $\mathscr{X}$, and $\mathscr{X}_{\mathrm{T}}$ are polytopes and $\ell$ is convex piecewise affine or convex quadratic, then linear programming or, respectively, quadratic programming, offers a relatively efficient way to solve these problems, compare

[Bemporad et al., 2002]. If $\mathcal{W}$ is a polytope, then the Pontryagin differences $\mathcal{X} \ominus \mathcal{F}_i$ and $\mathcal{U} \ominus K\mathcal{F}_i$ can be computed by linear programming, see [Kolmanovsky and Gilbert, 1998]. Terminal sets $\mathcal{X}_T$ satisfying the required assumptions can be computed using the algorithm described in [Kolmanovsky and Gilbert, 1995], compare [Chisci et al., 2001].

# 3. Linear Systems Perturbed by Bounded Disturbances

In this chapter, we consider the control of systems defined by

$$x_{t+1} = Ax_t + Bu_t + w_t \tag{3.1a}$$
$$y_t = Cx_t + v_t. \tag{3.1b}$$

as described in Section 2.2, where we assume bounds on the disturbance $w_t$ and the noise $v_t$ to be known. In particular, we assume $(w_t, v_t) \in \mathscr{W} \times \mathscr{V}$, $t \in \mathbb{N}$, where $\mathscr{W} \subseteq \mathbb{R}^{n_x}$ and $\mathscr{V} \subseteq \mathbb{R}^{n_y}$ are convex and compact sets containing the respective origins. Our goal is the stabilization of the system in terms of Definition 2.2, that is, we want to design the trigger mechanism and the controller in Figure 2.1 (depicting the state-feedback case) such that a certain subset of the state space is $\theta$-uniformly asymptotically stable and the rate of transmissions over the network is low. In the approaches presented in this chapter, we exploit the structure of linear systems subject to bounded additive disturbances in the design of aperiodic control schemes. In particular, we assume that a matrix $K \in \mathbb{R}^{n_u \times n_x}$ is known such that $A + BK$ is stable and employ known properties of systems described by

$$x_{t+1} = (A + BK)x_t + w_t \tag{3.2}$$

with $w_t$ constrained to a compact set.

In Section 3.2, we derive aperiodic controllers based on a direct application of Theorem 2.1, that is, by guaranteeing explicitly that a certain Lyapunov function decreases over time. As a defining feature of the approaches proposed in the thesis, this Lyapunov function will be defined in terms of a *set* that is asymptotically stable for any dynamical system generating $(x_t)_{t \in \mathbb{N}}$ for (3.2), in the sense of Definition 2.2; in earlier approaches, the Lyapunov function employed is usually defined for the undisturbed system $x_{t+1} = (A + BK)x_t$, that is, it can be used to prove the stability of the origin, or it is defined for system (3.2), but is used to prove input-to-state stability of the origin with respect to the disturbances or a finite gain (expressed as, for example, $\mathscr{L}_2$ stability) between the disturbances and some output of the system. Such approaches can, for example, be found in [Tabuada, 2007, Velasco et al., 2009, Wang and Lemmon, 2009, Lemmon, 2010, Anta and Tabuada, 2010, Eqtami et al., 2010, Wang and Lemmon, 2011, Heemels et al., 2012, Dimarogonas et al., 2012, Tiberi and Johansson, 2013, Di

Benedetto et al., 2013, Forni et al., 2014, Seuret et al., 2014, Borgers and Heemels, 2014, Tallapragada and Chopra, 2014, Girard, 2015, Postoyan et al., 2015, Fiter et al., 2015, Abdelrahim et al., 2017, Hetel et al., 2017] or the references therein (part of which have been added to the list above).

In Section 3.3, we present approaches that rely on ensuring set-membership conditions, using Theorem 2.3 in order to prove stability. These approaches are somewhat related to the earliest event-triggered schemes such as [Årzén, 1999, Åström and Bernhardsson, 2002], but also later ones such as [Heemels et al., 2008, Lunze and Lehmann, 2010, Heemels and Donkers, 2013], where the event-conditions are defined in terms of a certain quantity crossing a threshold, that is, leaving a ball of a certain size centered at the origin. In fact, we will show that the stability properties of a certain class of such control schemes can be analyzed within our framework, potentially leading to less conservative results. The approaches we develop in this section are based on disturbance reachable sets of linear systems, with the results in [Iino et al., 2009] and [Kögel and Findeisen, 2014] being most closely related, due to their employing set-theoretic methods established for linear systems. Similarly related aperiodic control schemes can also be found in the context of predictive control (to which the two works mentioned above also belong) and are discussed in Chapter 5.

The results in this chapter are, in parts, based on [Brunner and Allgöwer, 2016, Brunner et al., 2016c, Brunner et al., 2017b].

# 3.1. Preliminaries

In this thesis, the size of the stabilized set, which also serves as an asymptotic bound on the system state, is the main measure of control performance. However, also the value of $\theta$, in our notion of stability, is a meaningful indicator of the closed-loop system behavior, as we will see, where a lower value of $\theta$ may be desirable. Both $\theta$ and the size of the stabilized set will be explicit tuning parameters in the devised control schemes and we will demonstrate in numerical examples the trade-off between these parameters and the average communication rate in the closed-loop system. In both sections of this chapter, we first present results for the state-feedback setup where we assume that the state $x_t$ is available as a measurement at any point in time. At the end of each section we show how the results can be extended to the output feedback case.

For the state-feedback case, following the framework presented in Section 2.3, we consider event-triggered control laws of the form

$$u_t = \kappa(x_{t_i}, t - t_i), \qquad t \in \{t_i, \ldots, t_{i+1} - 1\}, \ i \in \{0, \ldots, i_{\max}\} \qquad (3.3a)$$

$$t_0 = 0 \qquad (3.3b)$$

$$t_{i+1} = \inf\{t \in \{t_i + 1, \ldots\} \mid \delta_t(x_0, x_1, \ldots, x_t, t_i) = 1\}, \qquad (3.3c)$$

and self-triggered control laws of the form

$$u_t = \kappa(x_{t_i}, t - t_i), \qquad t \in \{t_i, \dots, t_{i+1} - 1\}, \; i \in \{0, \dots, i_{\max}\} \tag{3.4a}$$
$$t_0 = 0 \tag{3.4b}$$
$$t_{i+1} = \mu_i(x_{t_0}, x_{t_1}, \dots, x_{t_i}, t_0, t_1, \dots, t_i). \tag{3.4c}$$

**Remark 3.1.** *For simplicity, we assume $\kappa$ to be of the same structure for the event-triggered and the self-triggered case. Note, however, that $t_{i+1}$ is known already at time $t_i$ in the self-triggered case (in contrast to the event-triggered setup), which suggests to make the input $u_t$ for $t \in \{t_i, \dots, t_{i+1} - 1\}$ additionally dependent on this information (see [Gommans et al., 2014] for an example of such a type of self-triggered control law). In Chapter 5 of this thesis, we consider optimization based control schemes where this additional degree of freedom in the self-triggered case is actually exploited.*

We assume the function $\kappa$ to be given with the restriction that

$$\kappa(x, 0) = Kx \tag{3.5}$$

for all $x \in \mathbb{R}^{n_x}$. This framework applies to several controllers from the literature on event-triggered and self-triggered control, such as *to-zero controllers*, where

$$\kappa(x, \tau) = \begin{cases} Kx, & \tau = 0, \; x \in \mathbb{R}^{n_x} \\ 0, & \text{otherwise,} \end{cases} \tag{3.6a}$$

*to-hold controllers*, where

$$\kappa(x, \tau) = Kx, \quad \tau \in \mathbb{N}, \; x \in \mathbb{R}^{n_x}, \tag{3.6b}$$

and *model-based controllers* [Lunze and Lehmann, 2010, Garcia and Antsaklis, 2013, Heemels and Donkers, 2013], where

$$\kappa(x, \tau) = K(A + BK)^\tau x, \quad \tau \in \mathbb{N}, \; x \in \mathbb{R}^{n_x}, \tag{3.6c}$$

see also [Schenato, 2009, Heemels et al., 2012, Gommans and Heemels, 2015], where controllers of these types are discussed. By Theorem 2.2, we know that with the control law $u_t = Kx_t$, the set $\mathscr{F}_\infty = \bigoplus_{j=0}^\infty (A + BK)^j \mathscr{W}$ is stabilized for the closed-loop system. Our goal will be to design the functions $\delta_t$ and $\mu_i$ such that for the resulting closed-loop system the set $\gamma\mathscr{F}_\infty$ is stabilized in the sense of Definition 2.2, where $\theta \in \{1, \dots\} \cup \{\infty\}$ and $\gamma \in [1, \infty)$ are design parameters. As we will demonstrate in numerical examples, it is possible to stabilize the set $\mathscr{F}_\infty$ without having to update the control law at every point in time for certain realizations of the disturbance sequence.

## 3.2. Lyapunov-based approach

In this section, we consider an approach for the design of event-triggered and self-triggered controllers based on a Lyapunov function for the set $\gamma\mathscr{F}_\infty$. The main idea is to ensure a decrease of this function under the worst-case future disturbances. Hereby, we relax the rate of decrease guaranteed under periodic control (given by $\lambda \in [0,1)$, as defined below), in that we only require a rate of decrease given by $\eta = \max\{\lambda, \nu\}$, with $\nu \in [0,1)$, of our choice. The main idea behind this relaxation is to make it easier for the aperiodic controller to satisfy the decrease condition, leading (hopefully) to longer inter-transmission times on average. A similar relaxation of required decrease rate of the Lyapunov function was also employed in [Mazo et al., 2010, Heemels et al., 2012]. See also [Kögel and Findeisen, 2015c] for a self-triggered scheme with similar exponentially decreasing bounds.

Hence, three design parameters for the closed-loop system are available for a trade-off against the transmission rate in the system, which want to reduce: the size of the set $\gamma\mathscr{F}_\infty$ that is stabilized, the rate of convergence, determined by the guaranteed rate of decrease $\nu$, and the value of $\theta$ in Definition 2.2, which describes the quality of stability we are achieving.

### 3.2.1. Theoretical results

In this subsection, we provide the theoretical framework for the proposed approach. Let $V$ be a function mapping $\mathbb{R}^{n_x}$ to $\mathbb{R}$, and let scalars $c_2, c_2, a \in (0, \infty)$, and $\lambda \in [0, 1)$ be given such that $c_1|x|^a_{\gamma\mathscr{F}_\infty} \leq V(x) \leq c_2|x|^a_{\gamma\mathscr{F}_\infty}$ and $V((A + BK)x + w) \leq \lambda V(x)$ for all $x \in \mathbb{R}^{n_x}$ and all $w \in \mathscr{W}$. Let finally $\nu \in [0, 1)$. We consider existence and construction of $V$, and approximations thereof, in Section 3.2.3.

#### Event-triggered control

In the event-triggered case, decisions whether to transmit information over the network are made at every point in time. Only the worst-case (unknown) disturbance acting currently on the system is taking into account, guaranteeing a decrease of the Lyapunov function at the very next time step.

Define the event-generating functions $\delta_t$, $t \in \{1, \ldots\}$, by

$$\delta_t(x_0, x_1, \ldots, x_t, t_i) := \begin{cases} 0 & \text{if } \exists \tau \in \{1, \ldots, \min\{t+1, \theta\}\}, \ \forall w \in \mathscr{W}, \\ & \quad V(Ax_t + B\kappa(x_{t_i}, t - t_i) + w) \leq \nu^\tau V(x_{t+1-\tau}) \\ 1 & \text{otherwise.} \end{cases} \quad (3.7)$$

We have the following result.

**Theorem 3.1.** *For any dynamical system generating* $(x_t)_{t\in\mathbb{N}}$ *for the closed loop consisting of (3.1a), (3.3), and (3.7), the set* $\gamma\mathscr{F}_\infty$ *is* $\theta$-*uniformly asymptotically stable with region of attraction* $\mathbb{R}^{n_x}$.

*Proof.* The event-triggered controller guarantees that $V$ satisfies the conditions for the Lyapunov function in Theorem 2.1 with $\eta = \max\{\lambda, \nu\}$: by $\kappa(x_t, 0) = Kx_t$ in the case of $\delta_t(x_0, \ldots, x_t, t_i) = 1$, and by definition in the case of $\delta_t(x_0, \ldots, x_t, t_i) = 0$. □

**Remark 3.2.** *For $\theta = \infty$, the event-triggered control law closely resembles the scheme proposed in [Wang and Lemmon, 2011], where the trigger rule is based on guaranteeing an exponential decrease of a Lyapunov function over time, starting at time $t = 0$. Compare also [Seyboth et al., 2013], where trigger rules based on exponentially decaying thresholds are employed. As we will show in a numerical example later in the chapter, the decrease of the Lyapunov function may not be monotonic, compare Figure 1 in [Wang and Lemmon, 2011], depicting such a non-monotonic decrease. Compare also [Linsenmayer et al., 2016], where non-monotonic Lyapunov functions were used for the design and analysis of an event-triggered control system.*

**Self-triggered control**

In the self-triggered case, decisions when to communicate are made only at the transmission instants themselves. A decrease of the Lyapunov function has to be guaranteed for all times between the current transmission instant and the next and multiple independent worst-case disturbances have to be taken into account.

Define the scheduling functions $\mu_i$, $i \in \mathbb{N}$, by

$$
\mu_i(x_{t_0}, x_{t_1}, \ldots, x_{t_i}, t_0, t_1, \ldots, t_i)
$$

$$
:= \max\left\{ t_i + 1, \; \sup\left\{ t' \in \{t_i + 2, \ldots\} \; \middle| \; \begin{array}{l} \forall k \in \{t_i + 2, \ldots, t'\}, \\ \exists \tau_k \in \{1, \ldots, \min\{k, \theta\}\}, \\ \forall w[l] \in \mathscr{W}, \; l \in \{t_j, \ldots, k-1\}, \\ V(x[k]_i) \leq \nu^{\tau_k} V(x[k - \tau_k]_i) \end{array} \right\} \right\},
$$

$$
\tag{3.8}
$$

with the supremum of the empty set being negative infinity. Here, we define for all $i \in \{0, \ldots, i_{\max}\}$,

$$
x[k]_i := A^{k-t_i} x_{t_i} + \sum_{l=t_i}^{k-1} A^{k-1-l}(B\kappa(x_{t_i}, l - t_i) + w[l]) \tag{3.9}
$$

for all $k \in \{t_i, \ldots\}$ and

$$
x[k - \tau_k]_i := A^{k-\tau_k-t_j} x_{t_j} + \sum_{l=t_j}^{k-\tau_k-1} A^{k-\tau_k-1-l}(B\kappa(x_{t_j}, l - t_j) + w[l]) \tag{3.10}
$$

with $j \in \mathbb{N}$ such that $k - \tau_k \in \{t_j, \ldots, t_{j+1} - 1\}$. The reasoning behind these definitions is the following. As we have assumed that measurements of the state $x_t$ are only available at the transmission instants in the self-triggered case, we have to replace $x_t$ with a worst-case estimate for times in between transmission instants if we want to compare the (worst-case) value of the Lyapunov function at these points in time. We chose to select the most recent transmission time before $k$ or, respectively, $k - \tau_k$, at which the actual state is known, and to evaluate the system dynamics under the assumed disturbance realization and the assumed (or known) input applied to the system to compute this estimate.

We have the following result.

**Theorem 3.2.** *For any dynamical system generating $(x_t)_{t \in \mathbb{N}}$ for the closed loop consisting of (3.1a), (3.3), and (3.8), the set $\gamma \mathscr{F}_\infty$ is $\theta$-uniformly asymptotically stable with region of attraction $\mathbb{R}^{n_x}$.*

*Proof.* The proof follows by the same arguments as the proof of Theorem 3.1 by noting that, due to all possible disturbance realizations being considered in the scheduling function, it is guaranteed that $V(x_k) \leq \max\{\nu, \lambda\}^{\tau_k} V(x_{k-\tau_k})$ for some $\tau_k \in \{1, \ldots, \min\{k, \theta\}\}$ for all $k \in \{t_i + 1, \ldots, t_{i+1}\}$ and all $i \in \{0, \ldots, i_{\max}\}$, where it is taken into account that $u_{t_i} = K x_{t_i}$ for all $i \in \{0, \ldots, i_{\max}\}$, such that a decrease of the Lyapunov function is always guaranteed for at least one time-step into the future. $\square$

**Remark 3.3.** *In the scheduling function, we assume $w_t \in \mathscr{W}$ in order to compute the worst-case values of unknown systems states in the past. Note that if two states at different times are known, say $x_{t_j}$ and $x_{t_i}$ for $t_j < t_i$, then one may infer tighter bounds on $w_t$ for $t \in \{t_j, \ldots, t_i - 1\}$ than merely $w_t \in \mathscr{W}$. In the simplest case, where $t_i = t_j + 1$, we have $w_j = x_{t_i} - A x_{t_j} - B \kappa(x_{t_j}, 0)$, that is, $w_{t_j}$ is known exactly. In order to not overcomplicate the discussion, we do not make use of this additional information in the thesis.*

### 3.2.2. Output feedback

Consider now the output-feedback case, that is, the control of system (3.1) where only $y_t$, but not $x_t$, is available as a measurement. We assume that the disturbance and noise acting on the system satisfy $(w_t, v_{t+1}) \in \mathscr{W} \times \mathscr{V}$, $t \in \mathbb{N}$, for known compact and convex sets $\mathscr{W} \subseteq \mathbb{R}^{n_x}$ and $\mathscr{V} \subseteq \mathbb{R}^{n_y}$. In general, it is not possible to stabilize the origin of system (3.1) (in the absence of disturbances) by a static output feedback controller, that is, by an input defined by a map $u_t = K_s y_t$, see, for example, [Syrmos et al., 1997]. However, it might still be possible to define $u_t$ as the output of a second dynamical system of which $y_t$ is an input, such that the origin of the *joint* state space of these two systems is stable. Here, we assume such a linear dynamic output-feedback controller to be given in the form

$$\hat{x}_{t+1} = A \hat{x}_t + B u_t + L(C(A \hat{x}_t + B u_t) - y_{t+1}) \qquad (3.11a)$$

$$u_t^{\text{lin}} = K \hat{x}_t, \qquad (3.11b)$$

with $L \in \mathbb{R}^{n_x \times n_y}$ where the matrices $A + BK$ and $A + LCA$ are stable. Define the estimation error $\tilde{x}_t := x_t - \hat{x}_t$ and the joint state $(\tilde{x}_t, \hat{x}_t)$. The dynamics of this joint state[1] are given by

$$\begin{bmatrix} \tilde{x}_{t+1} \\ \hat{x}_{t+1} \end{bmatrix} = \hat{A} \begin{bmatrix} \tilde{x}_t \\ \hat{x}_t \end{bmatrix} + \hat{B}(u_t - K\hat{x}_t) + \hat{w}_t \tag{3.12}$$

with $\hat{w}_t \in \hat{\mathscr{W}}$, where

$$\hat{A} := \begin{bmatrix} A + LCA & 0 \\ -LCA & A + BK \end{bmatrix}, \tag{3.13}$$

$$\hat{B} := \begin{bmatrix} 0 \\ B \end{bmatrix}, \tag{3.14}$$

$$\hat{w}_t := \begin{bmatrix} I + LC & L \\ -LC & -L \end{bmatrix} \begin{bmatrix} w_t \\ v_{t+1} \end{bmatrix}, \tag{3.15}$$

for $t \in \mathbb{N}$, and

$$\hat{\mathscr{W}} := \begin{bmatrix} I + LC & L \\ -LC & -L \end{bmatrix} (\mathscr{W} \times \mathscr{V}). \tag{3.16}$$

Define further

$$\hat{\mathscr{F}}_i = \bigoplus_{j=0}^{i-1} \hat{A}^j \hat{\mathscr{W}} \tag{3.17}$$

for $i \in \mathbb{N} \cup \{\infty\}$. Analogously to the state-feedback case, the set $\hat{\mathscr{F}}_\infty$ is 1-uniformly asymptotically stable for any dynamical system generating $((\tilde{x}_t, \hat{x}_t))_{t \in \mathbb{N}}$ for (3.1) in closed loop with (3.11) where $u_t = u_t^{\text{lin}}$. Our goal in the output-feedback case is to stabilize the set $\gamma \hat{\mathscr{F}}_\infty$ for a chosen $\gamma \in [1, \infty)$, in the sense of Definition 2.2.

At this point, we want to highlight a difference in state estimation between the event-triggered and the self-triggered case. In the event-triggered case, we assume that the output of the system is measured at each point in time and, hence, it is reasonable to assume that the state estimate $\hat{x}$ is updated according to (3.11a) at every time point $t \in \mathbb{N}$. In the self-triggered case, under the standing assumption that no measurements are taken between transmission instants, the estimate $\hat{x}_t$ must be updated without measurements at the time points $t \in \{t_i + 1, \ldots, t_{i+1} - 1\}$. We assume that these updates satisfy the equation $\hat{x}_t = A\hat{x}_{t-1} + Bu_t$. The aperiodic control loops for the event-triggered and self-triggered setups are depicted in Figure 3.1 and Figure 3.2, respectively, where a *set-valued* estimate $\tilde{\mathcal{X}}_t$—discussed below—is supplied by the estimator.

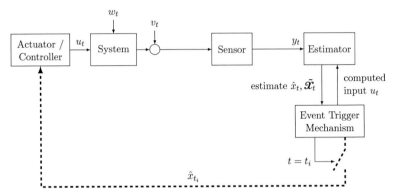

Figure 3.1.: Event-triggered control loop in the output-feedback case. The estimator receives measurement information at every point in time; further, it is assumed that the actual input applied to the plant is computed in the trigger mechanism and supplied to the estimator.

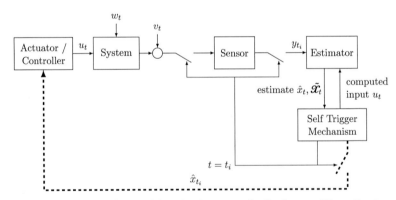

Figure 3.2.: Self-triggered control loop in the output-feedback case. The estimator receives measurement information only at the transmission instants.

Define the open-loop system for the event-triggered setup,

$$\begin{bmatrix} \tilde{x}_{t+1} \\ \hat{x}_{t+1} \end{bmatrix} = \underbrace{\begin{bmatrix} A + LCA & 0 \\ -LCA & A \end{bmatrix}}_{=:\tilde{A}_{et}} \begin{bmatrix} \tilde{x}_t \\ \hat{x}_t \end{bmatrix} + \hat{B}u_t + \hat{w}_t \tag{3.18}$$

and the open-loop system for the self-triggered setup

$$\begin{bmatrix} \tilde{x}_{t+1} \\ \hat{x}_{t+1} \end{bmatrix} = \underbrace{\begin{bmatrix} A & 0 \\ 0 & A \end{bmatrix}}_{=:\tilde{A}_{st}} \begin{bmatrix} \tilde{x}_t \\ \hat{x}_t \end{bmatrix} + \hat{B}u_t + \underbrace{\begin{bmatrix} I & I \\ 0 & 0 \end{bmatrix}}_{:=\bar{D}_{st}} \hat{w}_t \tag{3.19}$$

for the time points $t \in \{t_i, \ldots, t_i - 2\}$, $i \in \{0, \ldots, i_{\max}\}$. For the time points $t = t_{i+1} - 1$, $i \in \{0, \ldots, i_{\max} - 1\}$, we assume that (3.18) holds in both the event-triggered and the self-triggered case.

**Remark 3.4.**

*(i) A self-triggered scenario with measurements taken at every point in time is also feasible. For simplicity, and because availability of measurements at every point in time suggest the use of event-triggered control, we stick to the setup described above.*

*(ii) Similar to the discussion in Remark 3.1, one might conceive of estimation schemes —both for the event-triggered and the self-triggered case—where the estimator gain depends on the time since the last transmission instant, compare, for example, [Brunner et al., 2015a].*

Event-triggered and self-triggered controllers may be defined along similar lines as in the state-feedback case, based on a Lyapunov function $V : \mathbb{R}^{2n_x} \to \mathbb{R}$ for the set $\gamma \hat{\mathscr{F}}_\infty$. Here, the problem arises that in general the argument of the function $V$, that is $(\tilde{x}_t, \hat{x}_t)$, is not known at every time $t \in \mathbb{N}$ because $x_t$ is not assumed to be available as a measurement. We assume therefore that a set-valued estimator providing the sets $\tilde{\mathscr{X}}_t \subseteq \mathbb{R}^{(t+1)n_x}$ is available such that $(\tilde{x}_0, \ldots, \tilde{x}_t) \in \tilde{\mathscr{X}}_t$ for all $t \in \mathbb{N}$. A brief description of set-valued estimation is provided in Section 2.4. The event-generating functions in the event-triggered case and the scheduling functions in the self-triggered case are designed such that $t_i$ is a transmission instant if for any element in $\tilde{\mathscr{X}}_t$ a decrease of the function $V$ is otherwise not ensured, $t$ being the time at which the decision is made. At a transmission instant $t_i$, the input $u_{t_i} = K\hat{x}_{t_i}$ is applied to the system, guaranteeing a decrease of the function $V$. More precisely, we assume that the input to the system is given by

$$u_t = \kappa(\hat{x}_{t_i}, t - t_i), \qquad t \in \{t_i, \ldots, t_{i+1} - 1\}, \; i \in \{0, \ldots, i_{\max}\}, \tag{3.20}$$

---

[1]Note that we use $(\tilde{x}_t, \hat{x}_t)$ and $[\tilde{x}_t^{\mathsf{T}}, \hat{x}_t^{\mathsf{T}}]^{\mathsf{T}}$ interchangeably.

where the function $\kappa$ satisfies $\kappa(\hat{x}, 0) = K\hat{x}$ for all $\hat{x} \in \mathbb{R}^{n_x}$. Further, let $V$ be a function mapping $\mathbb{R}^{2n_x}$ to $\mathbb{R}$, and let scalars $c_2, c_2, a \in (0, \infty)$ and $\lambda \in [0, 1)$ be given such that $c_1 |z|^a_{\gamma\mathscr{F}_\infty} \leq V(z) \leq c_2 |z|^a_{\gamma\mathscr{F}_\infty}$ and $V(\hat{A}z + \hat{w}) \leq \lambda V(z)$ for all $z \in \mathbb{R}^{2n_x}$ and all $\hat{w} \in \mathscr{W}$. Let finally $\nu \in [0, 1)$.

For the event-triggered setup, define the event-generating functions $\delta_t$, $t \in \{1, \dots\}$, by

$$\delta_t \big(\hat{x}_0, \dots, \hat{x}_t, \tilde{\mathfrak{X}}_t, t_i\big)$$
$$:= \begin{cases} 0 & \text{if } \exists \tau \in \{1, \dots, \min\{t+1, \theta\}\}, \ \forall \hat{w} \in \mathscr{W}, \ \forall (\tilde{x}_0, \dots, \tilde{x}_t) \in \tilde{\mathfrak{X}}_t, \\ & \qquad V(\bar{A}_{\text{et}}(\tilde{x}_t, \hat{x}_t) + \hat{B}\kappa(\hat{x}_{t_i}, t - t_i) + \hat{w}) \leq \nu^\tau V((\tilde{x}_{t+1-\tau}, \hat{x}_{t+1-\tau})) & (3.21) \\ 1 & \text{otherwise.} \end{cases}$$

For the self-triggered setup, define the scheduling functions $\mu_i$, $i \in \mathbb{N}$, by

$$\mu_i\big(\hat{x}_0, \dots, \hat{x}_t, \tilde{\mathfrak{X}}_{t_i}, t_0, t_1, \dots, t_i\big)$$
$$:= \max \left\{ t_i + 1, \ \sup \left\{ t' \in \{t_i + 2, \dots\} \ \left| \ \begin{array}{c} \forall k \in \{t_i + 2, \dots, t'\}, \\ \exists \tau_k \in \{1, \dots, \min\{k, \theta\}\}, \\ \forall \hat{w}[l] \in \mathscr{W}, \ l \in \{k - \tau_k, \dots, k-1\}, \\ \forall (\tilde{x}_0, \dots, \tilde{x}_{t_i}) \in \tilde{\mathfrak{X}}_{t_i}, \\ V(z[k]_i) \leq \nu^{\tau_k} V(z[k - \tau_k]_i), \\ \text{if } k \in \{t_i + 1, \dots, t' - 1\}, \\ V(\bar{z}[t']_i) \leq \nu^{\tau_{t'}} V(z[t' - \tau_{t'}]_i) \end{array} \right. \right\} \right\}.$$
$$(3.22)$$

Here, for all $i \in \{0, \dots, i_{\max}\}$,

$$z[t]_i := \bar{A}_{\text{st}}^{t-t_i} \begin{bmatrix} \tilde{x}_{t_i} \\ \hat{x}_{t_i} \end{bmatrix} + \sum_{l=t_i}^{t-1} \bar{A}_{\text{st}}^{t-1-l} \left( \hat{B}\kappa(\hat{x}_{t_i}, l - t_i) + \bar{D}_{\text{st}} \hat{w}[l] \right), \qquad (3.23)$$

$$\bar{z}[t]_i := \bar{A}_{\text{et}} \bar{A}_{\text{st}}^{t-1-t_i} \begin{bmatrix} \tilde{x}_{t_i} \\ \hat{x}_{t_i} \end{bmatrix} + \bar{A}_{\text{et}} \sum_{l=t_i}^{t-2} \bar{A}_{\text{st}}^{t-2-l} \left( \hat{B}\kappa(\hat{x}_{t_i}, l - t_i) + \bar{D}_{\text{st}} \hat{w}[l] \right)$$
$$+ \hat{B}\kappa(\hat{x}_{t_i}, t - 1 - t_i) + \hat{w}[t-1] \quad (3.24)$$

for all $t \in \{t_i + 1, \dots\}$, and

$$z[t]_i := \begin{bmatrix} \tilde{x}_t \\ \hat{x}_t \end{bmatrix} \qquad (3.25)$$

for all $t \in \{0, \dots, t_i\}$. Analogously to the state-feedback case, the quantities $z[t]_i$ and $\bar{z}[t]_i$ constitute estimates of the (joint) system state at time $t$, employed at the transmission instant $t_i$. However, there are two notable differences: first, the dynamics are not the same for the whole time span $\{t_i, \ldots, t_{i+1} - 1\}$, due to the assumed estimation structure discussed above, which leads to the difference between $z[t]_i$ and $\bar{z}[t]_i$. Second, in contrast to the state-feedback case, the state $x_t$ cannot be assumed to be known exactly at *any* point in time and the estimator state $\hat{x}_t$ is known at *every* point in time, which is why we employ (3.25) for all states in the past.

The resulting closed-loop system is given by

$$x_{t+1} = Ax_t + Bu_t + w_t \tag{3.26a}$$

$$y_t = Cx_t + v_t \tag{3.26b}$$

$$\hat{x}_{t+1} = A\hat{x}_t + Bu_t + \sigma_t L(C(A\hat{x}_t + Bu_t) - y_{t+1}) \tag{3.26c}$$

$$u_t = \kappa(\hat{x}_{t_i}, t - t_i), \qquad t \in \{t_i, \ldots, t_{i+1} - 1\}, \; i \in \{0, \ldots, i_{\max}\}, \tag{3.26d}$$

where the transmission instants $t_i$ are generated by

$$t_0 = 0, \quad t_{i+1} = \inf\{t \in \{t_i + 1, \ldots\} \mid \delta_t(\hat{x}_0, \hat{x}_1, \ldots, \hat{x}_{t-1}, \hat{x}_t, \tilde{\boldsymbol{\mathcal{X}}}_t, t_i) = 1\} \tag{3.27}$$

in the event-triggered case and by

$$t_0 = 0, \quad t_{i+1} = \mu_i(\hat{x}_0, \hat{x}_1 \ldots, \hat{x}_{t_i-1}, \hat{x}_{t_i}, \tilde{\boldsymbol{\mathcal{X}}}_{t_i}, t_0, t_1, \ldots, t_i) \tag{3.28}$$

in the self-triggered case. The value of $\sigma_t \in \{0, 1\}$ in (3.26c) is given by $\sigma_t = 1$ in the event-triggered case and by $\sigma_t = 1 \Leftrightarrow \exists i \in \{0, \ldots, i_{\max}\} : t = t_i$ in the self-triggered case, compare [Sinopoli et al., 2004].

The statements of Theorem 3.1 and Theorem 3.2 apply analogously to the output-feedback case, ensuring $\theta$-uniform asymptotic stability of the set $\gamma \bar{\mathscr{F}}_\infty$ for the respective closed-loop systems. Note that no assumptions on the set-valued estimator providing the sets $\tilde{\boldsymbol{\mathcal{X}}}_t$ were made besides that it provided correct bounds for the unknown estimation errors.

### 3.2.3. Computational aspects

In this section, we show how the Lyapunov functions employed in the previous section can be defined, whose existence was only assumed up to this point. Further, we discuss how the functions can be approximated: a major obstacle to employing a Lyapunov function for $\gamma \bar{\mathscr{F}}_\infty$ is that this set is, in general, not explicitly available. Even if $\mathscr{W}$ is a polytope, it might not be possible to define $\bar{\mathscr{F}}_\infty$ as the intersection of a finite number of half spaces, compare [Kolmanovsky and Gilbert, 1998, Remark 4.2]. We overcome this problem by finding appropriate overapproximations and underapproximations of the Lyapunov function in question, which are in turn obtained by inner and outer approximations of $\bar{\mathscr{F}}_\infty$. Finally, we describe how the event-generating and scheduling functions can be evaluated and discuss the computational effort involved. All methods described for the state-feedback case work analogously for the output-feedback case.

## Lyapunov functions

Consider first the existence of the function $V$.

**Lemma 3.1.** *Let a continuous function $W : \mathbb{R}^n \to \mathbb{R}^n$, $\mathscr{K}_\infty$-functions $\alpha_1$, $\alpha_2$, and a scalar $\lambda \in [0,1)$ be given such that $\alpha_1(|x|) \leq W(x) \leq \alpha_2(|x|)$ and $W((A+BK)x) \leq \lambda W(x)$ for all $x \in \mathbb{R}^{n_x}$. Let $\gamma \in [1,\infty)$ and define $V(x) = \min_{y \in \gamma \mathscr{F}_\infty} W(x-y)$ for all $x \in \mathbb{R}^{n_x}$. Then it holds that $\alpha_1(|x|_{\gamma \mathscr{F}_\infty}) \leq V(x) \leq \alpha_2(|x|_{\gamma \mathscr{F}_\infty})$ and $V((A+BK)x+w) \leq \lambda V(x)$ for all $x \in \mathbb{R}^{n_x}$ and all $w \in \mathscr{W}$.*

The existence of the function $W$ is ensured by the stability of $A + BK$. We still face the problem that the set $\gamma \mathscr{F}_\infty$ might not be explicitly available, that is, determined by a finite number of inequalities. Hence, in general the function $V$ as defined above is hard to evaluate. Lower and upper approximations of $V$ can be defined as follows.

**Proposition 3.1.** *Let $V$ and $W$ be defined as in Lemma 3.1. Let outer and inner approximating compact sets $\Omega_1, \Omega_2 \subseteq \mathbb{R}^{n_x}$ be given such that $\Omega_1 \supseteq \mathscr{F}_\infty \supseteq \Omega_2$ and define $V_1(x) = \min_{y \in \gamma \Omega_1} W(x-y)$ and $V_2(x) = \min_{y \in \gamma \Omega_2} W(x-y)$ for all $x \in \mathbb{R}^{n_x}$. Then it holds that $V_1(x) \leq V(x) \leq V_2(x)$ for all $x \in \mathbb{R}^{n_x}$.*

The statement follows immediately from the definitions.

With these approximations, we replace all expressions of the form

$$V(z_1) \leq \nu^\tau V(z_2) \tag{3.29}$$

in the various trigger conditions by

$$V_2(z_1) \leq \nu^\tau V_1(z_2). \tag{3.30}$$

Clearly, in the case these inequalities hold (and, hence, no transmission time is scheduled for the time point in question), a decrease of the actual Lyapunov function $V$ is still ensured, such that the stability properties remain guaranteed. However, depending on the tightness of the approximation $V_1(x) \leq V(x) \leq V_2(x)$, the average transmission rate in the closed-loop system might increase.

If $\mathscr{W}$ is a polyhedron, there are ways to obtain polyhedral outer and inner approximations of the set $\mathscr{F}_\infty$, see [Raković and Kouramas, 2006a] and the references therein, especially [Hirata and Ohta, 2003]. The function $W$ in Lemma 3.1 can be chosen as a convex quadratic function, for example by solving a Lyapunov equation, or as a convex piecewise affine function, see, for example, [Lazar, 2010]. Hence, evaluating the functions $V_1$ and $V_2$ becomes equivalent to solving a quadratic program or a linear program. While these classes of optimization problems can be solved efficiently, in certain applications it might be desirable that the functions $V_1$ and $V_2$ are available in an explicit form[2].

---

[2]What constitutes an "explicit form" is a question of definition. We assume that rational powers count as explicit.

For any compact set $\mathscr{Y} \subseteq \mathbb{R}^{n_x}$ containing the origin in its interior define the Minkowski gauge function $\Phi_{\mathscr{Y}} : \mathbb{R}^{n_x} \to \mathbb{R}$ by $\Phi_{\mathscr{Y}}(x) = \min\{s \in [0, \infty) \mid x \in s\mathscr{Y}\}$ for all $x \in \mathbb{R}^{n_x}$. See, for example, [Raković and Lazar, 2014] and the references therein for an overview of the properties of the Minkowski gauge function.

We have the following result.

**Lemma 3.2.** *Let $\mathscr{Y} \subseteq \mathbb{R}^{n_x}$ be a compact and convex set containing the origin in its interior, let $\zeta \in \mathscr{K}_\infty$, and let $W : \mathbb{R}^{n_x} \to \mathbb{R}$, $x \mapsto \zeta(\Phi_{\mathscr{Y}}(x))$. Then it holds that*

$$\min_{y \in c\mathscr{Y}} W(x - y) = \zeta(\max\{0, \zeta^{-1}(W(x)) - c\}) \tag{3.31}$$

*for all $x \in \mathbb{R}^{n_x}$ and any $c \in (0, \infty)$.*

As the Minkowski gauge function is itself defined by an optimization problem, the benefits of Lemma 3.2 are not immediately obvious. However, convex quadratic functions and convex piecewise linear functions can be defined in terms of Minkowski gauge functions:

**Proposition 3.2.** *(i) Let $Q \in \mathbb{R}^{n_x \times n_x}$ be a symmetric positive definite matrix and define $V_Q : \mathbb{R}^{n_x} \to \mathbb{R}$, $x \mapsto x^\mathsf{T} Q x$. Then it holds that $V_Q(x) = (\Phi_{\mathcal{Q}}(x))^2$ for all $x \in \mathbb{R}^{n_x}$, where $\mathcal{Q} = \{x \in \mathbb{R}^{n_x} \mid x^\mathsf{T} Q x \leq 1\}$. (ii) Let $P_i \in \mathbb{R}^{1 \times n_x}$, $i \in \{1, \dots, p\}$, be such that the function $V_P : \mathbb{R}^{n_x} \to \mathbb{R}$, $x \mapsto \max\{P_i x \mid i \in \{1, \dots, p\}\}$ is positive definite. Then it holds that $V_P(x) = \Phi_{\mathscr{P}}(x)$ for all $x \in \mathbb{R}^{n_x}$, where $\mathscr{P} = \{x \in \mathbb{R}^{n_x} \mid P_i x \leq 1, \ i \in \{1, \dots, p\}\}$.*

In both cases, the statement follows readily from the definitions, compare also [Blanchini et al., 1995]. Hence, for convex quadratic or piecewise linear functions, the evaluation of $V_1$ and $V_2$ becomes especially simple if $\Omega_1$ and $\Omega_2$ are chosen as scaled versions of the sets $\mathcal{Q}$ and $\mathscr{P}$, respectively. If *any* polytopic sets $\bar{\Omega}_1, \bar{\Omega}_2 \subseteq \mathbb{R}^{n_x}$ containing the origin in their interiors are known with $\bar{\Omega}_1 \supseteq \mathscr{F}_\infty \supseteq \bar{\Omega}_2$, appropriate scalars $c_1, c_2 \in (0, \infty)$ such that $c_1 \mathcal{Q} \supseteq \mathscr{F}_\infty \supseteq c_2 \mathcal{Q}$ (or $c_1 \mathscr{P} \supseteq \mathscr{F}_\infty \supseteq c_2 \mathscr{P}$) can be found with semi-definite (or linear) programming.

**Event-generating and scheduling functions**

The way the event generating functions $\delta_t$ and the scheduling functions $\mu_i$ are defined in (3.7) and (3.8), the computational effort and storage requirements involved with their evaluation increases with the cardinality of the set $\{1, \dots, \min\{t + 1, \theta\}\}$ and is unbounded if $\theta = \infty$. In order to reduce the computational effort, one may check the inequalities in the trigger conditions only for a subset of $\{1, \dots, \min\{t + 1, \theta\}\}$ and generate a transmission instant if the respective inequality is violated for all members of this subset. While introducing conservatism, the stability guarantees are retained. In the following, we discuss several possible choices of the subset and their interpretations. In particular, assume $\{1, \dots, \min\{t + 1, \theta\}\}$ in the definition of $\delta_t$ for $t \in \{t_i + 1, \dots, t_{i+1} - 1\}$ to be replaced with $\mathscr{R}(t + 1, t_i)$ for a set-valued function $\mathscr{R} : \mathbb{N} \times \mathbb{N} \to 2^\mathbb{N}$. In the scheduling functions $\mu_i$ for $i \in \mathbb{N}$ we equivalently assume $\{1, \ldots, \min\{k, \theta\}\}$ to be replaced with $\mathscr{R}(k, t_i)$. Below, we only discuss the event-triggered case. We refer to the inequalities that have to be checked in both (3.7) and (3.8) as "trigger conditions". Further, we redefine $\theta$ to be equal to $\sup\{\mathscr{R}(t+1, t_i) \mid t \in \mathbb{N}, \ i \in \mathbb{N}\}$ in the following, which is consistent with the stability results. The special cases of the functions $\mathscr{R}$ below require only a vector of bounded dimension to be stored in the controller:

1. $\mathscr{R}(t+1, t_i) = \{1\}$.

   With this function, the controller requires only the state at the current point in time and ensures 1-uniform asymptotic stability of $\gamma \mathscr{F}_\infty$ in the closed-loop system.

2. $\mathscr{R}(t+1, t_i) = \{t+1\}$.

   This function requires only the initial state to be stored. Considering that the trigger conditions guarantee $V(x_{t+1}) \leq \nu^{t+1} V(x_0)$ for *any* choice of $\mathscr{R}$, this particular definition offers the most relaxed trigger conditions, if the overapproximation and underapproximation of $V$ is neglected. On the other hand, it only guarantees $\infty$-uniform asymptotic stability.

3. $\mathscr{R}(t+1, t_i) = \{\min\{t+1, \hat{\theta}\}\}$ for a $\hat{\theta} \in \{1, \ldots\}$.

   With this function, the storage requirements in the controller are not reduced, but the computational effort reduces to validating the inequality in the trigger conditions for a single $\tau$. A controller with this set guarantees $\hat{\theta}$-uniform asymptotic stability.

4. $\mathscr{R}(t+1, t_i) = \{t+1 - t_i\}$.

   With this function the controller requires only the states at the current time step and at the last transmission instant. Hence, both the storage requirements and the computational effort are considerably reduced. If $t_{i+1} - t_i \leq \Delta_{\max}$ is artificially enforced for an *a priori* chosen $\Delta_{\max} \in \{1, \ldots\} \cup \{\infty\}$, then the controller ensures $\Delta_{\max}$-uniform asymptotic stability.

5. $\mathscr{R}(t+1, t_i) = \{t+1-t_j \mid t_j \in \{t_0, t_1, \ldots, t_i\}, \ t+1-t_j \leq \hat{\theta}\}$ for a $\hat{\theta} \in \{1, \ldots\} \cup \{\infty\}$.

   This function provides a relaxed version of case 4 and guarantees $\hat{\theta}$-uniform asymptotic stability. The storage requirements and computational effort involved with evaluating this condition are higher than in case 4, but still only the states at transmission instants need to be stored.

We restrict the following discussion to the case where $\mathscr{W}$ is a polytope and the functions $V_1$ and $V_2$ are convex. Hence, in order to evaluate the event-generating and scheduling functions, inequalities of the form $\forall z \in \mathscr{Z}, f_2(y(z)) \leq f_1(x(z))$ have to be verified, where $f_1$ and $f_2$ are convex functions, $x$ and $y$ are affine functions and $\mathscr{Z}$ is a polytope. In the event-triggered case, the function $x$ is constant such that the inequality can be verified by evaluating $f_2(y(z))$ at the vertices of $\mathcal{Z}$, as a convex function attains its maximum on a polytope at one of its vertices. This statement also holds in the self-triggered case if $k - \tau_k = t_j$ for some $j \in \{0, \ldots, i\}$ in (3.8). This condition can be ensured by a careful choice of $\mathcal{R}$, for example case 4 and case 5 in the list above. The issue here is that we assumed that $x_t$ is only available as a measurement at the transmission instants in the self-triggered setting. If it does not hold that $t + 1 - \tau = t_j$ for some $j \in \{0, \ldots, i\}$, both $x$ and $y$ are functions of $z$, such that evaluating the trigger condition requires computing the maximum of a function that is defined as the difference between two convex functions, and, hence, is in general neither convex nor concave. As computing this maximum is a hard task, we propose the following convex relaxation to the problem. Instead of checking whether $\max_{z \in \mathcal{Z}} f_2(y(z)) - f_1(x(z)) \leq 0$, we may check whether $\max_{z_2 \in \mathcal{Z}} f_2(y(z_2)) - \min_{z_1 \in \mathcal{Z}} f_1(x(z_1)) \leq 0$. While introducing conservatism and still involving more computational effort than in the case where $x$ is constant, this approach allows the trigger condition to be evaluated with convex programming techniques.

In the event-triggered case, the computational complexity is determined by the number of vertices of the set $\mathcal{Z} = \mathcal{W}$, as this many function evaluations are required in order to verify the inequality. In the self-triggered case, assuming for the moment that $x$ is also a constant here, we have that $\mathcal{Z} = \bigoplus_{j=0}^{t-t_i} A^j \mathcal{W}$ for each $t \in \{t_i, \ldots\}$ that is checked, such that in the worst case the number of vertices of $\mathcal{Z}$ grows exponentially with $t$. Overapproximating $\mathcal{Z}$ with polytopes of lower complexity[3] could help in reducing the computational effort.

**Output feedback**

We assume that the set-valued estimates $\tilde{\mathcal{X}}_t$ are such that polytopic bounds for the arguments of the functions $V_1$ and $V_2$ in (3.21) and (3.22) are available for each point in time, and, in particular, that the vertices of these sets are known. Then, the same discussion as above applies, that is, the trigger conditions amount to checking whether $\max_{z \in \mathcal{Z}} f_2(y(z)) - f_1(x(z)) \leq 0$ for a polytope $\mathcal{Z}$ and convex functions $f_1$ and $f_2$. In contrast to the state-feedback case, we never have the case that the functions $x$ or $y$ are constant, except if the set-valued estimation returns a singleton set at some point in time. Hence, in the output feedback case one is always faced with the need to solve non-convex optimization problems or to use possibly conservative relaxations.

### 3.2.4. Numerical examples

In the following, we investigate the behavior of the closed-loop system with the proposed event-triggered and self-triggered controllers in the state-feedback case.[4] Consider a

---

[3]This introduces conservatism, but, by the convexity of the functions evaluated, still guarantees stability.

[4]Because of the difficulties arising from non-convexity—as discussed in Section 3.2.3—we postpone numerical examples for the output-feedback case to the next section, where these problems are

system as in (3.1a) with system matrices

$$A = \begin{bmatrix} 1 & 0.3 \\ 0 & 1 \end{bmatrix}, \ B = \begin{bmatrix} 0.045 \\ 0.3 \end{bmatrix}, \tag{3.32}$$

which correspond to a continuous-time double integrator system discretized with a sample time of 0.3. The disturbance bound was chosen to $\mathscr{W} = [-1, 1] \times [-1, 1]$. We computed the linear feedback gain LQ-optimal, using the weighting matrices $Q = \begin{bmatrix} 1 & 0 \\ 0 & 1 \end{bmatrix}$ and $R = 1$ to obtain $K = [-0.7719 \ -1.4628]$. The matrix describing the optimal infinite horizon cost function resulted in $P = \begin{bmatrix} 6.32 & 3.37 \\ 3.37 & 6.38 \end{bmatrix}$. We used the algorithm in [Raković et al., 2005] to obtain an outer bound $\Omega_1 \supseteq \mathscr{F}_\infty$ by selecting the parameter $\epsilon$ in the algorithm as 0.1. This resulted in $\Omega_1 = \frac{1}{1-\alpha} \bigoplus_{i=0}^{s-1} (A + BK)^i \mathscr{W}$ with $s = 20$ and $\alpha = 0.0083$. As an inner approximation, we chose $\Omega_2 = \bigoplus_{i=0}^{s-1} (A + BK)^i \mathscr{W}$. These approximating sets are polyhedrons described by 80 hyperplanes each. We defined the functions $V_1$ and $V_2$ by $V_i(x) = \min_{y \in \gamma \Omega_i} (x - y)^\mathsf{T} P(x - y)$ for $i \in \{1, 2\}$, as suggested by Proposition 3.1. We evaluated these functions by solving quadratic programs. In the scheduling functions for self-triggered control, we evaluated the maximum value for $V(x[k]_i)$ by evaluating the expression $V_2(A^{k-t_i} x_{t_i} + \sum_{j=0}^{k-t_i-1} A^{k-t_i-1-j} B\kappa(x_{t_i}, j) + \tilde{w})$ for every $\tilde{w}$ in the set of vertices of $\bigoplus_{j=0}^{k-t_i-1} A^k \mathscr{W}$. We enforced $t_{i+1} - t_i \leq \Delta_{\max}$ with $\Delta_{\max} = 20$ in the self-triggered case; the number of vertices of the set $\bigoplus_{j=0}^{19} A^j \mathscr{W}$ was 42. Further, in order to avoid the non-convexity issues described in Section 3.2.3, we chose the function $\mathscr{R}$ of type 5 as described therein for the self-triggered case, with $\hat{\theta} = \theta$; that is, we only compare the worst case future value of $V_2$ to values of $V_1$ at states that are known to the controller, more specifically, states at transmission instants. Thereby, we further restrict the time between transmission instants to be upperbounded by $\theta$. In the event-triggered case, we do not have non-convexity problems and allow $V_1$ to be evaluated at arbitrary states in the past, restricted by $\theta$. Because of possible numerical inaccuracies in evaluating $V_1$ and $V_2$, which become especially pronounced for system states close to or contained in $\gamma \mathscr{F}_\infty$, we artificially increased the right-hand side of all inequalities of the type $V_2(z_1) \leq \nu^\tau V_1(z_2)$ by 0.001. The case $\theta = \infty$ was treated by only considering $\tau = t + 1$ in the event-generating function (3.7) and $\tau_k = k$ in the scheduling function (3.8). Finally, in order to quantify the decrease rate of the Lyapunov function defined by $V(x) = \min_{y \in \gamma \mathscr{F}_\infty} (x - y)^\mathsf{T} P(x - y)$ under the all-time triggered control $u_t = Kx_t$, we solved the semi-definite program $\lambda = \min\{\lambda' \in [0, \infty) \mid (A + BK)^\mathsf{T} P(A + BK) \preceq \lambda' P\}$, resulting in $\lambda = 0.757$.

In Table 3.1 and Table 3.2 we report the average transmission rates (number of transmissions during the simulation divided by the number of time points in the simulation) for various values of $\theta$ and $\gamma$ and different choices of $\kappa$. The parameter $\nu$ was chosen to 0.8 in all simulations, where the system was initialized at random points in $[-1000, 1000] \times [-1000, 1000]$. The disturbances were generated by sampling $w_t$ uniformly and independently on the set $\mathscr{W}$. For every initial condition we sampled one

---

absent.

random realization of the disturbance sequence. We kept these sequences and initial conditions identical for the various parameter combinations. As can be seen from these results, the average transmission rate decreases with increasing $\theta$ and $\gamma$. In Table 3.3 and Table 3.4 we report average transmission rates for the system state being initialized at the origin. Note that if $x_0 \in \gamma\mathscr{F}_\infty$, then $V(x_t) = 0$ for all $t \in \mathbb{N}$ in the closed-loop system and, hence, the parameters $\nu$ and $\theta$ have no influence on the transmission rate (beyond the restriction $t_{i+1} - t_i \leq \theta$ in the self-triggered case). Hence, we only report results for $\theta = 5$. The comparison of the asymptotic and the transient behavior of the closed-loop systems by these results show that the influence of the parameter $\gamma$ on the transmission rate depends on the distance of the state to the set $\gamma\mathscr{F}_\infty$: if the state is far away from $\gamma\mathscr{F}_\infty$, the value of the functions $V_1$ and $V_2$ does not change much (relatively) with $\gamma$, reducing the influence of this parameter on the transmission rate.

In Table 3.5, we report average transmission rates for different choices of $\nu$ and an initial condition far away from $\gamma\mathscr{F}_\infty$. As expected, an increase of $\nu$ leads to a reduction of the average transmission rate. Note that this reduction can be achieved even for $\nu < \lambda$.

In Figure 3.3, we depict the values of $V_2(x_t)$ over time for a simulation with $\theta = \infty$, $\gamma = 1$, and $\nu = 0.99$, showing the non-monotonic behavior of the function over time.

Finally, we want to illustrate the influence of the parameter $\theta$ on the stability properties of the closed-loop system. By Definition 2.2, if $\theta = 1$ and $x_{t^0} \in \gamma\mathscr{F}_\infty$ for some $t^0 \in \mathbb{N}$ in the closed-loop system (for any of the controllers considered in this chapter so far), then it must also necessarily hold that $x_t \in \gamma\mathscr{F}_\infty$ for all $t \in \{t^0 + 1, \ldots\}$. In other words, the set $\gamma\mathscr{F}_\infty$ is positively invariant for the closed-loop system. For $\theta \in \{2, \ldots\} \cup \{\infty\}$, however, this is not necessarily the case. In Figure 3.4, we depict state trajectories from the simulations associated with Table 3.2 demonstrating this fact.

Table 3.1.: Average transmission rates in the closed-loop system with event-triggered control over 30 random simulations with 30 time points each and random initial states.

| | to-zero control | | | | | | | to-hold control | | | | | | | model-based control | | | | | |
|---|---|---|---|---|---|---|---|---|---|---|---|---|---|---|---|---|---|---|---|---|
| | $\gamma$ | | | | | | | $\gamma$ | | | | | | | $\gamma$ | | | | | |
| $\theta$ | 1 | 1.25 | 1.5 | 2 | 2.5 | 3 | $\theta$ | 1 | 1.25 | 1.5 | 2 | 2.5 | 3 | $\theta$ | 1 | 1.25 | 1.5 | 2 | 2.5 | 3 |
| 1 | 0.58 | 0.53 | 0.50 | 0.47 | 0.46 | 0.43 | 1 | 0.25 | 0.24 | 0.25 | 0.25 | 0.23 | 0.22 | 1 | 0.10 | 0.07 | 0.06 | 0.05 | 0.05 | 0.04 |
| 2 | 0.48 | 0.46 | 0.44 | 0.40 | 0.38 | 0.36 | 2 | 0.24 | 0.24 | 0.24 | 0.24 | 0.24 | 0.24 | 2 | 0.08 | 0.06 | 0.06 | 0.05 | 0.05 | 0.04 |
| 5 | 0.36 | 0.35 | 0.34 | 0.33 | 0.32 | 0.31 | 5 | 0.25 | 0.25 | 0.25 | 0.24 | 0.24 | 0.24 | 5 | 0.07 | 0.06 | 0.06 | 0.05 | 0.05 | 0.04 |
| 10 | 0.32 | 0.31 | 0.31 | 0.30 | 0.29 | 0.29 | 10 | 0.24 | 0.23 | 0.24 | 0.23 | 0.23 | 0.23 | 10 | 0.06 | 0.06 | 0.06 | 0.05 | 0.05 | 0.04 |
| $\infty$ | 0.30 | 0.29 | 0.29 | 0.28 | 0.28 | 0.27 | $\infty$ | 0.24 | 0.24 | 0.24 | 0.24 | 0.24 | 0.24 | $\infty$ | 0.04 | 0.04 | 0.04 | 0.03 | 0.03 | 0.03 |

Table 3.2.: Average transmission rates in the closed-loop system with self-triggered control over 30 random simulations with 30 time points each and random initial states.

to-zero control

| $\theta$ | $\gamma$ 1 | 1.25 | 1.5 | 2 | 2.5 | 3 |
|---|---|---|---|---|---|---|
| 1 | 1.00 | 1.00 | 1.00 | 1.00 | 1.00 | 1.00 |
| 2 | 0.73 | 0.69 | 0.66 | 0.63 | 0.61 | 0.59 |
| 5 | 0.45 | 0.44 | 0.44 | 0.42 | 0.40 | 0.40 |
| 10 | 0.37 | 0.36 | 0.36 | 0.35 | 0.34 | 0.33 |
| $\infty$ | 0.33 | 0.33 | 0.32 | 0.31 | 0.31 | 0.30 |

to-hold control

| $\theta$ | $\gamma$ 1 | 1.25 | 1.5 | 2 | 2.5 | 3 |
|---|---|---|---|---|---|---|
| 1 | 1.00 | 1.00 | 1.00 | 1.00 | 1.00 | 1.00 |
| 2 | 0.51 | 0.50 | 0.50 | 0.50 | 0.50 | 0.50 |
| 5 | 0.35 | 0.30 | 0.30 | 0.27 | 0.26 | 0.26 |
| 10 | 0.25 | 0.25 | 0.25 | 0.24 | 0.24 | 0.24 |
| $\infty$ | 0.24 | 0.24 | 0.24 | 0.24 | 0.24 | 0.24 |

model-based control

| $\theta$ | $\gamma$ 1 | 1.25 | 1.5 | 2 | 2.5 | 3 |
|---|---|---|---|---|---|---|
| 1 | 1.00 | 1.00 | 1.00 | 1.00 | 1.00 | 1.00 |
| 2 | 0.54 | 0.50 | 0.50 | 0.50 | 0.50 | 0.50 |
| 5 | 0.34 | 0.28 | 0.27 | 0.23 | 0.20 | 0.20 |
| 10 | 0.25 | 0.22 | 0.21 | 0.17 | 0.15 | 0.14 |
| $\infty$ | 0.09 | 0.08 | 0.08 | 0.08 | 0.07 | 0.07 |

Table 3.3.: Average transmission rates in the closed-loop system with event-triggered control over 10 random simulations with 300 time points each and initial states at the origin.

to-zero control

| $\theta$ | $\gamma$ 1 | 1.25 | 1.5 | 2 | 2.5 | 3 |
|---|---|---|---|---|---|---|
| 5 | 0.11 | 0.08 | 0.06 | 0.04 | 0.03 | 0.03 |

to-hold control

| $\theta$ | $\gamma$ 1 | 1.25 | 1.5 | 2 | 2.5 | 3 |
|---|---|---|---|---|---|---|
| 5 | 0.20 | 0.20 | 0.21 | 0.20 | 0.20 | 0.19 |

model-based control

| $\theta$ | $\gamma$ 1 | 1.25 | 1.5 | 2 | 2.5 | 3 |
|---|---|---|---|---|---|---|
| 5 | 0.08 | 0.05 | 0.04 | 0.03 | 0.03 | 0.02 |

Table 3.4.: Average transmission rates in the closed-loop system with self-triggered control over 10 random simulations with 300 time points each and initial states at the origin.

to-zero control

| $\theta$ | $\gamma$ 1 | 1.25 | 1.5 | 2 | 2.5 | 3 |
|---|---|---|---|---|---|---|
| 5 | 0.53 | 0.40 | 0.35 | 0.26 | 0.21 | 0.20 |

to-hold control

| $\theta$ | $\gamma$ 1 | 1.25 | 1.5 | 2 | 2.5 | 3 |
|---|---|---|---|---|---|---|
| 5 | 0.50 | 0.34 | 0.33 | 0.26 | 0.22 | 0.22 |

model-based control

| $\theta$ | $\gamma$ 1 | 1.25 | 1.5 | 2 | 2.5 | 3 |
|---|---|---|---|---|---|---|
| 5 | 0.50 | 0.34 | 0.33 | 0.24 | 0.20 | 0.20 |

Table 3.5.: Average transmission rate over 30 random simulations for a model-based controller, $\gamma = 1$, $\theta = 5$, $x_0 = (1000, 1000)$, and 30 time points in the simulation.

| $\nu$ | 0.5 | 0.6 | 0.7 | 0.8 | 0.9 | 0.99 |
|---|---|---|---|---|---|---|
| event-triggered | 0.53 | 0.36 | 0.18 | 0.06 | 0.06 | 0.06 |
| self-triggered | 0.76 | 0.60 | 0.41 | 0.30 | 0.30 | 0.29 |

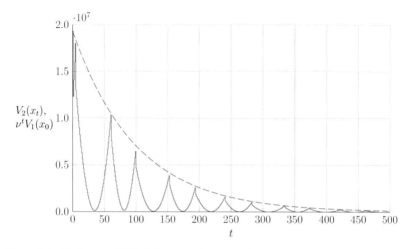

Figure 3.3.: Evolution of $V_2(x_t)$ for $x_0 = (1000, 1000)$, $\theta = \infty$, $\gamma = 1$, $\nu = 0.99$, and a "to-zero" controller (blue, solid). For comparison, $\nu^t V_1(x_0)$ is also depicted (orange, dashed). The trigger rule in this case is based on enforcing exactly $V_2(x_t) \leq \nu^t V_1(x_0)$.

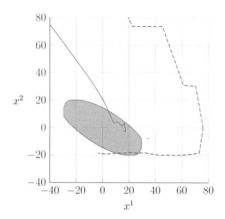

Figure 3.4.: Set $\gamma\Omega_1$ (outer approximation of $\gamma\mathscr{F}_\infty$) and state trajectories for the "to-zero" event-triggered controller with $\gamma = 3$, $\nu = 0.8$, $\theta = 1$ (blue, solid) and $\theta = \infty$ (violet, dashed). The initial condition and disturbance realizations were identical in both simulations.

41

# 3.3. Set-based approach

In this section, we provide a modification of the trigger conditions proposed in the last section, which were defined in terms of Lyapunov functions. Here, we instead define the conditions in terms of set inclusions, polytope inclusions in a particular case, which can be verified efficiently. In particular, we employ Theorem 2.3, which allows stability to be guaranteed by checking set-membership conditions for the sets $\gamma \mathscr{F}_i$, $i \in \mathbb{N}$. These sets have the properties described in Section 2.2 and, additionally, are convex as $\mathscr{W}$ is convex. The main benefit of this approach is that checking set memberships is expected to be computationally less demanding than checking the Lyapunov decrease conditions in Section 3.2. As we will see, the non-convexity issues appearing in the Lyapunov-based trigger conditions in the last section do not appear in the set-based approaches considered in this section.

In contrast to the previous section, we do not have access to the guaranteed rate of convergence; only the size of the stabilized set, in the form of $\gamma \in [1, \infty)$, and the uniformity parameter $\theta \in \{1, \dots\} \cup \{\infty\}$ in the stability definition are available for tuning.

## 3.3.1. State feedback

Consider the following event-generating functions and scheduling functions, defined analogously to the functions in Section 3.2. Instead of guaranteeing a decrease of the Lyapunov function for future points in time under the worst-case disturbance, we instead guarantee that a certain set-membership condition holds.

**Event-triggered control**

Define the event-generating functions $\delta_t$, $t \in \{1, \dots\}$, by

$$\delta_t(x_0, x_1, \dots, x_t, t_i) = \begin{cases} 0 & \text{if } \exists \tau \in \{1, \dots, \min\{t+1, \theta\}\},\ \forall w \in \mathscr{W}, \\ & \quad Ax_t + BK\kappa(x_{t_i}, t-t_i) + w \in \{(A+BK)^\tau x_{t+1-\tau}\} \oplus \gamma \mathscr{F}_\tau \\ 1 & \text{otherwise.} \end{cases}$$

(3.33)

We have the following result.

**Theorem 3.3.** *For any dynamical system generating* $(x_t)_{t \in \mathbb{N}}$ *for the closed loop consisting of* (3.1a), (3.3), *and* (3.33), *the set* $\gamma \mathscr{F}_\infty$ *is* $\theta$*-uniformly asymptotically stable with region of attraction* $\mathbb{R}^{n_x}$.

*Proof.* The statement follows from Theorem 2.3 by noting that the inclusion $x_{t+1} \in \{(A+BK)^\tau x_{t+1-\tau}\} \oplus \gamma \mathscr{F}_\tau$ is either verified explicitly (if $\delta_t(x_0, \dots, x_t, t_i) = 0$), or implicitly (otherwise), as then $u_t = Kx_t$ and, hence, $x_{t+1} \in \{(A+BK)x_t\} \oplus \mathscr{W} \subseteq \{(A+BK)x_t\} \oplus \gamma \mathscr{F}_1$, exploiting the fact that $\mathscr{W}$ is convex and contains the origin. $\square$

**Self-triggered control**

Define the scheduling functions $\mu_i$, $i \in \mathbb{N}$, by

$$\mu_i(x_{t_0}, x_{t_1}, \ldots, x_{t_i}, t_0, t_1, \ldots, t_i)$$

$$:= \sup \left\{ t' \in \{t_i+1, \ldots\} \ \middle| \ \begin{array}{l} \forall k \in \{t_i+1, \ldots, t'\}, \\ \exists \tau_k \in \{1, \ldots, \min\{k, \theta\}\}, \\ \forall w[l] \in \mathscr{W}, \ l \in \{t_j, \ldots, k-1\}, \\ x[k]_i \in \{(A+BK)^{\tau_k} x[k-\tau_k]_i\} \oplus \gamma \mathscr{F}_{\tau_k}) \end{array} \right\} \quad (3.34)$$

where, as in (3.8), for all $i \in \{0, \ldots, i_{\max}\}$,

$$x[k]_i := A^{k-t_i} x_{t_i} + \sum_{l=t_i}^{k-1} A^{k-1-l}(B\kappa(x_{t_i}, l - t_i) + w[l]) \quad (3.35)$$

for all $k \in \{t_i, \ldots\}$ and

$$x[k-\tau_k]_i := A^{k-\tau_k-t_j} x_{t_j} + \sum_{l=t_j}^{k-\tau_k-1} A^{k-\tau_k-1-l}(B\kappa(x_{t_j}, l - t_j) + w[l]) \quad (3.36)$$

with $j \in \mathbb{N}$ such that $k - \tau_k \in \{t_j, \ldots, t_{j+1} - 1\}$. We have the following result.

**Theorem 3.4.** *For any dynamical system generating $(x_t)_{t \in \mathbb{N}}$ for the closed loop consisting of (3.1), (3.3), and (3.34), the set $\gamma \mathscr{F}_\infty$ is $\theta$-uniformly asymptotically stable with region of attraction $\mathbb{R}^{n_x}$.*

The proof follows analogously to the proof of Theorem 3.3.

## 3.3.2. Analysis of given aperiodic schemes

The results in the previous section are of a synthetic nature, that is, they allow the design of controllers with a given worst-case asymptotic bound. Here, we show exemplarily how these methods can be used to obtain, *a posteriori*, worst-case asymptotic bounds for certain aperiodic control schemes.

Consider the event-triggered controller defined by

$$u_t = K(A+BK)^{t-t_i} x_{t_i}, \qquad t \in \{t_i, \ldots, t_{i+1} - 1\}, \ i \in \{0, \ldots, i_{\max}\} \quad (3.37a)$$

$$t_0 = 0 \quad (3.37b)$$

$$t_{i+1} = \inf\{t \in \{t_i+1, \ldots\} \mid x_t \notin \{(A+BK)^{t-t_i} x_{t_i}\} \oplus \mathscr{T}_{t-t_i}\}, \quad (3.37c)$$

with the *threshold sets* $\mathscr{T}_\tau \subseteq \mathbb{R}^{n_x}$ for $\tau \in \mathbb{N}$. This event-triggered control scheme is "model-based" and uses the function $\mathscr{R}$ of type 4 as described in Section 3.2.3. Schemes falling in this framework were, for example, proposed in [Lunze and Lehmann, 2010, Garcia and Antsaklis, 2013, Heemels and Donkers, 2013]. We have the following result.

**Theorem 3.5.** *Let $\mathcal{T}_0 = \{0\}$ and define the sets $\mathcal{H}_\tau \subseteq \mathbb{R}^{n_x}$ for $\tau \in \mathbb{N}$ with $\mathcal{H}_0 = \{0\}$ and*

$$\mathcal{H}_{\tau+1} = A(\mathcal{H}_\tau \cap \mathcal{T}_\tau) \oplus \mathcal{W} \tag{3.38}$$

*for $\tau \in \{1, \ldots\}$ with the convention that $\mathcal{H}_{\bar{\tau}} = \emptyset \Rightarrow \mathcal{H}_\tau = \emptyset$ for any $\bar{\tau}, \tau \in \mathbb{N}$ with $\tau > \bar{\tau}$. If $\gamma = \min\{c \in [1, \infty) \mid \mathcal{H}_\tau \subseteq c\mathcal{F}_\tau, \ \tau \in \mathbb{N}\}$, then the set $\gamma\mathcal{F}_\infty$ is $\theta$-uniformly asymptotically stable with region of attraction $\mathbb{R}^{n_x}$ for any dynamical system generating $(x_t)_{t \in \mathbb{N}}$ for the closed loop consisting of (3.1a) and (3.37), where $\theta = \sup\{t_{i+1} - t_i \mid i \in \{0, \ldots, i_{\max}\}\}$. Further, if $\mathcal{W}$ is convex then, for any $c \in [1, \gamma)$, the set $c\mathcal{F}_\infty$ is not asymptotically stable for the closed-loop system.*

**Remark 3.5.**

(i) *The bound obtained in Theorem 3.5 is similar to the bound proposed in [Heemels et al., 2008] in that it is defined by $\mathcal{F}_\infty$, scaled by a certain factor. The particular scaling, however, is defined in a different way.*

(ii) *The bound is still an overapproximation of the actual minimal robust positive invariant set for the closed-loop system under the event-triggered controller. Considering that the closed-loop system is nonlinear, we expect the task of obtaining the exact minimal robust positive invariant set to be difficult in general.*

**Example**

In [Grüne et al., 2010, Lehmann et al., 2013], an asymptotic bound on the system state for a closed-loop system of the type in (3.37) with $\mathcal{T}_\tau = \mathcal{T}$, $\tau \in \mathbb{N}$ was obtained by rewriting the closed-loop dynamics under event-triggered control as (reported here in our own notation)

$$x_{t+1} = (A + BK)x_t + w_t - BK(x_t - (A + BK)^{t-t_i}x_{t_i}),$$
$$t \in \{t_i, \ldots, t_{i+1} - 1\}, \ i \in \{0, \ldots, i_{\max}\} \tag{3.39}$$

the term $x_t - (A + BK)^{t-t_i}x_{t_i}$ was considered as an additional additive disturbance, which—by the trigger conditions—is bounded by

$$x_t - (A + BK)^{t-t_i}x_{t_i} \in A\mathcal{T} \oplus \mathcal{W}. \tag{3.40}$$

From this, it follows that the set

$$\mathcal{F}' := \bigoplus_{j=0}^{\infty}(A + BK)^j(\mathcal{W} \oplus BK(A\mathcal{T} \oplus \mathcal{W})) \tag{3.41}$$

is 1-uniformly asymptotically stable for the dynamical system generating $(x_t)_{t \in \mathbb{N}}$ for the closed loop. Note that in [Lehmann et al., 2013] an outer bound of this set is obtained in the form $r\mathscr{B}_{n_x}$ for an $r \in [0, \infty)$.

Consider now the parameters

$$A = \begin{bmatrix} 1 & 2 \\ 0 & 3 \end{bmatrix}, \ B = \begin{bmatrix} 0.5 \\ 1 \end{bmatrix}, \tag{3.42}$$

$K = [-0.2603 \ -2.9269]$ (LQ optimal gain for identity weighting matrices), $\mathscr{W} = [-1, 1] \times [-1, 1]$, and $\mathscr{T} = [-1, 1] \times [-1, 1]$. We computed an upper bound $\bar{\gamma}$ on $\gamma$ in Theorem 3.5 by replacing $\mathscr{H}_\tau$ with $A\mathscr{T} \oplus \mathscr{W}$ and $\mathscr{F}_\tau$ with $\mathscr{F}_{10}$ for $\tau \geq 10$ (leading to an optimization problem with a finite number of constraints). This yielded $\bar{\gamma} = 3$. In Figure 3.5, we depict an *inner* approximation of $\mathscr{F}'$, namely $\bigoplus_{j=0}^{9}(A + BK)^j(\mathscr{W} \oplus BK(A\mathscr{T} \oplus \mathscr{W}))$ and an *outer* approximation of $\bar{\gamma} \bigoplus_{j=0}^{\infty}(A + BK)^j\mathscr{W}$ (computed with the algorithm in [Raković et al., 2005] for $\epsilon = 0.1$), demonstrating the reduction of conservatism achieved by our method. Presumably, the way the open-loop matrix $A$ enters the approximation in [Grüne et al., 2010, Lehmann et al., 2013] is the main source of this conservatism; in Figure 3.6 we depict the same sets, but this time computed for

$$A = \begin{bmatrix} 0.1 & 0.2 \\ 0 & 0.3 \end{bmatrix} \tag{3.43}$$

and $K = [-0.0222 \ -0.1820]$. For these parameters, the method proposed here loses its advantage, leading to an asymptotic bound of roughly the same size.

**Remark 3.6.** *We need to stress that the comparison here is only meaningful in terms of the asymptotic bound, but not necessarily in terms of stability: the set $\mathscr{F}'$ is 1-uniformly asymptotically stable, whereas the set $\gamma\mathscr{F}_\infty$ is only $\theta$-uniformly asymptotically stable for some $\theta \in \{1, \ldots\} \cup \{\infty\}$.*

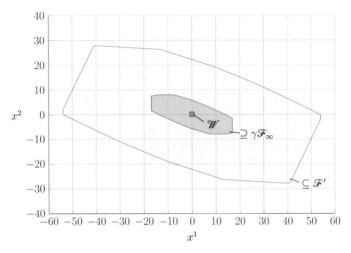

Figure 3.5.: Inner approximation of $\mathscr{F}'$ (yellow) and outer approximation of $\gamma\mathscr{F}_\infty$ (green) for an unstable open loop matrix $A$. The set $\mathscr{W}$ is depicted in red.

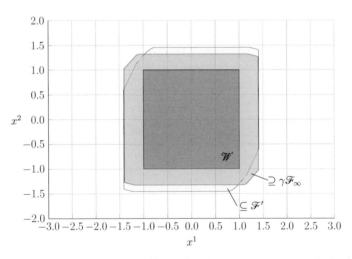

Figure 3.6.: Inner approximation of $\mathscr{F}'$ (yellow) and outer approximation of $\gamma\mathscr{F}_\infty$ (green) for a stable open loop matrix $A$. The set $\mathscr{W}$ is depicted in red.

### 3.3.3. Output feedback

We consider an output feedback scheme analogous to the one proposed in Section 3.2.2, where the event-generating functions $\delta_t$ and the scheduling functions are modified to

$$\delta_t\big(\hat{x}_0,\ldots,\hat{x}_t,\tilde{\boldsymbol{\mathcal{X}}}_t,t_i\big)$$
$$= \begin{cases} 0 & \text{if } \exists\tau \in \{1,\ldots,\min\{t+1,\theta\}\},\ \forall\hat{w} \in \hat{\mathscr{W}},\ \forall(\tilde{x}_0',\ldots,\tilde{x}_t') \in \tilde{\boldsymbol{\mathcal{X}}}_t, \\ \qquad \bar{A}_{\mathrm{et}}(\tilde{x}_t',\hat{x}_t) + \hat{B}\kappa(\hat{x}_{t_i},t-t_i) + \hat{w} \in \{\hat{A}^\tau(\tilde{x}_{t+1-\tau}',\hat{x}_{t+1-\tau})\} \oplus \gamma\hat{\mathscr{F}}_\tau & (3.44) \\ 1 & \text{otherwise} \end{cases}$$

and

$$\mu_i\big(\hat{x}_0,\ldots,\hat{x}_t,\tilde{\boldsymbol{\mathcal{X}}}_{t_i},t_0,t_1,\ldots,t_i\big)$$
$$:= \sup\left\{t' \in \{t_i+1,\ldots\} \left| \begin{array}{l} \forall k \in \{t_i+1,\ldots,t'\}, \\ \exists\tau_k \in \{1,\ldots,\min\{k,\theta\}\}, \\ \forall\hat{w}[l] \in \hat{\mathscr{W}},\ l \in \{k-\tau_k,\ldots,k-1\}, \\ \forall(\tilde{x}_0,\ldots,\tilde{x}_{t_i}) \in \tilde{\boldsymbol{\mathcal{X}}}_{t_i}, \\ z[k]_i \in \{\hat{A}^{\tau_k}z[k-\tau_k]_i\} \oplus \gamma\hat{\mathscr{F}}_{\tau_k}, \\ \text{if } k \in \{t_i+1,\ldots,t'-1\}, \\ \bar{z}[t']_i \in \{\hat{A}^{\tau_{t'}}z[t'-\tau_{t'}]_i\} \oplus \gamma\hat{\mathscr{F}}_{\tau_{t'}} \end{array} \right.\right\}. \quad (3.45)$$

where, as in (3.22), for all $i \in \{0,\ldots,i_{\max}\}$,

$$z[t]_i := \bar{A}_{\mathrm{st}}^{t-t_i}\begin{bmatrix}\tilde{x}_{t_i} \\ \hat{x}_{t_i}\end{bmatrix} + \sum_{l=t_i}^{t-1}\bar{A}_{\mathrm{st}}^{t-1-l}\left(\hat{B}\kappa(\hat{x}_{t_i},l-t_i) + \bar{D}_{\mathrm{st}}\hat{w}[l]\right), \quad (3.46)$$

$$\bar{z}[t]_i := \bar{A}_{\mathrm{et}}\bar{A}_{\mathrm{st}}^{t-1-t_i}\begin{bmatrix}\tilde{x}_{t_i} \\ \hat{x}_{t_i}\end{bmatrix} + \bar{A}_{\mathrm{et}}\sum_{l=t_i}^{t-2}\bar{A}_{\mathrm{st}}^{t-2-l}\left(\hat{B}\kappa(\hat{x}_{t_i},l-t_i) + \bar{D}_{\mathrm{st}}\hat{w}[l]\right)$$
$$+ \hat{B}\kappa(\hat{x}_{t_i},t-1-t_i) + \hat{w}[t-1] \quad (3.47)$$

for all $t \in \{t_i+1,\ldots\}$, and

$$z[t]_i := \begin{bmatrix}\tilde{x}_t \\ \hat{x}_t\end{bmatrix} \quad (3.48)$$

for all $t \in \{0,\ldots,t_i\}$.

Stability results analogous to the state-feedback case hold.

### 3.3.4. Computational aspects

The discussion in Section 3.2.3 concerning the computational effort involved with checking a condition for every member of the set $\{1, \ldots, \min\{t+1, \theta\}\}$ in the event-generating and scheduling functions applies also here; one may replace $\{1, \ldots, \min\{t+1, \theta\}\}$ with the sets $\mathscr{R}(t+1, t_i)$ as proposed in that section.

In the evaluation of the event-generating function (3.33), one has to check, for a given $\tau \in \mathbb{N}$, whether

$$\forall w \in \mathscr{W}, \; Ax_t + B\kappa(x_{t_i}, t - t_i) + w \in \{(A + BK)^\tau x_{t+1-\tau}\} \oplus \gamma \mathscr{F}_\tau, \tag{3.49}$$

which is, using the Pontryagin difference, equivalent to checking whether

$$Ax_t + B\kappa(x_{t_i}, t - t_i) - (A + BK)^\tau x_{t+1-\tau} \in \gamma \mathscr{F}_\tau \ominus \mathscr{W}. \tag{3.50}$$

The left-hand side of the set-membership condition in (3.50) can be easily computed from quantities known to the trigger mechanism at time $t$. The right-hand side only depends on $\tau$. At least in the case of polytopic sets $\mathscr{W}$, the expression $\gamma \mathscr{F}_\tau \ominus \mathscr{W}$ can be computed and is again a polytope, compare [Kolmanovsky and Gilbert, 1998]. It is straight-forward to see that replacing $\gamma \mathscr{F}_\tau \ominus \mathscr{W}$ in (3.50) with any subset thereof does not deteriorate the stability guarantees of the closed-loop system (triggering, however, becomes more likely). Further, as it holds that $\mathscr{F}_i \subseteq \mathscr{F}_{i+1}$ for all $i \in \mathbb{N}$ (due to $0 \in \mathscr{W}$), one may replace $\mathscr{F}_\tau$ with $\mathscr{F}_{\bar{\tau}}$ in the conditions for all $\tau \in \{\bar{\tau}, \ldots, \}$ and a chosen $\bar{\tau} \in \mathbb{N}$. Considering that the complexity of the set $\mathscr{F}_\tau$, determined by the number of hyperplanes required for its representation, grows with $\tau$, this substitution might even become necessary, especially if $\theta = \infty$.

For the evaluation of the scheduling function (3.34) in the self-triggered case, we need to highlight again the issue of $x_t$ only being measured at the transmission instants. Considering that the input $u_t$, however, *is* known at every point in time, we obtain the bounds

$$x_t \in \left\{ A^{t-t_j} x_{t_j} + \sum_{k=0}^{t-t_j-1} A^{t-t_j-1-k} B\kappa(x_{t_j}, k) \right\} \oplus \mathscr{G}_{t-t_j} \tag{3.51}$$

for $t \in \{t_j, \ldots, t_{j+1}\}$, $j \in \mathbb{N}$, where

$$\mathscr{G}_i := \bigoplus_{k=0}^{i-1} A^k \mathscr{W} \tag{3.52}$$

for all $i \in \mathbb{N}$. With (3.51), we may replace the set-membership conditions in the definition of the scheduling function (3.34) for a given $\tau_k \in \mathbb{N}$ with

$$\left( A^{k-t_i} x_{t_i} + \sum_{l=t_i}^{k-1} A^{k-1-l} B\kappa(x_{t_i}, l-t_i) \right)$$

$$- (A+BK)^{\tau_k} \left( A^{k-\tau_k-t_j} x_{t_j} + \sum_{l=t_j}^{k-\tau_k-1} A^{k-\tau_k-1-l} B\kappa(x_{t_j}, l-t_j) \right)$$

$$\in \gamma \mathscr{F}_{\tau_k} \ominus \left( \mathscr{G}_{k-t_i} \oplus (-(A+BK)^{\tau_k}) \mathscr{G}_{k-\tau_k-t_j} \right) \quad (3.53)$$

for $k - \tau_k \in \{t_j, \ldots, t_{j+1} - 1\}$ and $j < i$, and with

$$(A^{\tau_k} - (A+BK)^{\tau_k}) \left( A^{k-\tau_k-t_i} x_{t_i} + \sum_{l=t_i}^{k-\tau_k-1} A^{k-\tau_k-1-l} B\kappa(x_{t_i}, l-t_i) \right)$$

$$+ \sum_{l=k-\tau_k}^{k-1} A^{k-1-l} B\kappa(x_{t_i}, l-t_i) \in \gamma \mathscr{F}_{\tau_k} \ominus \mathscr{G}_{\tau_k} \ominus (A^{\tau_k} - (A+BK)^{\tau_k}) \mathscr{G}_{k-\tau_k-t_i} \quad (3.54)$$

for $k - \tau_k \in \{t_i, \ldots\}$.

In the second expression, we additionally used the information that

$$x_k = A^{\tau_k} x_{k-\tau_k} + \sum_{l=k-\tau_k}^{k-1} A^{k-1-l}(B\kappa(x_{t_i}, l-t_i) + w_l), \quad (3.55)$$

implying that employing (3.54) is less conservative than employing (3.53) for the case $k - \tau_k \geq t_i$. The same discussion as for the event-triggered case regarding the use of inner approximating substitutes of the sets in the right-hand sides of (3.53) and (3.54) applies.

**Output feedback**

For the following discussion, it is assumed that $\mathscr{W}$ and $\mathscr{V}$ are polytopes. We consider three different approaches for the output-feedback case: in the first approach, we assume that the sets $\tilde{\mathscr{X}}_t$ are given in the form $\mathscr{E}_0 \times \mathscr{E}_1 \times \cdots \times \mathscr{E}_t$ for *a priori* given polytopes $\mathscr{E}_t \subseteq \mathbb{R}^{n_x}$, as discussed in Section 2.4.1. This setup is only reasonable for the event-triggered case, however, as in the self-triggered case the estimator dynamics are time-varying and, moreover, depend on the realization of the transmission times. Analogously to the state-feedback case, for a fixed $\tau \in \mathbb{N}$, with this choice of $\tilde{\mathscr{X}}_t$, the condition to be checked in the event-generating function (3.44) is equivalent to

$$\bar{A}_{\text{et}} \begin{bmatrix} 0 \\ \hat{x}_t \end{bmatrix} + \hat{B}\kappa(\hat{x}_{t_i}, t-t_i) - \hat{A}^{\tau} \begin{bmatrix} 0 \\ \hat{x}_{t+1-\tau} \end{bmatrix} \in \gamma \hat{\mathscr{F}}_{\tau} \ominus \hat{\mathscr{W}} \ominus (\bar{A}_{\text{et}} - \hat{A})(\mathscr{E}_t \times \{0\}), \quad (3.56)$$

for $\tau = 1$ and to

$$\bar{A}_{\mathrm{et}} \begin{bmatrix} 0 \\ \hat{x}_t \end{bmatrix} + \hat{B}\kappa(\hat{x}_{t_i}, t - t_i) - \hat{A}^\tau \begin{bmatrix} 0 \\ \hat{x}_{t+1-\tau} \end{bmatrix}$$
$$\in \gamma \hat{\mathscr{F}}_\tau \ominus \hat{\mathscr{W}} \ominus \bar{A}_{\mathrm{et}}(\mathscr{E}_t \times \{0\}) \ominus (-\hat{A}^\tau)(\mathscr{E}_{t+1-\tau} \times \{0\}), \quad (3.57)$$

for $\tau \in \{2, \ldots\}$. In both cases the right-hand side, although time-varying, may be determined *a priori*, at least if $\tau$ is bounded (that is, $\theta$ is finite).

The second and the third approach are based on determining $\tilde{\mathscr{X}}_t$ at runtime depending on the realization of the measurements and transmission times, which make them applicable to the self-triggered case. The application to event-triggered control follows straight forward if the self-triggered case has been established and is omitted here.

For the second approach, we parameterize $\tilde{\mathscr{X}}_t$ as $c_0 \mathscr{E} \times c_1 \mathscr{E} \times \ldots c_t \mathscr{E}$, where $\mathscr{E} \subseteq \mathbb{R}^{n_x}$ is a fixed polytope and $c_t$, $t \in \mathbb{N}$, are scalars determined at time $t$, respectively. In particular, we follow the method proposed in [Brunner et al., 2015a] and compute these scalars as solutions to optimization problems. Compare [Kögel and Findeisen, 2015b] for a similar approach. In particular, we assume $c_0 \in [0, \infty)$ to be given such that $\tilde{x}_0 \in c_0 \mathscr{E}$ and define

$$c_t := \inf \left\{ c \in [0, \infty) \middle| \tilde{\mathscr{X}}_t \subseteq c\mathscr{E} \right\}, \quad (3.58a)$$

where

$$\tilde{\mathscr{X}}_t := \left\{ \tilde{x}_t' \in \mathbb{R}^{n_x} \middle| \begin{array}{l} \tilde{x}_i' \in c_i \mathscr{E}, \\ \tilde{x}_{i+1}' = A\tilde{x}_i' + w_i' + \sigma_{i+1}(LCA\tilde{x}_i' + LCw_i' + Lv_{i+1}'), \\ w_i' \in \mathscr{W}, \\ \quad i \in \{\max\{0, t - M\}, \ldots, t - 1\}, \\ v_j' \in \mathscr{V}, \\ y_j = C(\hat{x}_j + \tilde{x}_j') + v_j', \\ \quad j \in \{j' \in \{\max\{0, t - M\} + 1, \ldots, t\} \mid \sigma_{j'} = 1\} \end{array} \right\} \quad (3.58b)$$

for all $t \in \{1, \ldots\}$, where $M \in \{1, \ldots\}$ is the estimation horizon and $\sigma_t \in \{0, 1\}$ is defined by $\sigma_t = 1 \Leftrightarrow \exists i \in \{0, \ldots, i_{\max}\} : t = t_i$ as in Section 3.2.2. As $\mathscr{W}$, $\mathscr{V}$, and $\mathscr{E}$ were assumed to be polytopes, the optimization problem in (3.58) can be solved with linear programming. With $c_t$ obtained in this manner, the condition $z[k]_i \in \{\hat{A}^{\tau_k} z[k - \tau_k]_i\} \oplus \gamma \mathscr{F}_{\tau_k}$ in (3.45) can be checked for a given $\tau_k \in \mathbb{N}$ by evaluating

$$\bar{A}_{\mathrm{st}}^{k-t_i} \begin{bmatrix} 0 \\ \hat{x}_{t_i} \end{bmatrix} + \sum_{l=t_i}^{k-1} \bar{A}_{\mathrm{st}}^{k-l} \hat{B}\kappa(\hat{x}_{t_i}, l - t_i) - \hat{A}^{\tau_k} \begin{bmatrix} 0 \\ \hat{x}_{k-\tau_k} \end{bmatrix}$$
$$\in \gamma \hat{\mathscr{F}}_{\tau_k} \ominus \hat{\mathscr{G}}_{k-t_i} \ominus \bar{A}_{\mathrm{st}}^{k-t_i}(c_{t_i}\mathscr{E} \times \{0\}) \ominus (-\hat{A}^{\tau_k})(c_{k-\tau_k}\mathscr{E} \times \{0\}) \quad (3.59)$$

for $k - \tau_k \in \{0, \ldots, t_i - 1\}$ where

$$\hat{\mathscr{G}}_i := \left( \bigoplus_{j=0}^{i-1} \bar{A}_{\mathrm{st}}^j \bar{D}_{\mathrm{st}} \hat{\mathscr{W}} \right) \tag{3.60}$$

for all $i \in \mathbb{N}$ and by evaluating

$$\left( \bar{A}_{\mathrm{st}}^{\tau_k} - \hat{A}^{\tau_k} \right) \left( \bar{A}_{\mathrm{st}}^{k-\tau_k-t_i} \begin{bmatrix} 0 \\ \hat{x}_{t_i} \end{bmatrix} + \sum_{l=t_i}^{k-\tau_k-1} \bar{A}_{\mathrm{st}}^{k-\tau_k-1-l} \hat{B} \kappa(\hat{x}_{t_i}, l - t_i) \right) + \sum_{l=k-\tau_k}^{k-1} \bar{A}_{\mathrm{st}}^{k-1-l} \hat{B} \kappa(\hat{x}_{t_i}, l - t_i)$$

$$\in \gamma \hat{\mathscr{F}}_{\tau_k} \ominus \hat{\mathscr{G}}_{\tau_k} \ominus \left( \bar{A}_{\mathrm{st}}^{\tau_k} - \hat{A}^{\tau_k} \right) \left( \bar{A}_{\mathrm{st}}^{k-\tau_k-t_i} (c_{t_i} \mathscr{E} \times \{0\}) \oplus \hat{\mathscr{G}}_{k-\tau_k-t_i} \right) \tag{3.61}$$

for $k - \tau_k \in \{t_i, \ldots\}$.

Similarly, the condition $\bar{z}[t']_i \in \{\hat{A}^{\tau_{t'}} z[t' - \tau_{t'}]_i\} \oplus \gamma \hat{\mathscr{F}}_{\tau_{t'}}$ can be checked for a given $\tau_{t'} \in \mathbb{N}$ by evaluating

$$\bar{A}_{\mathrm{et}} \left( \bar{A}_{\mathrm{st}}^{t'-1-t_i} \begin{bmatrix} 0 \\ \hat{x}_{t_i} \end{bmatrix} + \sum_{l=t_i}^{t'-2} \bar{A}_{\mathrm{st}}^{t'-2-l} \hat{B} \kappa(\hat{x}_{t_i}, l - t_i) \right) + \hat{B} \kappa(\hat{x}_{t_i}, t' - 1 - t_i) - \hat{A}^{\tau_{t'}} \begin{bmatrix} 0 \\ \hat{x}_{t'-\tau_{t'}} \end{bmatrix}$$

$$\in \gamma \hat{\mathscr{F}}_{\tau_{t'}} \ominus \left( \bar{A}_{\mathrm{et}} \hat{\mathscr{G}}_{t'-1-t_i} \oplus \hat{\mathscr{W}} \right) \ominus \bar{A}_{\mathrm{et}} \bar{A}_{\mathrm{st}}^{t'-1-t_i} (c_{t_i} \mathscr{E} \times \{0\}) \ominus (-\hat{A}^{\tau_{t'}})(c_{t'-\tau_{t'}} \mathscr{E} \times \{0\}) \tag{3.62}$$

for $t' - \tau_{t'} \in \{0, \ldots, t_i - 1\}$ and by evaluating

$$\left( \bar{A}_{\mathrm{et}} \bar{A}_{\mathrm{st}}^{\tau_{t'}-1} - \hat{A}^{\tau_{t'}} \right) \left( \bar{A}_{\mathrm{st}}^{t'-\tau_{t'}-t_i} \begin{bmatrix} 0 \\ \hat{x}_{t_i} \end{bmatrix} + \sum_{l=t_i}^{t'-\tau_{t'}-1} \bar{A}_{\mathrm{st}}^{t'-\tau_{t'}-1-l} \hat{B} \kappa(\hat{x}_{t_i}, l - t_i) \right)$$

$$+ \bar{A}_{\mathrm{et}} \sum_{l=t'-\tau_{t'}}^{t'-2} \bar{A}_{\mathrm{st}}^{t'-2-l} \hat{B} \kappa(\hat{x}_{t_i}, l - t_i) + \hat{B} \kappa(\hat{x}_{t_i}, t' - 1 - t_i)$$

$$\in \gamma \hat{\mathscr{F}}_{\tau_{t'}} \ominus \left( \bar{A}_{\mathrm{et}} \hat{\mathscr{G}}_{\tau_{t'}-1} \oplus \hat{\mathscr{W}} \right)$$

$$\ominus \left( \bar{A}_{\mathrm{et}} \bar{A}_{\mathrm{st}}^{\tau_{t'}-1} - \hat{A}^{\tau_{t'}} \right) \left( \bar{A}_{\mathrm{st}}^{t'-\tau_{t'}-t_i} (c_{t_i} \mathscr{E} \times \{0\}) \oplus \hat{\mathscr{G}}_{t'-\tau_{t'}-t_i} \right) \tag{3.63}$$

for $t' - \tau_{t'} \in \{t_i, \ldots\}$.

While the set-membership conditions in (3.59) to (3.63) are analogous to those in the state-feedback case in that the left-hand side can be simply calculated at time $t_i$, the question remains how to obtain the right-hand side, considering that the quantities $c_t$ are not known *a priori*. However, as shown in [Raković et al., 2012b], for polytopic sets $\mathscr{X} = \{x \in \mathbb{R}^n \mid H_j x \le h_j, j \in \{1, \ldots, q\}\}$ and $\mathscr{Y} = \mathrm{convh}\left( \bigcup_{i=1}^p \{r_i\} \right)$ with $H_j \in \mathbb{R}^{1 \times n}$, $h_j \in \mathbb{R}$, $j \in \{1, \ldots, q\}$, $r_i \in \mathbb{R}^n$, $i \in \{1, \ldots, p\}$ it holds that

$$\mathscr{X} \ominus c\mathscr{Y} = \left\{ x \in \mathbb{R}^n \,\middle|\, H_j x \le h_j - c \max_{i \in \{1, \ldots, p\}} H_j r_i, j \in \{1, \ldots, q\} \right\}, \tag{3.64}$$

and, hence, checking whether $z \in \mathcal{X} \ominus c\mathcal{Y}$ amounts to checking whether

$$\max_{j \in \{1,\dots,q\}} H_j z - h_j + c\tilde{h}_j \geq 0, \tag{3.65}$$

where the quantities $h_j := \max_{i \in \{1,\dots,p\}} H_j r_i$ can be computed *a priori*.

**Remark 3.7.** *Note that the quality of the set-valued estimates, defined here by the value of $c_t$, does not deteriorate the stability guarantees in the closed-loop system, as long as $\tilde{x}_t \in c_t \mathcal{E}$ (which is ensured by the constraints in (3.58)). Even $c_t = \infty$ poses no other problem than a high likelihood of a transmission occurring at the respective point in time.*

Finally, for the third approach, we employ a set-valued moving-horizon estimator along the lines of Section 2.4.2; that is, we define for all $t \in \mathbb{N}$

$$\tilde{\mathcal{X}}_t := \left\{ (\tilde{x}_0', \dots, x_t') \in \mathbb{R}^{(t+1)n_x} \left| \begin{array}{c} w_k' \in \mathcal{W}, \ v_{k+1}' \in \mathcal{V}, \\ \tilde{x}_{k+1}' = (A + LCA)\tilde{x}_k' + (I + LC)w_k' + Lv_{k+1}', \\ C(A\hat{x}_k + Bu_k) - y_{k+1} = -CA\tilde{x}_k' - Cw_k' - v_{k+1}', \\ k \in \{\max\{0, t - M\}, \dots, t - 1\}, \end{array} \right. \right\} \tag{3.66}$$

for a fixed $M \in \mathbb{N}$. Note that we accounted for the fact that no measurements or inputs for times before $t = 0$ are available. Considering (3.62), the condition in (3.45) for a given $\tau_{t'} \in \mathbb{N}$ with $t' - \tau_{t'} \in \{t_j, \dots, t_{j+1} - 1\}$ and $j < i$, can thus be checked by evaluating whether for all $(\tilde{x}_0', \dots, \tilde{x}_{t_i}') \in \tilde{\mathcal{X}}_{t_i}$ it holds that

$$\bar{A}_{\text{et}} \left( \bar{A}_{\text{st}}^{t'-t_i} \begin{bmatrix} \tilde{x}_{t_i}' \\ \hat{x}_{t_i} \end{bmatrix} + \sum_{l=t_i}^{t'-2} \bar{A}_{\text{st}}^{t'-2-l} \hat{B}\kappa(\hat{x}_{t_i}, l - t_i) \right) + \hat{B}\kappa(\hat{x}_{t_i}, t' - 1 - t_i) - \hat{A}^{\tau_{t'}} \begin{bmatrix} \tilde{x}_{t'-\tau_{t'}}' \\ \hat{x}_{t'-\tau_{t'}} \end{bmatrix}$$
$$\in \gamma \hat{\mathcal{F}}_{\tau_{t'}} \ominus \left( \bar{A}_{\text{et}} \hat{\mathcal{G}}_{t'-1-t_i} \oplus \hat{\mathcal{W}} \right). \tag{3.67}$$

Here, the right-hand side of the set membership is a known polytope, which we assume for the moment to admit a representation of the form $\{x \in \mathbb{R}^{2n_x} \mid Gx \leq g\}$ for a matrix $G \in \mathbb{R}^{q \times 2n_x}$ and a vector $g \in \mathbb{R}^q$. Further, by (3.66), there exists known matrices $H \in \mathbb{R}^{2n_x \times r}$, $F \in \mathbb{R}^{p \times r}$, and a vector $f \in \mathbb{R}^p$ such that $(\tilde{x}_0', \dots, \tilde{x}_t') \in \tilde{\mathcal{X}}_t$ implies $(\tilde{x}_{t+1-\tau}', \tilde{x}_t') \in \{He \mid Fe \leq f\}$. Hence, the condition in (3.67) is equivalent to checking whether

$$\forall e \in \{e \in \mathbb{R}^r \mid Fe \leq f\}, \ EHe + \bar{x} \in \{x \in \mathbb{R}^{2n_x} \mid Gx \leq g\}, \tag{3.68}$$

where $E \in \mathbb{R}^{2n_x \times 2n_x}$ is some given matrix and $\bar{x} \in \mathbb{R}^{2n_x}$ is known. The main result in [Hennet, 1989] implies that (3.68) is equivalent to the existence of a matrix $P \in \mathbb{R}^{q \times p}$

satisfying

$$P \geq 0, \tag{3.69a}$$

$$PF = GEH \tag{3.69b}$$

$$Pf \leq g - GEH\bar{x}. \tag{3.69c}$$

Hence, (3.67) can be checked by solving a linear feasibility problem. The same holds, by analogous derivations, for the case $t' + 1 - \tau_{t'} \geq t_i$; the cases involving $\tau_k \in \{1, \ldots, t' - 1\}$ lead to similar conditions.

### 3.3.5. Numerical examples

**State feedback**

We consider the same system as in Section 3.2.4, that is, we have

$$A = \begin{bmatrix} 1 & 0.3 \\ 0 & 1 \end{bmatrix}, \ B = \begin{bmatrix} 0.045 \\ 0.3 \end{bmatrix}, \tag{3.70}$$

$\mathscr{W} = [-1, 1] \times [-1, 1]$, and $K = [-0.7719 \ -1.4628]$.

We simulated the closed-loop systems with both the event-triggered and the self-triggered controllers proposed in this section with the same parameters as in Section 3.2.4 with the following difference: first, we lifted the restrictions on $\tau$ in the scheduling function (3.34) that we implemented in the Lyapunov-based approaches in order to circumvent non-convexity. Further, we treated the case $\theta = \infty$ by only considering $\tau = t + 1$ in the event-generating function (3.33) and $\tau_k = k$ in the scheduling function (3.34). In order to keep the complexity of the involved sets bounded, we replaced every occurrence of $\mathscr{F}_\tau$ with $\mathscr{F}_{20}$ for $\tau \geq 20$.

In Table 3.6 and Table 3.7 we report average transmission rates for various values of $\theta$ and $\gamma$, where we chose the same initial conditions and disturbance realizations as in Section 3.2.4. In Table 3.8 and Table 3.9 we report average transmission rates for initial system states at the origin. As can be seen from this data, the effect of $\theta$ and $\gamma$ on the average transmission rates is the same as for the Lyapunov-based approaches, although the transmission rates are higher here. This coincides with our assessment that the set-based approaches lead to more conservative trigger conditions.

Table 3.6.: Average transmission rates in the closed-loop system with event-triggered control over 30 random simulations with 30 time points each and random initial states.

| | to-zero control | | | | | | | to-hold control | | | | | | | model-based control | | | | | |
| | γ | | | | | | | γ | | | | | | | γ | | | | | |
| $\theta$ | 1 | 1.25 | 1.5 | 2 | 2.5 | 3 | $\theta$ | 1 | 1.25 | 1.5 | 2 | 2.5 | 3 | $\theta$ | 1 | 1.25 | 1.5 | 2 | 2.5 | 3 |
|---|---|---|---|---|---|---|---|---|---|---|---|---|---|---|---|---|---|---|---|---|
| 1 | 1.00 | 0.87 | 0.78 | 0.73 | 0.69 | 0.66 | 1 | 1.00 | 0.84 | 0.73 | 0.64 | 0.59 | 0.54 | 1 | 1.00 | 0.47 | 0.23 | 0.12 | 0.09 | 0.07 |
| 2 | 0.81 | 0.76 | 0.72 | 0.67 | 0.64 | 0.61 | 2 | 0.77 | 0.66 | 0.60 | 0.54 | 0.49 | 0.46 | 2 | 0.46 | 0.21 | 0.13 | 0.09 | 0.07 | 0.06 |
| 5 | 0.73 | 0.69 | 0.66 | 0.62 | 0.59 | 0.56 | 5 | 0.62 | 0.55 | 0.50 | 0.45 | 0.43 | 0.41 | 5 | 0.16 | 0.11 | 0.09 | 0.06 | 0.05 | 0.05 |
| 10 | 0.70 | 0.66 | 0.63 | 0.59 | 0.57 | 0.54 | 10 | 0.55 | 0.51 | 0.47 | 0.43 | 0.41 | 0.39 | 10 | 0.12 | 0.09 | 0.08 | 0.06 | 0.05 | 0.04 |
| ∞ | 0.69 | 0.64 | 0.62 | 0.58 | 0.56 | 0.52 | ∞ | 0.53 | 0.50 | 0.47 | 0.42 | 0.41 | 0.39 | ∞ | 0.11 | 0.08 | 0.07 | 0.05 | 0.05 | 0.04 |

Table 3.7.: Average transmission rates in the closed-loop system with self-triggered control over 30 random simulations with 30 time points each and random initial states.

| | to-zero control | | | | | | | to-hold control | | | | | | | model-based control | | | | | |
| | γ | | | | | | | γ | | | | | | | γ | | | | | |
| $\theta$ | 1 | 1.25 | 1.5 | 2 | 2.5 | 3 | $\theta$ | 1 | 1.25 | 1.5 | 2 | 2.5 | 3 | $\theta$ | 1 | 1.25 | 1.5 | 2 | 2.5 | 3 |
|---|---|---|---|---|---|---|---|---|---|---|---|---|---|---|---|---|---|---|---|---|
| 1 | 1.00 | 1.00 | 1.00 | 0.89 | 0.83 | 0.78 | 1 | 1.00 | 1.00 | 1.00 | 0.84 | 0.77 | 0.69 | 1 | 1.00 | 1.00 | 1.00 | 0.50 | 0.33 | 0.33 |
| 2 | 1.00 | 1.00 | 0.91 | 0.83 | 0.76 | 0.71 | 2 | 1.00 | 1.00 | 0.86 | 0.75 | 0.65 | 0.58 | 2 | 1.00 | 1.00 | 0.50 | 0.50 | 0.33 | 0.27 |
| 5 | 0.93 | 0.84 | 0.79 | 0.73 | 0.68 | 0.65 | 5 | 0.96 | 0.87 | 0.80 | 0.69 | 0.60 | 0.54 | 5 | 0.96 | 0.84 | 0.50 | 0.33 | 0.27 | 0.20 |
| 10 | 0.85 | 0.80 | 0.75 | 0.69 | 0.64 | 0.60 | 10 | 0.87 | 0.81 | 0.75 | 0.67 | 0.58 | 0.52 | 10 | 0.84 | 0.78 | 0.46 | 0.33 | 0.26 | 0.20 |
| ∞ | 0.83 | 0.77 | 0.73 | 0.68 | 0.63 | 0.60 | ∞ | 0.75 | 0.65 | 0.60 | 0.52 | 0.46 | 0.44 | ∞ | 0.58 | 0.41 | 0.37 | 0.27 | 0.23 | 0.20 |

Table 3.8.: Average transmission rates in the closed-loop system with event-triggered control over 10 random simulations with 300 time points each and initial states at the origin.

| | to-zero control | | | | | | | to-hold control | | | | | | | model-based control | | | | | |
| | γ | | | | | | | γ | | | | | | | γ | | | | | |
| $\theta$ | 1 | 1.25 | 1.5 | 2 | 2.5 | 3 | $\theta$ | 1 | 1.25 | 1.5 | 2 | 2.5 | 3 | $\theta$ | 1 | 1.25 | 1.5 | 2 | 2.5 | 3 |
|---|---|---|---|---|---|---|---|---|---|---|---|---|---|---|---|---|---|---|---|---|
| 1 | 1.00 | 0.53 | 0.30 | 0.14 | 0.09 | 0.06 | 1 | 1.00 | 0.50 | 0.28 | 0.19 | 0.17 | 0.17 | 1 | 1.00 | 0.46 | 0.21 | 0.09 | 0.06 | 0.04 |
| 2 | 0.44 | 0.24 | 0.16 | 0.09 | 0.07 | 0.05 | 2 | 0.44 | 0.25 | 0.19 | 0.19 | 0.19 | 0.18 | 2 | 0.42 | 0.17 | 0.11 | 0.06 | 0.04 | 0.04 |
| 5 | 0.19 | 0.13 | 0.09 | 0.06 | 0.05 | 0.03 | 5 | 0.22 | 0.20 | 0.19 | 0.20 | 0.20 | 0.20 | 5 | 0.13 | 0.08 | 0.06 | 0.04 | 0.03 | 0.03 |
| 10 | 0.14 | 0.09 | 0.07 | 0.05 | 0.03 | 0.03 | 10 | 0.21 | 0.21 | 0.20 | 0.20 | 0.20 | 0.20 | 10 | 0.09 | 0.06 | 0.05 | 0.03 | 0.03 | 0.03 |
| ∞ | 0.11 | 0.08 | 0.06 | 0.04 | 0.03 | 0.03 | ∞ | 0.21 | 0.21 | 0.21 | 0.20 | 0.20 | 0.19 | ∞ | 0.07 | 0.05 | 0.04 | 0.03 | 0.03 | 0.03 |

Table 3.9.: Average transmission rates in the closed-loop system with self-triggered control over 10 random simulations with 300 time points each and initial states at the origin.

| | to-zero control | | | | | | | to-hold control | | | | | | | model-based control | | | | | |
|---|---|---|---|---|---|---|---|---|---|---|---|---|---|---|---|---|---|---|---|---|
| | $\gamma$ | | | | | | | $\gamma$ | | | | | | | $\gamma$ | | | | | |
| $\theta$ | 1 | 1.25 | 1.5 | 2 | 2.5 | 3 | $\theta$ | 1 | 1.25 | 1.5 | 2 | 2.5 | 3 | $\theta$ | 1 | 1.25 | 1.5 | 2 | 2.5 | 3 |
| 1 | 1.00 | 1.00 | 1.00 | 0.60 | 0.46 | 0.36 | 1 | 1.00 | 1.00 | 1.00 | 0.51 | 0.46 | 0.36 | 1 | 1.00 | 1.00 | 1.00 | 0.50 | 0.33 | 0.33 |
| 2 | 1.00 | 1.00 | 0.66 | 0.50 | 0.36 | 0.30 | 2 | 1.00 | 1.00 | 0.56 | 0.50 | 0.35 | 0.30 | 2 | 1.00 | 1.00 | 0.50 | 0.50 | 0.33 | 0.25 |
| 5 | 0.86 | 0.59 | 0.50 | 0.36 | 0.28 | 0.22 | 5 | 0.87 | 0.60 | 0.48 | 0.34 | 0.27 | 0.24 | 5 | 0.86 | 0.59 | 0.47 | 0.33 | 0.25 | 0.20 |
| 10 | 0.62 | 0.51 | 0.39 | 0.29 | 0.23 | 0.20 | 10 | 0.61 | 0.50 | 0.39 | 0.29 | 0.24 | 0.22 | 10 | 0.60 | 0.49 | 0.37 | 0.27 | 0.22 | 0.19 |
| $\infty$ | 0.54 | 0.41 | 0.35 | 0.26 | 0.21 | 0.18 | $\infty$ | 0.51 | 0.34 | 0.34 | 0.26 | 0.22 | 0.21 | $\infty$ | 0.51 | 0.34 | 0.34 | 0.24 | 0.20 | 0.17 |

**Output feedback**

As the output-feedback control schemes are structurally almost identical to the state-feedback schemes, we focus here on the main difference, which is the required set-valued estimation. In particular, we employed the set-valued estimator defined by (3.58) and investigated the impact of the quality of estimation—determined by the set $\mathscr{E}$ and the length of the estimation horizon $M$—on the average transmission rate.

We consider a system as in (3.1) with matrices

$$A = \begin{bmatrix} 1 & 0.3 \\ 0 & 1 \end{bmatrix}, B = \begin{bmatrix} 0.045 \\ 0.3 \end{bmatrix}, C = \begin{bmatrix} 1 & 0 \end{bmatrix}, \tag{3.71}$$

which correspond to a continuous-time double integrator system discretized with a sample time of 0.3. The disturbance bounds were chosen to $\mathscr{W} = [-0.05, 0.05] \times [-0.05, 0.05]$ and $\mathscr{V} = [-0.1, 0.1]$. We computed both the linear feedback and the estimator gain LQ-optimal, using the weighting matrices $Q = \begin{bmatrix} 1 & 0 \\ 0 & 1 \end{bmatrix}$ and $R = 1$ to obtain $K = [-0.7719 \quad -1.4628]$. The gain $L$ was computed as the transpose of the LQ-optimal feedback gain for the matrix pair $(A^\mathsf{T}, A^\mathsf{T} C^\mathsf{T})$ and weighting matrices $Q = \begin{bmatrix} 1 & 0 \\ 0 & 100 \end{bmatrix}$ and $R = 0.3$, resulting in $L = [-0.9758 \quad -2.8374]^\mathsf{T}$. Thus, the eigenvalues of $A + BK$ are (both) at 0.7719 and the eigenvalues of $A + LCA$ (both) at 0.1554, which is consistent with the common advice to choose the dynamics of the estimator four to five times as fast as the closed-loop dynamics under state feedback.

For the set-valued estimator, we considered two choices of the set $\mathscr{E}$; that is $\mathscr{E} = \mathscr{E}_{\text{box}} = [-1, 1] \times [-1, 1]$ which can be represented by the intersection of 4 half-spaces and $\mathscr{E} = \mathscr{E}_{\text{inv}}$ where $\mathscr{E}_{\text{inv}}$ is the 0.1-outer approximation of the minimal robust positively invariant set for the estimation error dynamics $e_{t+1} = (A + LCA)e_t + (I + LC)w_t + Lv_{t+1}$, computed with the algorithm in [Raković et al., 2005]. The set $\mathscr{E}_{\text{inv}}$ is represented by the intersection of 18 half-spaces. Our reasoning was that employing the set $\mathscr{E}_{\text{inv}}$ should lead to a tighter set-valued state estimate, but at the price of an increased computational burden as the linear program defining $c_t$ becomes more complicated with the number of half-spaces required to represent $\mathscr{E}$. Both $\mathscr{E}_{\text{box}}$ and $\mathscr{E}_{\text{inv}}$ are depicted in Figure 3.7.

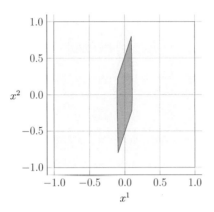

Figure 3.7.: The sets $\mathscr{E}_{\text{box}}$ and $\mathscr{E}_{\text{inv}}$ used as the set $\mathscr{E}$ in the set-valued estimator.

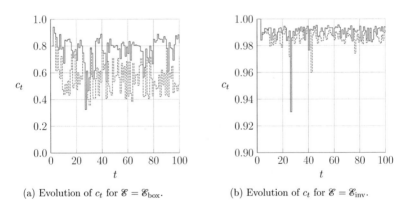

(a) Evolution of $c_t$ for $\mathscr{E} = \mathscr{E}_{\text{box}}$.          (b) Evolution of $c_t$ for $\mathscr{E} = \mathscr{E}_{\text{inv}}$.

Figure 3.8.: Evolutions of $c_t$ for the event-triggered case, $\gamma = 2$, $M = 2$ (blue, solid) and $M = 10$ (orange, dotted). Note the different scaling of the vertical axes.

Table 3.10.: Average transmission rates in the closed-loop system in the output feedback case over 30 random simulations with 100 time points each.

event-triggered control, $\mathscr{E} = \mathscr{E}_{\mathrm{box}}$

| $M$ | $\gamma$ | | |
|---|---|---|---|
| | 2 | 5 | 10 |
| 2 | 0.813 | 0.333 | 0.051 |
| 3 | 0.813 | 0.332 | 0.048 |
| 4 | 0.813 | 0.332 | 0.046 |
| 5 | 0.813 | 0.333 | 0.046 |
| 10 | 0.813 | 0.330 | 0.044 |

event-triggered control, $\mathscr{E} = \mathscr{E}_{\mathrm{inv}}$

| $M$ | $\gamma$ | | |
|---|---|---|---|
| | 2 | 5 | 10 |
| 2 | 0.813 | 0.333 | 0.048 |
| 3 | 0.813 | 0.333 | 0.048 |
| 4 | 0.813 | 0.333 | 0.048 |
| 5 | 0.813 | 0.333 | 0.048 |
| 10 | 0.813 | 0.333 | 0.048 |

self-triggered control, $\mathscr{E} = \mathscr{E}_{\mathrm{box}}$

| $M$ | $\gamma$ | | |
|---|---|---|---|
| | 2 | 5 | 10 |
| 2 | 1.000 | 0.998 | 0.908 |
| 3 | 1.000 | 0.996 | 0.621 |
| 4 | 1.000 | 0.994 | 0.559 |
| 5 | 1.000 | 0.993 | 0.525 |
| 10 | 1.000 | 0.993 | 0.492 |

self-triggered control, $\mathscr{E} = \mathscr{E}_{\mathrm{inv}}$

| $M$ | $\gamma$ | | |
|---|---|---|---|
| | 2 | 5 | 10 |
| 2 | 1.000 | 0.612 | 0.468 |
| 3 | 1.000 | 0.593 | 0.423 |
| 4 | 1.000 | 0.593 | 0.414 |
| 5 | 1.000 | 0.593 | 0.413 |
| 10 | 1.000 | 0.593 | 0.412 |

In Table 3.10 we report average transmission rates for different combinations of the parameters $\gamma$ and $M$, random initial states and random realizations of the disturbance and noise sequences. We employed the same realization for every parameter combination. We chose the "model-based" controller type for $\kappa$ and $\theta = 4$. Further, we replaced $\hat{\mathscr{F}}_\tau$ with $\hat{\mathscr{F}}_4$ for $\tau > 4$. In the self-triggered case, we enforced $t_{i+1} - t_i \leq 20$ for all $i \in \mathbb{N}$. For $\mathscr{E} = \mathscr{E}_{\mathrm{box}}$, we set $c_0 = 0.8$; for $\mathscr{E} = \mathscr{E}_{\mathrm{inv}}$, we set $c_0 = 1$, in order to compensate for the difference in size of the sets. In all simulations, we initialized $\hat{x}_0 = x_0$. As can be seen from the data, both increasing $M$, as well as choosing a—presumably—more appropriate shape of $\mathscr{E}$ leads to a decrease in the average transmission rate. The effect of increasing $M$ is higher for the case $\mathscr{E} = \mathscr{E}_{\mathrm{box}}$, which can also be seen in the exemplary evolutions of $c_t$ depicted in Figure 3.8, which show that increasing $M$ has a greater effect on the quality of the state estimate in this case.

## 3.4. Summary

We have proposed event-triggered and self-triggered controllers that stabilize certain compact sets in the state space. In the first part of the chapter, the respective event-generating functions and scheduling functions were based on guaranteeing—explicitly— the decrease of a Lyapunov function over time. In the second part, these decrease conditions were replaced by set-membership conditions, which, in turn, imply the decrease of a certain Lyapunov function. We treated the output-feedback case by designing first a stabilizing linear output-feedback controller updated at every point in time and then posing the Lyapunov-decrease-conditions and set-membership conditions, respectively, in the extended state space composed of the system state and the estimator state. Additionally, we have shown that our set-based event-triggered approaches are a generalization of certain existing threshold based schemes and suggested a way of refining the *a posteriori* stability analysis for these methods.

The approaches based on set-membership conditions introduce, conceptually, a certain amount of conservatism when compared to the Lyapunov-based approaches. In the numerical examples, this conservatism was reflected in a generally higher average transmission rate for the set-based approaches. On the other hand, the Lyapunov-based methods are expected to entail a higher computational cost than the set-based approaches, due to the general necessity of solving optimization problems in order to obtain the values of the Lyapunov functions. Moreover, in the self-triggered setup and in all output-feedback cases, these optimization problems are, in general, non-convex. Finally, since the minimal robustly positive invariant set defining the Lyapunov functions does not admit a finite representation in general, only upper and lower bounds on the function values can be computed, such that the Lyapunov function-based approaches themselves are, in general, not free of conservatism.

# 4. Stochastic Threshold Design in Event-triggered Control

In the previous chapter, aperiodic control schemes were designed based on a worst case bound on the disturbances only. In the numerical examples, it could be demonstrated that for certain randomly sampled disturbance sequences the required amount of communication in the control system could be reduced. However, neither were the stochastic properties of the disturbance used in the design of the controllers, nor was the closed-loop system analyzed based on this additional information. In this chapter, we design aperiodic controllers with guarantees on the time between communication instant based on the knowledge of the probability density function governing the disturbance acting on the system. In particular, we focus on event-triggered controllers where the arguments of the event-generating function consist only of the state at the most recent transmission instant, the current state, and the time since the last transmission instant. This restriction allows us to exploit a Markov-like property of the resulting control system, in the sense that the probability of a certain point in time being a communication instant will only depend on the disturbances realized in the time span since the last communication instant.

For threshold-based event-triggered control schemes as described in Section 3.3, it is in general hard to compute the average event rate as an explicit function of the threshold sets, except for special cases as, for example, considered in [Åström and Bernhardsson, 2002, Rabi and Johansson, 2009, Blind and Allgöwer, 2012, Blind and Allgöwer, 2013]. The reason for this is that the complexity of the involved probability density functions grows with time, due to convolution and conditioning. This even holds in the most simple—and presumably benign—case of disturbances that are identically and independently distributed according to normal densities [Cervin and Henningsson, 2008], compare the discussion in [Han et al., 2015, Wu et al., 2016a] and in Section 4.2 later in the chapter. In [Wu et al., 2013, Ebner and Trimpe, 2016] approximative approaches for the analytic computation of the communication rate in the case of Gaußian disturbances were proposed; see also [Sijs et al., 2010], where posterior distributions—conditioned on the trigger conditions—are approximated by Gaußian distributions.

In Section 4.1, we consider the state-feedback case for disturbances distributed according to arbitrary bounded probability density functions, where the event-generating functions are similar to those investigated in Section 3.3. That is, the trigger conditions reduce to checking whether the error between the actual system state and a predicted nominal state crosses a certain threshold. We show that such thresholds can be designed for any desired probability distribution of the time between events and suggest numerical methods for their computation. In the case of bounded disturbances, stability follows from the results established in the previous chapter. Related results can be found for example in [Battistelli et al., 2012] in the context of event-based estimation, where trigger thresholds are designed for a special class of probability distributions.

In Section 4.2, we restrict the class of probability density functions to Gaußian functions. The thresholds employed here are random variables themselves which allows us to obtain the trigger probabilities as analytic functions of the parameters, greatly simplifying the design procedure. Similar trigger rules were proposed in [Han et al., 2015, Shi et al., 2016, Wu et al., 2016a] in the context of state estimation; the main idea is to ensure that the conditional distribution of the variable entering the trigger rule remains Gaußian if no event is generated—an impossibility if the trigger rules are based on deterministic thresholds. In [Demirel et al., 2017b], a similar event-triggered scheme as presented in this thesis was proposed for scalar systems, based on a dead-beat control law. General stochastic trigger rules were considered in [Chen et al., 2015, Huang et al., 2016]. Other probabilistic scheduling techniques have been proposed for example in [Xu and Hespanha, 2004a], where transmissions are scheduled randomly according to a doubly stochastic Poisson process and the instantaneous rate increases with the current error. Similarly, in [Mamduhi et al., 2013] a randomized transmission protocol for a network of multiple systems is proposed, where the probability of an individual subsystem gaining access to the network depends on the relative size of its associated error. See also [Tabbara and Nešić, 2008, Donkers et al., 2012, Antunes et al., 2012] for examples of purely random scheduling policies without dependence on the current state (compare [Mamduhi et al., 2013] for these references).

Another type of trigger rules exploiting stochastic properties of the disturbances was proposed in [Trimpe and Andrea, 2014], where the current variance of the state estimation error determines whether an event occurs. In contrast to most other trigger rules employed in event-based control and estimation with probabilistic disturbances, the variables on which the events are based in [Trimpe and Andrea, 2014] are not random, leading to a deterministic trigger behavior.

In the literature on event-triggered control where the stochasticity of the disturbances is explicitly considered, the performance is usually defined in terms of an integral performance index, sometimes also including a term weighting the cost of communication, see, for example, [Xu and Hespanha, 2004b, Imer and Başar, 2006, Cogill, 2009, Henningsson et al., 2008, Molin and Hirche, 2009, Molin and Hirche, 2013, Rabi et al., 2012, Meng and Chen, 2012, Lipsa and Martins, 2013, Antunes and Heemels, 2014, Gommans et al., 2014]. The controller designs studied in the present thesis aim chiefly at stabilization, using the size of the guaranteed asymptotic bound on the system state as a measure of control performance. This measure is only meaningful if the disturbances are bounded. In Section 4.2—where the disturbances are unbounded—we instead are satisfied with providing a bound for the asymptotic variance of the system state.

The content of Section 4.1 is partly based on [Brunner et al., 2015b, Brunner et al., 2017c]. Section 4.2 is, in parts, based on [Brunner et al., 2017a].

# 4.1. Threshold design for arbitrarily distributed disturbances

Consider the system

$$x_{t+1} = Ax_t + Bu_t + w_t, \tag{4.1}$$

with state $x_t \in \mathbb{R}^{n_x}$, input $u_t \in \mathbb{R}^{n_u}$, and disturbance $w_t \in \mathbb{R}^{n_x}$ at time $t \in \mathbb{N}$. The disturbance is considered to be a random variable with the associated probability density function $\rho_w : \mathbb{R}^{n_x} \to [0, \infty)$, where $w_t$ is assumed to be the realization of a random variable[1] which is identically and independently distributed at each time point $t \in \mathbb{N}$. Further, we assume that $\rho_w$ is bounded, which, in particular, excludes the case of discretely distributed disturbances.

The input $u_t$ is determined[2] by an event-triggered scheme as described in Section 3.3.2, that is,

$$u_t = K(A + BK)^{t-t_i}x_{t_i}, \qquad t \in \{t_i, \dots, t_{i+1} - 1\}, \ i \in \{0, \dots, i_{\max}\} \tag{4.2a}$$

$$t_0 = 0 \tag{4.2b}$$

$$t_{i+1} = \inf\{t \in \{t_i + 1, \dots\} \mid x_t \notin \{(A + BK)^{t-t_i}x_{t_i}\} \oplus \mathscr{T}_{t-t_i}\}, \tag{4.2c}$$

where $\mathscr{T}_\tau \subseteq \mathbb{R}^{n_x}$ for $\tau \in \mathbb{N}$. Our goal is to find the threshold sets $\mathscr{T}_\tau$ such that the random variable $\Delta_i := t_{i+1} - t_i$ in the closed-loop systems is distributed according to a desired probability mass function $p_\Delta : \{1, \dots\} \to [0, 1]$. In order to simplify the theoretical analysis and, later, the numerical determination of the threshold sets, we impose the constraint $\mathscr{T}_N = \emptyset$ for some $N \in \mathbb{N}$. This choice immediately implies $p_\Delta(i) = 0$ for $i \in \{N + 1, \dots\}$. As a consequence, if our goal is achieved, the expected value of the inter-event times $\mathbb{E}[\Delta_i] = \sum_{j=1}^{\infty} jp_\Delta(j) = \sum_{j=1}^{N} jp_\Delta(j)$ is guaranteed to exist. Without loss of generality, we assume $\mathscr{T}_0 = \{0\}$. The structure of the chosen event-triggered controller implies that

$$x_t - (A + BK)^{t-t_i}x_{t_i} = \sum_{j=0}^{t-1} A^{t-1-j}w_{t_i+j} \tag{4.3}$$

for all $t \in \{t_i, \dots, t_{i+1}\}$ and all $i \in \{0, \dots, i_{\max}\}$. Hence, the quantity $x_t - (A+BK)^{t-t_i}x_{t_i}$ only depends on the realization of the disturbances in the time span $\{t_i, \dots, t_{i+1} - 1\}$.

---

[1]For brevity of exposition, we do not distinguish between random variables and their realization if the meaning is clear from the context.

[2]Consequently, $u_t$ is a function of random variables. Whenever functions of random variables occur in this chapter, we assume implicitly that the functions im question are Baire functions.

As the disturbances were assumed to be identically and independently distributed, the same holds for the random variable $\Delta_i$ (where the identity and independence of the distribution are with respect to $i$). In the following, we hence omit or reassign the indices of the variables $\Delta_i$ and $w_t$ if it simplifies exposition. Different indices of the variable $w_t$ now only indicate that these are two different random variables (sampled at different points in time), but we do not associate the variable $t$ with specific points in time anymore. The indices of the threshold sets $\mathscr{T}_\tau$ retain their meaning.

Define $P_i$ to be the probability of an event being generated $i$ time steps after the last transmission instant and define $\bar{P}_i$ to be the probability that no event has been generated for the first $i$ time steps since the last transmission instant. That is,

$$P_i = \mathbb{P}(\Delta = i) = \mathbb{P}\left( \sum_{k=0}^{j-1} A^{j-1-k} w_k \in \mathscr{T}_j,\ j \in \{1, \ldots, i-1\},\ \sum_{k=0}^{i-1} A^{i-1-k} w_k \notin \mathscr{T}_i, \right)$$

(4.4)

and

$$\bar{P}_i = \mathbb{P}(\Delta > i) = \mathbb{P}\left( \sum_{k=0}^{j-1} A^{j-1-k} w_k \in \mathscr{T}_j,\ j \in \{1, \ldots, i\} \right)$$

(4.5)

for $i \in \mathbb{N}$. With $\mathscr{T}_0 = \{0\}$ and $\mathscr{T}_N = \emptyset$, it holds that $P_i = 0$ for $i \in \{0\} \cup \{N+1, \ldots\}$ and, hence, that $\bar{P}_i = 1 - \sum_{j=1}^{i} P_j = \sum_{j=i+1}^{N} P_j$ for $i \in \mathbb{N}$. We also have $P_i = \bar{P}_{i-1} - \bar{P}_i$ for $i \in \{1, \ldots\}$.

We can state the following result.

**Lemma 4.1.** *Let $P_i$ and $\bar{P}_i$ be the probabilities as defined in (4.4) and (4.5), associated with thresholds sets $\mathscr{T}_i \in \mathbb{R}^{n_x}$ for $i \in \{1, \ldots, N\}$. Similarly, let $P_i'$ and $\bar{P}_i'$ be the probabilities associated with the threshold sets $\mathscr{T}_j' \in \mathbb{R}^{n_x}$ for $j \in \{1, \ldots, N\}$. If $\mathscr{T}_j \subseteq \mathscr{T}_j'$ for $j \in \{0, \ldots, N\}$ it follows that $\bar{P}_i \leq \bar{P}_i'$ for all $i \in \{0, \ldots, N\}$ and $\sum_{i=1}^{N} i P_i \leq \sum_{i=1}^{N} i P_i'$.*

In the following sections we will exploit this monotonicity in order to assign desired values to the probabilities $P_i$ by parameterizing the sets $\mathscr{T}_j$ in an appropriate fashion. In order to simplify the analysis, we introduce the change of coordinates defined by

$$v_j = \sum_{k=0}^{j-1} A^{j-1-k} w_k,\ j \in \{1, \ldots, N\}.$$

(4.6)

Equivalently, with $\boldsymbol{v}_i := (v_1^\mathsf{T}, v_2^\mathsf{T}, \ldots, v_i^\mathsf{T})^\mathsf{T}$ and $\boldsymbol{w}_i := (w_0^\mathsf{T}, w_1^\mathsf{T}, \ldots, w_{i-1}^\mathsf{T})^\mathsf{T}$, we have $\boldsymbol{v}_i = T_i \boldsymbol{w}_i$ for $i \in \{1, \ldots, N\}$, where

$$T_i = \begin{bmatrix} I & 0 & & \cdots & 0 \\ A & I & 0 & \cdots & \\ A^2 & A & I & \cdots & 0 \\ \vdots & \vdots & & \ddots & \\ A^{i-1} & A^{i-2} & \cdots & & I \end{bmatrix}.$$

(4.7)

It holds that $\det(T_i) = 1$, $i \in \{1, \ldots, N\}$, such that the probability density functions $\rho_{\mathsf{w}}^i$ and $\rho_{\mathsf{v}}^i$ associated with the random variables $\boldsymbol{w}_i$ and $\boldsymbol{v}_i$ satisfy $\rho_{\mathsf{v}}^i(\boldsymbol{v}_i) = \rho_{\mathsf{w}}^i(T^{-1}\boldsymbol{v}_i)$ for all $\boldsymbol{v}_i \in \mathbb{R}^{in_{\mathsf{x}}}$, $i \in \{1, \ldots, N\}$, see for example [Anderson, 1958]. The functions $\rho_{\mathsf{w}}^i$, using the assumed independence of disturbances at different points in time, are defined by $\rho_{\mathsf{w}}^i(\boldsymbol{w}_i) = \prod_{j=1}^i \rho_{\mathsf{w}}(w_{j-1})$ for all $\boldsymbol{w}_i \in \mathbb{R}^{in_{\mathsf{x}}}$, $i \in \{1, \ldots, N\}$.

## 4.1.1. Probability assigment

Let $\mathscr{T}$ be any compact and convex set containing the origin in its interior and define $\mathscr{T}_j = r_j\mathscr{T}$ with $r_j \in [0, \infty)$ for $j \in \{1, \ldots, N-1\}$. Given desired values for the probabilities $P_i$, the probability assignment problem reduces now to finding the appropriate scalars $r_j$. Using the probability density functions and the change of variables introduced above, we obtain

$$
\begin{aligned}
P_i &= \int_{\{\boldsymbol{w} \in \mathbb{R}^{in} \mid \sum_{k=0}^{j-1} A^{j-1-k}w_k \in r_j\mathscr{T}, \; j \in \{1,\ldots,i-1\}, \; \sum_{k=0}^{i-1} A^{i-1-k}w_k \notin r_i\mathscr{T}\}} \rho_{\mathsf{w}}^i(\boldsymbol{w}) \, \mathrm{d}\boldsymbol{w} \\
&= \int_{\{\boldsymbol{v} \in \mathbb{R}^{in} \mid v_j \in r_j\mathscr{T}, \; j \in \{1,\ldots,i-1\}, \; v_i \notin r_i\mathscr{T}\}} \rho_{\mathsf{v}}^i(\boldsymbol{v}) \, \mathrm{d}\boldsymbol{v} \\
&= \int_{r_1\mathscr{T} \times r_2\mathscr{T} \times \cdots \times (\mathbb{R}^{n_{\mathsf{x}}} \setminus r_i\mathscr{T})} \rho_{\mathsf{v}}^i(\boldsymbol{v}) \, \mathrm{d}\boldsymbol{v}
\end{aligned}
\tag{4.8}
$$

for $i \in \{1, \ldots, N-1\}$. One could, in principle, analytically solve the equations in (4.8) for the unknown variables $r_j$, $j \in \{1, \ldots, N-1\}$. We expect this task to be cumbersome or, depending on the functions $\rho_{\mathsf{w}}$, even intractable. Instead, approximate numerical methods will have to be applied to solve this problem in general. Below, we propose a heuristic method that only requires the integrals in (4.8) to be (approximately) *evaluated* for given $r_j$. In a first step, however, we establish that the problem is well-posed in the sense that for any given desired values for the probabilities $P_1, \ldots, P_N$ (or, equivalently, for the quantities $\bar{P}_1, \ldots, \bar{P}_N$), there exist appropriate scalars $r_1, \ldots, r_{N-1}$ such that (4.8) holds. In particular, considering that for any choice of the probabilities $P_i$ it holds that $P_i \in [0, 1 - \sum_{j=1}^{i-1} P_j]$ for $i \in \{1, \ldots, N-1\}$, the following statement implies that appropriate scalars $r_1, \ldots, r_{N-1}$ can be found sequentially, starting with $r_1$.

**Lemma 4.2.** *For all $i \in \{1, \ldots, N-1\}$, $P_i$ is a monotonously non-increasing continuous function of $r_i$ with $P_i = 1 - \sum_{j=1}^{i-1} P_j$ for $r_i = 0$. Further, $\lim_{r_i \to \infty} P_i = 0$.*

Note that the case $P_i = 0$ can be achieved by simply never triggering an event $i$ steps after the last event, without checking any set-memberships. Similarly, $P_i = 1 - \sum_{j=1}^{i-1} P_j$ can be achieved by enforcing an event $i$ steps after the last event.

**Stochastic approximation**

Recursive algorithms based on stochastic difference equations are generally known under the name "stochastic approximation", see for example [Kushner and Yin, 2003] which contains many references to the literature on the topic and provides a theoretical analysis and multiple applications of stochastic approximation methods.

Here, we propose a simple heuristic based on one of the earliest stochastic approximation methods, as described in [Robbins and Monro, 1951]. Therein, an algorithm is proposed which approximates the (unique) scalar $\theta \in \mathbb{R}$ such that $M(\theta) = \alpha$, where $M : \mathbb{R} \to \mathbb{R}$ and $\alpha \in \mathbb{R}$ is a desired value. It is assumed that the evaluation of the function $M$ is corrupted by noise, such that for a sequence $(\theta_n)_{n \in \mathbb{N}}$ only a sequence of random variables $y_n$ with $\mathbb{E}[y_n] = M(\theta_n)$ is available. The proposed algorithm is to define[3] $\theta_{n+1} = \theta_n + a_n(y_n - \alpha)$ for all $n \in \mathbb{N}$, where $(a_n)_{n \in \mathbb{N}}$ is an *a priori* chosen sequence with $\lim_{n \to \infty} a_n = 0$.

Note that in that work the authors do provide a theoretical analysis of their algorithm, together with conditions that imply convergence. Instead of showing that these conditions hold for the setup presented in this subsection (which seems unlikely, in general), we instead perform an *a posteriori* analysis of the scalars $r_1, \ldots, r_{N-1}$ we obtained and provide confidence intervals for the associated probabilities $P_1, \ldots, P_N$. Algorithm 1 describes the stochastic approximation method for finding $r_1, \ldots, r_{N-1}$, where we use a Monte Carlo method to approximate the value of the probabilities $P_i$. In particular, for given $r_1, \ldots, r_i$, a certain number of trials is performed where disturbances $w_0, \ldots, w_{i-1}$ are sampled according to $\rho_w$. The sample mean of the random variable

$$p_i = \begin{cases} 1 & \text{if } \sum_{k=0}^{j-1} A^{j-1-k} w_k \in r_j \mathscr{T}, \; j \in \{1, \ldots, i-1\}, \; \sum_{k=0}^{i-1} A^{i-1-k} w_k \notin r_i \mathscr{T} \\ 0 & \text{otherwise} \end{cases} \quad (4.9)$$

over all trials is used as an approximation $P_i'$ of $P_i$. The sequence $(a_n)_{n \in \mathbb{N}}$ is chosen as $a_n = \frac{1}{n}$, $n \in \mathbb{N}$. The *a posteriori* analysis is performed according to Algorithm 2.

**Remark 4.1.** *One should be careful not to misinterpret the results of the algorithms presented here, especially if they are executed repeatedly—while varying the parameters or not—until satisfactory results are obtained: As the confidence intervals output by Algorithm 2 are random variables, one is likely to obtain a seemingly satisfactory outcome simply by running the procedure often enough, whether or not the scalars $r_i$ actually correspond to the desired trigger probabilities.*

### Stability analysis

If the probability density function $\rho_w$ has bounded support, then it holds that $w_t \in \mathscr{W}$, $t \in \mathbb{N}$, for a compact and convex set $\mathscr{W}$ containing the origin. Consequently, the results of Chapter 3, in particular Section 3.3.2, apply, allowing compact subsets of the state space to be determined which are asymptotically stable for the closed-loop system.

---

[3]We reverse the sign of the right-hand here side compared to [Robbins and Monro, 1951], as the function $M$ in our case is decreasing instead of increasing.

---

**Algorithm 1** Stochastic Approximation

---

**Input:** desired probabilities $P_1, \ldots, P_{N-1}$, number of iterations $n_{\max}$, number of trials $T$

**Output:** scalars $r_1, \ldots, r_{N-1}$

---

1: **for** $i \in \{1, \ldots, N-1\}$ **do**
2:    $r_i \leftarrow 1$
3:    **for** $n \in \{1, \ldots, n_{\max}\}$ **do**
4:      **for** $s \in \{1, \ldots, T\}$ **do**
5:        **for** $k \in \{0, \ldots, i-1\}$ **do**
6:          Sample $w_k$ according to $\rho_{\mathrm{w}}$
7:        **end for**
8:        **if** $\sum_{k=0}^{j-1} A^{j-1-k} w_k \in r_j \mathscr{T}$ for all $j \in \{1, \ldots, i-1\}$ and $\sum_{k=0}^{i-1} A^{i-1-k} w_k \notin r_i \mathscr{T}$ **then**
9:          $p_i^s \leftarrow 1$
10:        **else**
11:          $p_i^s \leftarrow 0$
12:        **end if**
13:      **end for**
14:      $P_i' \leftarrow \frac{1}{T} \sum_{s=1}^{T} p_j^s$
15:      $r_i \leftarrow r_i + \frac{1}{n}(P_i' - P_i)$
16:    **end for**
17: **end for**

---

**Numerical example**

Consider system (4.1) with

$$A = \begin{bmatrix} 1 & 0.2 \\ 0 & 1 \end{bmatrix}, \ B = \begin{bmatrix} 0.02 \\ 0.2 \end{bmatrix}, \tag{4.10}$$

and $\rho_{\mathrm{w}}$ corresponding to a uniform distribution on $\mathscr{W} = [-1, 1] \times [-1, 1]$. We chose $\mathscr{T} = [-1, 1] \times [-1, 1]$ as a base shape for the thresholds, $N = 5$, and executed Algorithm 1 with the inputs $P_1 = 0.1$, $P_2 = 0.2$, $P_3 = 0.4$, $P_4 = 0.2$, $n_{\max} = 10^6$, and $T = 1$. The resulting values for $r_i$ were $r_1 = 0.9488$, $r_2 = 1.2833$, $r_3 = 0.8628$ and $r_4 = 0.6511$. The *a posteriori* evaluation with Algorithm 2 yielded the 95% confidence intervals $\mathscr{C}_1 = [0.0989, 0.1026]$, $\mathscr{C}_2 = [0.1992, 0.2042]$, $\mathscr{C}_3 = [0.4007, 0.4067]$, $\mathscr{C}_4 = [0.1951, 0.2000]$, and $\mathscr{C}_5 = [0.0999, 0.1036]$.

Next, we evaluated the closed-loop behavior with the resulting thresholds. We computed the feedback gain as the LQ-optimal controller for the weighting matrices $Q = \begin{bmatrix} 1 & 1 \\ 1 & 1 \end{bmatrix}$ and $R = 0.001$, leading to $K = [-4.6342 \ -5.3658]$. We simulated the closed-loop system consisting of (4.1) and (4.2) with $\mathscr{T}_i = r_i \mathscr{T}$, $i \in \{1, \ldots, N-1\}$ and $\mathscr{T}_N = \emptyset$.

---

**Algorithm 2** A posteriori analysis

**Input:** scalars $r_1, \ldots, r_{N-1}$, confidence level $c$, number of trials $T$.
**Output:** confidence intervals $\mathscr{C}_1, \ldots, \mathscr{C}_N$ for the probabilities $P_1, \ldots, P_N$

1: **for** $i \in \{1, \ldots, N\}$ **do**
2:    **for** $s \in \{1, \ldots, T\}$ **do**
3:       **for** $k \in \{0, \ldots, i-1\}$ **do**
4:          Sample $w_k$ according to $\rho_w$
5:       **end for**
6:       **if** $\sum_{k=0}^{j-1} A^{j-1-k} w_k \in r_j \mathscr{T}$ for all $j \in \{1, \ldots, i-1\}$ and $\sum_{k=0}^{i-1} A^{i-1-k} w_k \notin r_i \mathscr{T}$ **then**
7:          $p_i^s \leftarrow 1$
8:       **else**
9:          $p_i^s \leftarrow 0$
10:       **end if**
11:    **end for**
12:    $P_i' \leftarrow \frac{1}{T} \sum_{s=1}^{T} p_i^s$
13:    $\delta_i' \leftarrow \frac{t^*(1-\frac{c}{2}, T-1)}{\sqrt{N}} \sqrt{\frac{1}{N-1} \sum_{s=1}^{T} (p_i^s - P_i')^2}$,
     where $t^*(a, b)$ is the value of the inverse cumulative distribution function of Student's t distribution with $b$ degrees of freedom evaluated at $a$.
14:    $\mathscr{C}_i \leftarrow [P_i' - \delta_i', P_i' + \delta_i']$
15: **end for**

---

In Table 4.1 we list the observed frequencies of the inter-transmission time spans for 1000 simulations (each with $T_{\mathrm{sim}} = 1000$ time points) with random initial conditions (uniformly sampled on $[-1, 1] \times [-1, 1]$) and random realizations of the disturbance sequence (according to $\rho_w$); these frequencies are consistent with the specified probabilities $P_i$, and the confidence intervals obtained in the *a posteriori* analysis.

The general goal of event-triggered control is to achieve a better trade-off between the transmission rate and performance than periodic control. With a given desired probability distribution of the inter-transmission times, the appropriate sampling scheme for comparison, however, is one where the transmission times $t_{i+1} - t_i$ are generated randomly, according to this distribution, and independent of the system state. The input in such a comparison scheme should be generated, as in the event-triggered case, according to (4.2a). We simulated the closed-loop with such a scheme for the same random initial conditions and disturbance realization as described above, and computed

Table 4.1.: Observed relative frequencies of the inter-transmission time spans over 1000 simulations.

| $t_{i+1} - t_i$ | 1 | 2 | 3 | 4 | 5 |
|---|---|---|---|---|---|
| frequency | 0.1002 | 0.2003 | 0.3995 | 0.1984 | 0.1017 |

the average value of the performance index

$$J_{\text{performance}} = \sum_{t=0}^{T_{\text{sim}}-1} x_t^{\mathsf{T}} Q x_t + u_t^{\mathsf{T}} R u_t \tag{4.11}$$

over all simulations for both the event-triggered and the purely randomly triggered scheme. Here, the average performance index of the event-triggered scheme was 30.6% lower than the index for the purely randomly triggered scheme (minimal improvement 16.7%, maximal improvement 40.3%, and median improvement 30.8% over all simulations).

## 4.1.2. Expected value assignment

In this subsection, we assume from the start that $\rho_w$ has bounded support and, more specifically, that $w_t \in \mathcal{W}$, $t \in \mathbb{N}$, for a compact and convex set $\mathcal{W}$ containing the origin in its (nonempty) interior. As already mentioned in the last subsection, and discussed in more detail in Section 3.3.2, the assumption of bounded disturbances allows an asymptotic bound to be computed for given threshold sets. Here, we assume a certain asymptotic bound to be given and design thresholds yielding the lowest average communication rate under this constraint.

Consider again the sets $\mathcal{H}_\tau \subseteq \mathbb{R}^{n_x}$, introduced in Section 3.3.2, where $\mathcal{H}_0 = \{0\}$ and $\mathcal{H}_{\tau+1} = A(\mathcal{H}_\tau \cap \mathcal{T}_\tau) \oplus \mathcal{W}$ for $\tau \in \mathbb{N}$. These sets have the property that replacing the sets $\mathcal{T}_\tau$ with $\mathcal{H}_\tau \cap \mathcal{T}_\tau$ would not change the event-triggered scheme in (4.2) in the sense that the same realization of the disturbances would lead to the same sequence $(t_i)_{i \in \mathbb{N}}$. This fact follows immediately from the first part of the proof of Theorem 3.5, where it is established that $x_{t_i+\tau} - (A + BK)^\tau x_{t_i} \in \mathcal{H}_\tau \cap \mathcal{T}_\tau$ if $t_i + \tau \in \{t_i, \ldots, t_{i+1} - 1\}$. Define the *effective threshold sets* $\mathcal{T}_\tau^{\text{eff}} = \mathcal{H}_\tau \cap \mathcal{T}_\tau$ for all $\tau \in \mathbb{N}$. Intuitively, for a given asymptotic bound $\gamma \mathcal{F}_\infty$ with $\mathcal{F}_\tau = \bigoplus_{j=0}^{\tau-1} (A + BK)^j \mathcal{W}$ for $\tau \in \mathbb{N} \cup \{\infty\}$ and $\gamma \in [1, \infty)$ (as defined in Theorem 3.5), the effective threshold sets $\mathcal{T}_\tau^{\text{eff}}$ allowing the greatest amount of communication to be saved are the largest sets (defined by set inclusion) $\mathcal{T}_\tau \subseteq \mathbb{R}^{n_x}$ satisfying $A\mathcal{T}_\tau \oplus \mathcal{W} \subseteq \gamma \mathcal{F}_{\tau+1}$, for $\tau \in \mathbb{N}$. This is formally expressed in the following statement.

**Theorem 4.1.** *Let $\gamma \in [1, \infty)$ be given and define $\mathscr{F}_\tau := \bigoplus_{j=0}^{\tau-1}(A + BK)^j \mathscr{W}$ for $\tau \in \mathbb{N}$. Let $P_i'$ be the probabilities as defined in (4.4), associated with the threshold sets*

$$\mathscr{T}_\tau' = A^{-1}(\gamma \mathscr{F}_{\tau+1} \ominus \mathscr{W}), \tag{4.12}$$

*$\tau \in \{1, \ldots, N-1\}$, and $\mathscr{T}_\tau' = \emptyset$ for $\tau \in \{N, \ldots\}$. Let further threshold sets $\mathscr{T}_\tau$ for $\tau \in \{1, \ldots, N-1\}$ be given where $\mathscr{T}_\tau = \emptyset$ for $\tau \in \{N, \ldots\}$ and define $\mathscr{H}_0 = \{0\}$ and $\mathscr{H}_{\tau+1} = A(\mathscr{H}_\tau \cap \mathscr{T}_\tau) \oplus \mathscr{W}$ for $\tau \in \mathbb{N}$. Then, if $\mathscr{H}_\tau \subseteq \gamma \mathscr{F}_\tau$ for all $\tau \in \mathbb{N}$, it holds that*

$$\sum_{i=1}^{N} i P_i \leq \sum_{i=1}^{N} i P_i' \tag{4.13}$$

*where $P_i$ are the probabilities as defined in (4.4) and associated with the threshold sets $\mathscr{T}_\tau$.*

The question remains whether an arbitrary assignment, similar to Section 4.1.1, of the expected value of the inter-event times $\sum_{i=1}^{N} i P_i$ is possible by choosing the scalar $\gamma$ appropriately. We can answer in the affirmative if certain additional assumptions hold:

**Lemma 4.3.** *Let $A$ be nonsingular and let $\mathscr{W}$ be a polytope. Let $\mathscr{T}_\tau = A^{-1}(r \mathscr{F}_{\tau+1} \ominus \mathscr{W})$ for $\tau \in \{1, \ldots, N-1\}$ and $\mathscr{T}_\tau = \emptyset$ for $\tau \in \{N, \ldots\}$, where $\mathscr{F}_\tau = \bigoplus_{j=0}^{\tau-1}(A + BK)^j \mathscr{W}$ and $r \in [0, \infty)$. Then, $\sum_{i=1}^{N} i P_i$ is a monotonously non-decreasing continuous function of $r$ where $\sum_{i=1}^{N} i P_i = 1$ for $r = 0$. Further, $\lim_{r \to \infty} \sum_{i=1}^{N} i P_i = N$.*

### Numerical example

We considered the same system parameters as in the numerical example of the previous subsection, that is

$$A = \begin{bmatrix} 1 & 0.2 \\ 0 & 1 \end{bmatrix}, \ B = \begin{bmatrix} 0.02 \\ 0.2 \end{bmatrix}, \tag{4.14}$$

and $\rho_{\mathrm{w}}$ corresponding to a uniform distribution on $\mathscr{W} = [-1, 1] \times [-1, 1]$. Further, we chose the feedback gain $K$ as the LQ-optimal controller for the weighting matrices $Q = \begin{bmatrix} 1 & 1 \\ 1 & 1 \end{bmatrix}$ and $R = 0.001$, that is, $K = [-4.6342 \ -5.3658]$.

In the example in the previous subsection, we assigned the probability mass function of the inter-event times, which satisfied $\sum_{i=1}^{N} P_i = 3$. Here, we employed a Stochastic

Table 4.2.: Observed relative frequencies of the inter-transmission time spans over 1000 simulations with the thresholds $A^{-1}(\gamma \mathscr{F}_{i+1} \ominus \mathscr{W})$.

| $t_{i+1} - t_i$ | 1 | 2 | 3 | 4 | 5 |
|---|---|---|---|---|---|
| frequency | 0.2105 | 0.2340 | 0.1694 | 0.1191 | 0.2671 |

Approximation algorithm similar to Algorithm 1 in order to compute an appropriate value for $\gamma$ such that $\sum_{i=1}^{N} P_i = 3$ also holds (approximately) for the thresholds defined by $\mathscr{T}_\tau = A^{-1}(\gamma \mathscr{F}_{\tau+1} \ominus \mathscr{W})$, $\tau \in \{1, \ldots, N-1\}$ and $\mathscr{T}_N = \emptyset$. This resulted in $\gamma = 1.5817$ and, using an analogue of Algorithm 2, a 95% confidence interval of $[2.9847, 3.0034]$ for the expected value $\sum_{i=1}^{N} iP_i$ of the inter-event times.

Performing the same closed-loop evaluation as in the previous subsection, we obtained the frequencies of the inter-transmission time spans reported in Table 4.2. The average time between transmission instants was 2.998, close to the desired value. Considering the performance index in (4.11), the closed-loop system based on the thresholds in this subsection achieved a reduction of 19.3% when compared to the event-triggered scheme based on the thresholds $r_i\mathscr{T}$. Note, however, that neither threshold design is based on optimizing that performance index.

On the other hand, the thresholds $A^{-1}(\gamma \mathscr{F}_{\tau+1} \ominus \mathscr{W})$ *are* designed to minimize a guaranteed asymptotic bound of the form $r\mathscr{F}_\infty$ for the closed-loop system. If we compute the minimal bound of this form based on Theorem 3.5 for the thresholds $r_i\mathscr{T}$ employed in the example of the previous subsection, we arrive at the bound $\bar{\gamma}\mathscr{F}_\infty$ with $\bar{\gamma} = 2.3627$, which is considerably larger than $\gamma = 1.5817$. This shows that the thresholds of the form $A^{-1}(\gamma \mathscr{F}_{\tau+1} \ominus \mathscr{W})$, while leading to the same average transmission rate, provide a quantitatively better stability guarantee than the thresholds $r_i\mathscr{T}$.

# 4.2. Stochastic thresholds for Gaußian noise disturbances

In Section 4.1, arbitrary probability density functions governing the disturbances were considered and appropriate threshold sets could only be approximated numerically. The question arises whether event conditions could be found analytically for a desired distributions of the inter-event times if the probability density function was of a particular type, for example, Gaußian. However, if the trigger conditions are based on threshold sets, as in the other parts of this thesis, then the probability density functions one faces cease to be Gaußian. As a simple example, consider the scalar system $x_{t+1} = ax_t + w_t + u_t$ where the disturbances $w_t$ are sampled independently from a normal distribution. Let an event be generated at time $t$ if $|x_t| > c$, for a $c \geq 0$. Let further $u_t = 0$ if no event occurs at time $t$ and let $u_t = -ax_t$ if an event occurs at time $t$, which leads to a system similar to the one studied in [Åström and Bernhardsson, 2002]. The probability of an event occurring one time step after the last event is obtained as

$$\mathbb{P}(|x_t| > c \mid |x_{t-1}| > c) = \frac{\mathbb{P}(|w_{t-1}| > c \wedge |x_{t-1}| > c)}{\mathbb{P}(|x_{t-1}| > c)} = \mathbb{P}(|w_{t-1}| > c) \qquad (4.15)$$

where we used the independence of the disturbances. This probability can be computed from the cumulative distribution function of the normal distribution. For the probability of an event occurring two time steps after the last event, we have to evaluate the expression

$$\mathbb{P}(|x_t| > c \mid |x_{t-1}| \le c \wedge |x_{t-2}| > c) = \mathbb{P}(|aw_{t-2} + w_{t-1}| > c \mid |w_{t-2}| \le c), \quad (4.16)$$

which is already a lot more complicated. The task of finding an appropriate value for $c$ for a given desired probability is even harder. The complexity of the distribution which is required for the computation of these probabilities grows with the inter-event time, making the computation of inter-event-times an intractable task in general, just as in the case of general probability distributions considered in Section 4.1. In Figure 4.1, we illustrate this problem by depicting the shapes of the involved probability density functions. Compare also [Wu et al., 2016a] and the references therein for a similar discussion of this issue. In [Han et al., 2015], the authors propose *stochastic* trigger conditions to circumvent this problem. As we will see, making the event threshold a random variable (with a particular probability distribution) itself, preserves the Gaußian nature of the involved probability distributions, greatly simplifying the necessary computations. This technique was used in [Han et al., 2015, Shi et al., 2016, Wu et al., 2016a] for the design of event-triggered estimation schemes. Here, we focus on event-triggered control.

### 4.2.1. State-Feedback

Consider a system of the form

$$x_{t+1} = Ax_t + Bu_t + w_t \quad (4.17)$$

where $A \in \mathbb{R}^{n_x \times n_x}$ and $B \in \mathbb{R}^{n_x \times n_u}$. The disturbances $w_t$ are independently, but not necessary identically, distributed according to the normal distribution with mean zero and covariance $0 \ne Q_t \in \mathbb{R}^{n_x \times n_x}$ for all $t \in \mathbb{N}$. We assume a feedback matrix $K \in \mathbb{R}^{n_u \times n_x}$ to be given such that $A + BK$ is stable. The proposed event-triggered controller has the form

$$u_t = K\bar{x}_t \quad (4.18a)$$

$$\bar{x}_t = \begin{cases} (A + BK)\bar{x}_{t-1} & t \in \{t_i + 1, \ldots, t_{i+1} - 1\}, \ i \in \{0, \ldots, i_{\max}\} \\ x_t & \text{otherwise} \end{cases} \quad (4.18b)$$

$$t_0 = 0 \quad (4.18c)$$

$$t_{i+1} = \inf\{t \in \{t_i + 1, \ldots\} \mid \delta(x_t, \bar{x}_{t-1}, r_t, t, t_i) = 1\}. \quad (4.18d)$$

Here, $\delta : \mathbb{R}^{n_x} \times \mathbb{R}^{n_x} \times \mathbb{R} \times \mathbb{N} \times \mathbb{N}$ is the event-generating function and $r_t$ is an independent random variable distributed uniformly on the interval $[0, 1]$. As in Section 4.1, $p_\Delta : \{1, \ldots\} \to [0, 1]$ describes the desired probability mass function for the time span between events $t_{i+1} - t_i$, with the difference that here, for the moment, we allow the support of $p_\Delta$ to be unbounded. We have the following result for the event-generating function $\delta$.

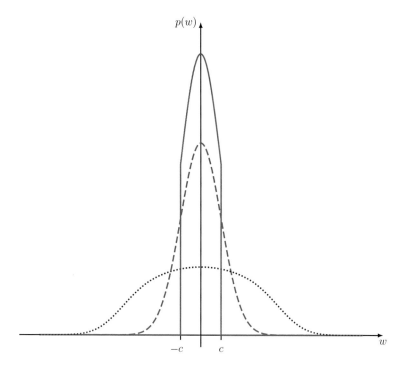

Figure 4.1.: Density of $w_{t-1}$ (blue, dashed); density of $w_{t-1}$, conditioned on the event $|w_{t-1}| \leq c$ (orange, solid); density of $aw_{t-1} + w_t$, conditioned on the event $|w_{t-1}| \leq c$ (violet, dotted). Note that neither of the two latter functions is Gaußian (compare Figure 2 in [Wu et al., 2016a]).

**Lemma 4.4.** *Let*

$$\delta_t := \delta(x_t, \bar{x}_{t-1}, r_t, t, t_i) = \begin{cases} 0 & \frac{1}{2}z_t^\mathsf{T}\Sigma_t^\dagger z_t \leq -\frac{1}{\lambda_t}\ln(r_t) \\ 1 & \text{otherwise} \end{cases} \tag{4.19}$$

*where* $z_t = x_t - (A + BK)\bar{x}_{t-1}$,

$$\Sigma_t = \begin{cases} \frac{1}{1+\lambda_{t-1}}A\Sigma_{t-1}A^\mathsf{T} + Q_{t-1} & t \in \{t_i + 2, \dots, t_{i+1}\}, \ i \in \{0, \dots, i_{\max}\} \\ Q_{t-1} & \text{otherwise,} \end{cases} \tag{4.20}$$

*and*

$$\lambda_t = \left( \frac{1 - \sum_{j=1}^{t-t_i} p_\Delta(j)}{1 - \sum_{j=1}^{t-t_i-1} p_\Delta(j)} \right)^{-\frac{2}{\mathrm{rk}(\Sigma_t)}} - 1 \tag{4.21}$$

*for all* $x_t, \bar{x}_{t-1} \in \mathbb{R}^{n_x}$, $r_t \in (0,1]$, $t \in \{t_i + 1, \dots, t_{i+1}\}$, *and* $i \in \{1, \dots, i_{\max}\}$. *Then it holds that* $\mathbb{P}(t_{i+1} - t_i = \tau) = p_\Delta(\tau)$ *for all* $i \in \mathbb{N}$ *and all* $\tau \in \{1, \dots\}$ *for the closed-loop system consisting of* (4.17) *and* (4.18).

**Remark 4.2.** *For the evaluation of* (4.21), *we use the conventions* $0/0 = 0$, $0^q = \infty$ *for* $q \in (-\infty, 0)$, *and* $\infty - 1 = \infty$. *The case* $\lambda_t = 0$, *which occurs if* $p_\Delta(t - t_i) = 0$, *is handled by enforcing* $\delta(x_t, \bar{x}_{t-1}, r_t, t, t_i) = 0$. *Similarly, the case* $\lambda_t = \infty$, *which occurs if* $\sum_{j=1}^{t-t_i-1} p_\Delta(j) = 1$, *is handled by enforcing* $\delta(x_t, \bar{x}_{t-1}, r_t, t, t_i) = 1$.

**Remark 4.3.** *The event generating function in* (4.19) *is equivalent to the condition proposed in [Han et al., 2015] for an appropriate choice of* $z_t$, $\Sigma_t$ *and* $\lambda_t$.

Of course, a desired distribution of the inter-event times can also be achieved by simply deciding randomly whether to trigger an event at a certain point in time, independently of the current state. The intuition behind the event-triggered scheme considered here is that events are more likely to be triggered if the state is farther away from the nominal trajectory, thereby introducing feedback to the system, compare the discussion in [Han et al., 2015, Remark 3]. A more structured approach is to compute the thresholds as the minimizer of the expected cost over some horizon, as was proposed in [Wu et al., 2016a] and in [Brunner et al., 2017a] (where this contribution is due to the second author of that work).

In the following, we investigate the stability of the closed-loop system under certain assumptions on the system and design parameters. We have the following result in terms of the second moment. Compare [Tarn and Rasis, 1976] for this notion of stability.

**Theorem 4.2.** *We consider the closed-loop system consisting of* (4.17) *and* (4.18) *where the event-generating function is defined as in Lemma 4.4. Let there exist a symmetric positive definite matrix* $P \in \mathbb{R}^{n_x \times n_x}$ *and a scalar* $\mu \in [0, 1)$ *such that* $(A + BK)^\mathsf{T}P(A +$

$BK) \preceq \mu P$ *for all* $t \in \mathbb{N}$. *Let further* $S \in \mathbb{R}^{n_x \times n_x}$ *be a positive semi-definite matrix such that* $g(\lambda_t, \mathrm{rk}(\Sigma_t))\Sigma_t \preceq S - (A + BK)S(A + BK)^\mathsf{T}$ *for all* $t \in \{1, \ldots\}$, *where*

$$g(\lambda, k) = \begin{cases} 1 + \frac{2}{k} & \lambda = 0 \\ \frac{1 - (1+\lambda)^{-(1+\frac{k}{2})}}{1 - (1+\lambda)^{-\frac{k}{2}}} & \lambda \in (0, \infty) \\ 1 & \lambda = \infty \end{cases} \tag{4.22}$$

*for all* $\lambda \in (0, \infty)$ *and all* $k \in \{1, \ldots\}$.

*Then it holds that*

$$\mathbb{E}[x_t^\mathsf{T} x_t] \le \mathrm{tr}(S) + \mu^t \frac{\lambda_{\max}(P)}{\lambda_{\min}(P)} x_0^\mathsf{T} x_0 \tag{4.23}$$

*for all* $x_0 \in \mathbb{R}^{n_x}$ *and all* $t \in \mathbb{N}$.

**Remark 4.4.**

(i) *The stability result we obtained is conservative as we disregard the individual probabilities of the realizations* $(t_i)_{i \in \{0, \ldots, i_{\max}\}}$ *of transmission instants in our proof.*

(ii) *The assumption that* $g(\lambda_t, \mathrm{rk}(\Sigma_t))\Sigma_t \preceq S - (A+BK)S(A+BK)^\mathsf{T}$ *for* $t \in \mathbb{N}$ *requires that the covariance matrices* $\Sigma_t$ *remain bounded. This can, for example, be achieved by (a) assuming that the origin of the open-loop system is asymptotically stable, (b) choosing a probability mass function for the inter-event times with a bounded support, or (c) ensuring that the values of* $\lambda_t$ *are large enough (which influences the choice of the probability mass function for the inter-event times). All three possibilities additionally require the existence of a bound on* $Q_t$.

**Heuristically minimized sampling rate for a given variance bound**

Based on point (ii)(c) in Remark 4.4 we propose an even-triggered control scheme that attempts to minimize the average rate of events in the closed-loop system while guaranteeing a certain given bound on the asymptotic variance of the system state. For simplicity, we assume that $Q_t = Q$ for all $t \in \mathbb{N}$. Let $\gamma \in [1, \infty)$ be a chosen scaling factor for the permitted asymptotic variance and define

$$\lambda_t := \inf \left\{ \lambda \in (0, \infty) \ \middle| \ \begin{array}{l} \dfrac{1}{1+\lambda} A\Sigma_t A^\mathsf{T} + Q \preceq \gamma \displaystyle\sum_{j=0}^{t-t_i} (A+BK)^j Q \left((A+BK)^j\right)^\mathsf{T} \\[2ex] g(\lambda, \mathrm{rk}(\Sigma_t))\Sigma_t \preceq \gamma \displaystyle\sum_{j=0}^{t-1-t_i} (A+BK)^j Q \left((A+BK)^j\right)^\mathsf{T} \end{array} \right\} \tag{4.24}$$

73

for all $t \in \{t_i + 1, \ldots, t_{i+1}\}$, $i \in \{0, \ldots, i_{\max}\}$, where $g$ is defined as in (4.22). As $g$ is continuous and monotonically decreasing in its first argument, the infimization problem in (4.24) can be solved with a bisection approach. The case $\lambda_t = \infty$ corresponds to forcing an event at time $t$. The stability result for this scheme follows readily from the proof of Theorem 4.2, the proof of which is given in the appendix:

**Corollary 4.1.** *Let there exist a symmetric positive definite matrix $P \in \mathbb{R}^{n_x \times n_x}$ and a scalar $\mu \in [0, 1)$ such that $(A + BK)^{\mathsf{T}} P (A + BK) \preceq \mu P$ for all $t \in \mathbb{N}$. Let further $\bar{S} \in \mathbb{R}^{n_x \times n_x}$ be a positive semi-definite matrix such that $(A + BK)\bar{S}(A + BK)^{\mathsf{T}} + Q = \bar{S}$. Let the closed-loop system consist of (4.17) and (4.18) where the event-generating function is defined as in Lemma 4.4 with $\lambda_t$ chosen as in (4.24). Then it holds that*

$$\mathbb{E}[x_t^{\mathsf{T}} x_t] \leq \gamma \operatorname{tr}(\bar{S}) + \mu^t \frac{\lambda_{\max}(P)}{\lambda_{\min}(P)} x_0^{\mathsf{T}} x_0 \tag{4.25}$$

*for all $x_0 \in \mathbb{R}^{n_x}$ and all $t \in \mathbb{N}$.*

**Remark 4.5.**

(i) *The idea behind this design of trigger conditions is similar to the threshold design in Section 4.1.2, where the thresholds were maximized under the condition that a certain asymptotic bound on the system state, depending on a scaling factor $\gamma$, could still be guaranteed. For the case that $\delta_t = 1$, $t \in \mathbb{N}$, it follows by simple calculations that $\mathbb{E}[x_t^{\mathsf{T}} x_t] \leq \operatorname{tr}(\bar{S}) + \mu^t \frac{\lambda_{\max}(P)}{\lambda_{\min}(P)} x_0^{\mathsf{T}} x_0$ for all $t \in \mathbb{N}$ and all $x_0 \in \mathbb{R}^{n_x}$.*

(ii) *Since $\lambda_t$ only depends on the time $t - t_i$ since the last event, its values can be computed offline, up to a desired upper bound on the time between events after which an event is simply forced (equivalent to choosing $\lambda_t = \infty$ at this point in time).*

(iii) *The dynamics of the variables $\Sigma_t$ resemble those studied in [Trimpe and Andrea, 2014], where an event trigger rule based on only monitoring the current variance of a certain variable was proposed, without taking into account the variable in question itself in the trigger decision. For a scalar example the authors showed that this leads to periodic trigger behavior. In our case, due to the stochastic nature of the event conditions and the explicit dependence on the realization of the disturbances, the triggering does not necessarily become periodic.*

**Numerical example**

We consider a system as in (4.17), with

$$A = \begin{bmatrix} 1.5 & 2 \\ -1 & 1.2 \end{bmatrix}, \ B = \begin{bmatrix} 0 \\ 1 \end{bmatrix}, \tag{4.26}$$

and

$$Q_t = Q = \begin{bmatrix} 1 & 0 \\ 0 & 1 \end{bmatrix} \tag{4.27}$$

for all $t \in \mathbb{N}$. First, we assigned the values of the desired probability mass function $p_\Delta$ according to a binomial distribution $B(9, 0.5)$, that is,

$$p_\Delta(i) = \binom{9}{i-1} 0.5^{i-1} 0.5^{9-i+1} \tag{4.28}$$

for $i \in \{1, \ldots, 10\}$ and $p_\Delta(i) = 0$ for $i \in \{11, \ldots\}$. This resulted in $\operatorname{tr}(S) = 18053$ in Theorem 4.2, where we computed $S$ as the minimizer of a semidefinite program with objective function $\operatorname{tr}(S)$ and constraints $g(\lambda_t, \operatorname{rk}(\Sigma_t))\Sigma_t \preceq S - (A + BK)S(A + BK)^\mathsf{T}$ for all $t \in \{1, \ldots, 10\}$. Second, we computed $\lambda_t$ according to (4.24) where we set $\gamma = \operatorname{tr}(S)/\operatorname{tr}(\bar{S})$ with $(A+BK)\bar{S}(A+BK)^\mathsf{T}+Q = \bar{S}$, leading to the probability mass function $\bar{p}_\Delta$ in this second case, with values given by $\bar{p}_\Delta(i) = 0$, $i \in \{1, \ldots, 5\}$, $\bar{p}_\Delta(6) = 0.6162$, $\bar{p}_\Delta(7) = 0.2665$, $\bar{p}_\Delta(8) = 0.0905$, $\bar{p}_\Delta(89) = 0.0267$, and $\bar{p}_\Delta(i) = 0$ for $i \in \{10, \ldots\}$. Hence, following the theoretical results in this section, event-triggered control based on either of the two probability mass functions should guarantee an asymptotic bound on the trace of the second moment of the state of 18053, while the average inter-transmission time should be 5.5 for the first scheme and 6.5278 for the second scheme.

The feedback gain $K$ was chosen LQ optimal for the weighting matrices $\begin{bmatrix} 1 & 0 \\ 0 & 1 \end{bmatrix}$ and 1, that is, $K = [0.2023 \; -2.1512]$. We simulated the closed-loop system consisting of (4.17) and (4.18) for both desired probability mass functions for $N_{\text{sim}} = 1000$ random initial conditions (uniformly distributed on $[-1, 1] \times [-1, 1]$) and disturbance realizations, each for $T_{\text{sim}} = 1000$ simulation time points. The observed relative frequencies of the inter-transmission times are reported in Table 4.3 and validate the theoretical claims. The resulting average inter-transmission time was 5.4953 for the first desired probability mass function, and 6.5256 for the second probability mass function, consistent with the predictions by the *a priori* computations (5.5 and 6.5278, respectively). Further, we obtained the sample second moment of the state at the end of the simulation time over all simulations, whose trace was 1682 for the first probability mass function and 3556 for the second. This shows the conservatism discussed in Remark 4.4, as both values are considerably lower than the predicted 18053.

Considering the discussion after Remark 4.3, we expect the event-triggered controller to perform better than a controller where the inter-transmission times are simply generated randomly according to the desired probability mass function. In order to examine this claim for these numerical experiments, we computed the performance index

$$J_{\text{performance}} = \sum_{t=0}^{T_{\text{sim}}-1} x_t^\mathsf{T} \begin{bmatrix} 1 & 0 \\ 0 & 1 \end{bmatrix} x_t + u_t^\mathsf{T} u_t \tag{4.29}$$

Table 4.3.: Observed relative frequencies of the inter-transmission time spans over 1000 simulations.

| $t_{i+1} - t_i$ | 1 | 2 | 3 | 4 | 5 | 6 | 7 | 8 | 9 | 10 |
|---|---|---|---|---|---|---|---|---|---|---|
| desired frequency ($p_\Delta$) | 0.0020 | 0.0176 | 0.0703 | 0.1641 | 0.2461 | 0.2461 | 0.1641 | 0.0703 | 0.0176 | 0.0020 |
| observed frequency | 0.0021 | 0.0175 | 0.0711 | 0.1650 | 0.2456 | 0.2459 | 0.1627 | 0.0708 | 0.0176 | 0.0018 |
| desired frequency ($\bar{p}_\Delta$) | 0 | 0 | 0 | 0 | 0 | 0.6162 | 0.2665 | 0.0905 | 0.0267 | 0 |
| observed frequency | 0 | 0 | 0 | 0 | 0 | 0.6174 | 0.2665 | 0.0893 | 0.0268 | 0 |

for both the closed-loop system under event-triggered control as well as the closed-loop system where the transmission times $t_i$ are purely randomly generated, according to the same probability distribution that we prescribed for the event-triggered control loop. For the event-triggered controller with the probability mass function in (4.28), we reported a performance index that was, on average, 77.3% lower than the performance index for the purely randomly triggered controller (minimal improvement 42.4%, maximal improvement 96.8%, and median improvement 78.0% over all simulations).

### 4.2.2. Output feedback

In the event-triggered state-feedback schemes we proposed in this chapter so far, we constructed the event-generating functions such that at the time of an event a "reset" occurs. Specifically, the disturbances affecting the system at times before an event had no influence on the trigger behavior after the event. In the output-feedback case, however, we are faced with the problem that the information that we have about the system state at any time depends, in general, on the realization of *all* past disturbances, with no reset occurring at the event times. This fact makes it difficult to design event-generating functions with guarantees on the transmission rate or the covariance of the system state.

In this subsection, we consider one special case where at least the *uncertainty* about the system state is independent of the past realizations of the disturbances. In particular, we consider a linear system subject to white-noise disturbances on the output and state, for which the Kalman filter provides an estimate of the state with just this property. Let the system being controlled be given by

$$x_{t+1} = Ax_t + Bu_t + w_t$$
$$y_t = Cx_t + v_t \tag{4.30}$$

where $A \in \mathbb{R}^{n_x \times n_x}$, $B \in \mathbb{R}^{n_x \times n_u}$ and $C \in \mathbb{R}^{n_y \times n_x}$. The disturbances $w_t$ and $v_t$ are independently, but not necessary identically, distributed according to the normal distributions with mean zero and covariances $Q_t \in \mathbb{R}^{n_x \times n_x}$ and $R_t \in \mathbb{R}^{n_y \times n_y}$, respectively, for all $t \in \mathbb{N}$. We assume that $R_t$ is positive definite for all $t \in \mathbb{N}$. An estimate $\hat{x}_t$ of the state $x_t$ is obtained at each time point $t \in \mathbb{N}$ according to the Kalman filter equations [Kalman, 1960, Stengel, 1994]

$$\hat{x}_{t+1} = A\hat{x}_t + Bu_t + L_{t+1}(y_{t+1} - C(A\hat{x}_t + Bu_t)) \tag{4.31a}$$

$$L_{t+1} = (AP_tA^\mathsf{T} + Q_t)C^\mathsf{T}(C(AP_tA^\mathsf{T} + Q_t)C^\mathsf{T} + R_{t+1})^{-1} \tag{4.31b}$$

$$P_{t+1} = (I - L_{t+1}C)(AP_tA^\mathsf{T} + Q_t), \tag{4.31c}$$

where we assume the unknown initial state $x_0$ to be distributed according to the normal distribution with known mean $\hat{x}_0$ and known covariance $P_0$. The following event-triggered output-feedback scheme is based on controlling (4.31a) with exactly the same approach as proposed for the state-feedback case in Section 4.2.1. An important property of the estimator in (4.31) is that the random variables $y_{t+1} - C(A\hat{x}_t + Bu_t)$ are independent for all $t \in \mathbb{N}$ and distributed according to the normal distribution with mean 0 and known covariance $C(AP_tA^\mathsf{T} + Q_t)C^\mathsf{T} + R_{t+1}$. Considering that $L_{t+1}$ is independent of the realizations of the disturbances, $\hat{w}_t := L_{t+1}(y_{t+1} - C(A\hat{x}_t + Bu_t))$ is also normally distributed with mean 0 and known covariance $\hat{Q}_t := L_{t+1}(C(AP_tA^\mathsf{T} + Q_t)C^\mathsf{T} + R_{t+1})L_{t+1}^\mathsf{T}$. Hence, (4.31a) satisfies the same assumptions as (4.17), that is, we have

$$\hat{x}_{t+1} = A\hat{x}_t + Bu_t + \hat{w}_t \tag{4.32}$$

where the disturbances $\hat{w}_t$ are independent random variables distributed according to normal distributions with mean zero and covariances[4] $\hat{Q}_t \neq 0$. Hence, we may employ the same event-triggered controller as in the state-feedback case, that is, (4.18) and (4.19) to (4.21), where every occurrence of $x_t$ is replaced with $\hat{x}_t$ and $Q_t$ is replaced with $\hat{Q}_t$ for all $t \in \mathbb{N}$. By the discussion above, Lemma 4.4 applies *mutatis mutandis* to the output feedback case. Further, it holds that

$$\begin{aligned}
\mathbb{E}[x_tx_t^\mathsf{T}] &= \mathbb{E}[(\hat{x}_t + (x_t - \hat{x}_t)(\hat{x}_t + (x_t - \hat{x}_t))^\mathsf{T}] \\
&= \mathbb{E}[\hat{x}_t\hat{x}_t^\mathsf{T}] + \mathbb{E}[\hat{x}_t(x_t - \hat{x}_t)^\mathsf{T}] + \mathbb{E}[(x_t - \hat{x}_t)\hat{x}_t^\mathsf{T}] + \mathbb{E}[(x_t - \hat{x}_t)(x_t - \hat{x}_t)^\mathsf{T}]. \tag{4.33}
\end{aligned}$$

The second and third addend in the second line are zero due to the orthogonality of the estimate and the estimation error produced by the Kalman filter, see [Kalman, 1960], if $\hat{x}_t$ is a square integrable Baire function of the random variables $\hat{w}_0, \ldots, \hat{w}_{t-1}$ and $r_1, \ldots, r_{t-1}$, compare [Doob, 1953, page 76]. The last addend in (4.33) is equal to $P_t$ in (4.31), that is, it is a quantity solely dependent on the system parameters and in particular independent of the input sequence $(u_t)_{t\in\mathbb{N}}$ and of $\hat{x}_0$. Sufficient conditions under which $P_t$ is bounded are, for example, discussed in [De Nicolao, 1992]. Hence, equivalent results as stated in Theorem 4.2 and Corollary 4.1 can also be obtained *mutatis mutandis*.

---

[4]We make the tacit assumption that $L_{t+1} \neq 0$ for all $t \in \mathbb{N}$, that is, that the output always contains information that can be used to improve the state estimate.

**Remark 4.6.** *The straight-forward extension of the results for the state-feedback case to the output-feedback case was the original motivation for considering time-varying non-identically distributed disturbances in Section 4.2.1. Note that the covariance of $\hat{w}_t$ is time-varying in general, even if the covariance matrices $Q_t$, and $R_t$ are constant. We want to emphasize here that the property of $y_{t+1} - C(A\hat{x}_t + Bu_t)$, $t \in \mathbb{N}$, being independently distributed with zero mean is a particular property of the Kalman filter, meaning that an arbitrary linear estimator cannot be employed to the same effect in this context.*

## 4.3. Summary

In this chapter, we have presented event-triggered control schemes that explicitly take information about the stochasticity of the disturbances into account. This allowed us to design the distribution of inter-transmission times *a priori*. For general probability distributions governing the disturbances, we relied on approximate numerical methods to design appropriate trigger thresholds. For the special case of normal distributions, we determined the trigger rules—based on random variables themselves—analytically. Exploiting the properties of the Kalman filter, we derived equivalent results for the output-feedback case.

# 5. Aperiodic Model Predictive Control of Constrained Linear Systems

In Chapter 3 and Chapter 4, the goal was to achieve stability in the closed-loop system while reducing the average communication rate. In this chapter and the next, we additionally impose constraints on the input and state in the form of set memberships. That is, we require $(x_t, u_t) \in \mathcal{X} \times \mathcal{U}$, $t \in \mathbb{N}$, for given compact sets $\mathcal{X} \subseteq \mathbb{R}^{n_x}$ and $\mathcal{U} \subseteq \mathbb{R}^{n_u}$. In general, it is not possible to achieve this goal for arbitrary initial conditions $x_0 \in \mathcal{X}$; counterexamples are readily obtained for unstable systems. Instead, we will be satisfied if the goal is achieved for initial conditions in a subset of $\mathcal{X}$.

In this chapter, we consider the state-feedback case, that is, the control of a system defined by

$$x_{t+1} = Ax_t + Bu_t + w_t \tag{5.1}$$

as described in Section 2.2, where $w_t \in \mathcal{W}$, $t \in \mathbb{N}$, for a known compact and convex set $\mathcal{W}$ containing the origin. As in the previous chapters, we assume that there exists a matrix $K \in \mathbb{R}^{n_u \times n_x}$ such that $A + BK$ is stable. In addition to the goal of satisfying the state and input constraints, we want to stabilize the set $\gamma \mathscr{F}_\infty$ for a chosen $\gamma \in [1, \infty)$. If we allow $u_t$ to be an explicit function of $x_t$ for all $t$, then robust MPC offers a solution to this problem; Section 2.5 gives a brief overview of the field.

Various MPC-based approaches to aperiodic control under constraints can be found in the literature. They can be roughly divided into three categories: event-triggered and self-triggered controllers with a structure as assumed in Section 2.3, and scheduling approaches, where the allocation of network resources over a certain time-span is decided online based on the solution of optimal control problems. Results of the first two types can, for example, be found in [Varutti et al., 2010, Maniatopoulos et al., 2012, Eqtami et al., 2011b, Eqtami et al., 2013, Eqtami, 2013, Lehmann et al., 2013, Liu et al., 2013, Li and Shi, 2014, Hashimoto et al., 2015a, Gommans and Heemels, 2015, Gommans, 2016, Mi and Li, 2016, Cerf et al., 2016, Zou et al., 2016, Incremona et al., 2017]. These approaches guarantee—explicitly—that the MPC value function decreases over time (whether communication occurs at a given point in time or not), trigger an event if the actual system state deviates too far from the predicted nominal system state or use a combination thereof, possibly with additional conditions enforced. An approach where the time until the predicted state is contained in a certain set around the origin is taken into account in the trigger conditions was proposed in [Hashimoto et al., 2015b]. Set-based approaches can be found in [Grüne and Müller, 2009], where events are defined in terms of a predefined partition of the state space; in [Iino et al., 2009], the scheduling function is defined in terms of set-membership constraints related to controllable sets; in [Zhe et al., 2015], robust invariant sets and robust reach sets determine the event-generating function; a self-triggered predictive control scheme based on homothetic sets was proposed in [Aydiner, 2014]. Event-triggered approaches based on the piecewise affinity of the optimal solution of the MPC problem were proposed in [Jost et al., 2015, Berner and Mönnigmann, 2016]. A scheme where events are only generated if the system state is far away from the origin can be found in [Demirel et al., 2017a]. Predictive controllers where events are considered in a network scheduling sense are, for example, investigated in [Hu et al., 2007, Varutti et al., 2009]. For schemes where MPC—or other finite horizon optimal control—techniques are employed for the allocation of network resources, also termed "scheduling", see [Zhang et al., 2007, Molin and Hirche, 2009, Lješnjanin et al., 2014, Henriksson et al., 2015]. Related to these contributions are also the *rollout*-approaches in [Antunes and Heemels, 2014, Gommans et al., 2017]. While not motivated necessarily from a networked control perspective, approaches which aim to achieve control signals that are only infrequently non-zero or only change their value infrequently are related to the goal of reducing communication in a feedback loop, and are mostly also based on optimal control concepts. See, for example, [Brockett, 1997, Gallieri and Maciejowski, 2012, Gallieri, 2016, Donkers et al., 2014, Nagahara et al., 2016, Gommans, 2016, Gommans et al., 2017] and the references therein (which are partly reported in this list).

In this thesis, we provide aperiodic extensions of the control scheme proposed in [Chisci et al., 2001], which is described, in detail, in Section 2.5. The algorithms developed here are based on the theoretical results obtained in Chapter 3; in particular, we propose event-triggered and self-triggered MPC schemes based on three different methods on guaranteeing stability of $\gamma \mathscr{F}_\infty$. First, we consider an approach based on a Lyapunov function for the closed-loop system under periodic MPC, that is, the approach in [Chisci et al., 2001]. As in Section 3.2, stability is guaranteed by, explicitly, guaranteeing that a certain function decreases over time in the closed-loop system. This approach is faced with the same difficulties as the one described in Section 3.2: as the set $\gamma \mathscr{F}_\infty$ is, in general, not explicitly available, neither is a Lyapunov function defined in terms of this set—one has to resort to upper and lower approximations. The second approach remedies this problem by employing only the value function of the MPC problem (which is not a Lyapunov function, in general), in combination with set-membership conditions equivalent to those described in Section 3.3. The third approach relies only on set-membership conditions, considerably reducing the computational effort involved, but requiring to restrict the constraints in the underlying MPC problems. For the self-triggered case, we distinguish, in each of the three approaches, between two different assumptions. In the first—and simpler—case, we assume the future inputs to the system to be given in the form a function $\kappa : \mathbb{R}^{n_x} \times \mathbb{N}$ as in Chapter 3 and merely check whether certain conditions on the worst case future states are met. In the second case, we include the future inputs as decision variables in the MPC problem solved at a given transmission time and include the conditions on the worst case future states as constraints.

The approaches in this chapter are, in parts, based on the results contained also in [Brunner et al., 2014, Brunner et al., 2015b, Brunner et al., 2016a, Brunner et al., 2016d, Brunner et al., 2017c].

## 5.1. Lyapunov-based approach

In this section, we pursue an approach mirroring the Lyapunov-based schemes proposed in Section 3.2. That is, we guarantee stability in the closed-loop system by having the controller explicitly check whether a Lyapunov function, defined for the closed-loop system under periodic control, decreases in time. We employ the robust MPC scheme presented in Section 2.5, based on [Chisci et al., 2001], as a reference controller defining this Lyapunov function.

### 5.1.1. A Lyapunov function for robust MPC

In a first step, we have to find a Lyapunov function for the controller proposed in [Chisci et al., 2001]. Here it is important to note that the MPC value function (2.19) is in general *not* positive definite with respect to the set $\mathscr{F}_\infty$, which is stabilized by the controller; in fact it holds that $\{x \in \mathbb{R}^{n_x} \mid J(x) = 0\} = \bar{\mathscr{X}}_0$, where

$$\bar{\mathscr{X}}_i := \{x \in \mathbb{R}^{n_x} \mid (A + BK)^{j-i}x \in \mathscr{X} \ominus \mathscr{F}_j,$$
$$K(A + BK)^{j-i}x \in \mathscr{U} \ominus K\mathscr{F}_j, \ j \in \{i, \ldots, N-1\}, \ (A + BK)^{N-i}x \in \mathscr{X}_\mathrm{T}\} \quad (5.2)$$

for $i \in \{0, \ldots, N-1\}$. The set $\bar{\mathscr{X}}_0$ is a subset of the set where the MPC problem is feasible: in particular, the set results from adding the additional constraints $u[i] = Kx[i]$, $i \in \{0, \ldots, N-1\}$ to $\mathscr{D}(x)$ in (2.16). Depending on the constraint sets $\mathscr{X}$ and $\mathscr{U}$, $\bar{\mathscr{X}}_0$ might be much larger than $\mathscr{F}_\infty$. Therefore, we define the *augmented value function* $V_\gamma : \mathbb{R}^{n_x} \to \mathbb{R}$ by $V_\gamma(x) := V_\gamma^\mathrm{s}(x) + J(x)$, where $V_\gamma^\mathrm{s}(x) := \min_{y \in \gamma \mathscr{F}_\infty} \bar{V}^\mathrm{s}(x - y)$, $\bar{V}^\mathrm{s} : \mathbb{R}^{n_x} \to \mathbb{R}$, and $\gamma \in [1, \infty)$. Further, we define

$$\mathscr{Y}^\gamma(x) := \{y \in \gamma \mathscr{F}_\infty \mid \bar{V}^\mathrm{s}(x - y) = V_\gamma^\mathrm{s}(x)\} \quad (5.3)$$

for all $x \in \mathbb{R}^{n_x}$ and require the following assumption to hold.

**Assumption 5.1.** *The function $\bar{V}^\mathrm{s}$ is continuous and there exist constants $\bar{c}_1$, $\bar{c}_2$, $\bar{c}_3$, $a \in (0, \infty)$ such that for all $x \in \mathbb{R}^{n_x}$ and all $v \in \mathbb{R}^{n_u}$ it holds that*

$$\bar{c}_1 |x|^a \leq \bar{V}^\mathrm{s}(x) \leq \bar{c}_2 |x|^a \quad (5.4a)$$
$$\bar{V}^\mathrm{s}(x) - \bar{V}^\mathrm{s}((A + BK)x + Bv) + \ell(v) \geq \bar{c}_3 |x|^a. \quad (5.4b)$$

**Remark 5.1.** *For quadratic $\ell$ and $a = 2$ or convex piecewise linear $\ell$ and $a = 1$, a suitable function $\bar{V}^s$ can always be constructed by multiplying a Lyapunov function of the form $x^\mathsf{T} Px$ or, respectively, $|Px|$, for the system $x_{t+1} = (A + BK)x_t$ with a sufficiently small positive scalar, compare Lemma A.7 in the appendix.*

The augmented value function indeed has the properties of a Lyapunov function:

**Lemma 5.1.** *There exists a $c_1 \in (0, \infty)$ such that $c_1 |x|^a_{\gamma \mathcal{F}_\infty} \leq V_\gamma(x)$ for all $\gamma \in [1, \infty)$ and all $x \in \mathcal{X}_\mathrm{f}$. Further, let $x_0 \in \mathcal{X}_\mathrm{f}$ and let $u_t \in \kappa_{\mathrm{MPC}}(x_t)$ for all $t \in \mathbb{N}$ for system (5.1). Then it holds that $V_\gamma(x_{t+1}) \leq V_\gamma(x_t) - c_3 |x_t|^a_{\gamma \mathcal{F}_\infty}$ for all $t \in \mathbb{N}$ and some $c_3 \in (0, \infty)$. Finally, if there exists an $\epsilon > 0$ such that $\gamma \mathcal{F}_\infty \oplus \epsilon \mathcal{B}_{n_x} \subseteq \bar{\mathcal{X}}_0$, then there exists a constant $c_2 \in (0, \infty)$ such that $V_\gamma(x) \leq c_2 |x|^a_{\gamma \mathcal{F}_\infty}$ for all $x \in \mathcal{X}_\mathrm{f}$.*

Lemma 5.1, together with Theorem 2.1, immediately leads us to the stability result below.

**Theorem 5.1.** *Let $x_0 \in \mathcal{X}_\mathrm{f}$ and let $u_t \in \kappa_{\mathrm{MPC}}(x_t)$ for all $t \in \mathbb{N}$ for system (5.1). It holds that $\lim_{t \to \infty} |x_t|_{\gamma \mathcal{F}_\infty} = 0$. If additionally there exists an $\epsilon > 0$ such that $\gamma \mathcal{F}_\infty \oplus \epsilon \mathcal{B}_{n_x} \subseteq \bar{\mathcal{X}}_0$, then $\gamma \mathcal{F}_\infty$ is 1-uniformly asymptotically stable for any dynamical system generating $(x_t)_{t \in \mathbb{N}}$ with $\mathcal{X}_\mathrm{f}$ belonging to its region of attraction.*

**Remark 5.2.** *The main idea in the stability proof is to augment the original value function with a term penalizing the distance to the set $\gamma \mathcal{F}_\infty$. This approach is related to robust predictive control schemes where the distance to a set is explicitly included in the cost function that is optimized online, see for example [Scokaert and Mayne, 1998], [Raković et al., 2012a], [Raković, 2012, Section 3.3], and the references in the latter work. An important difference in the approach here is that we do not optimize over this distance online, but only use it in the a priori stability analysis, similar to the analysis in [Ghaemi et al., 2008]. However, in some of the approaches in Section 5.1, we will optimize over future worst case values of the augmented cost function, resulting in optimization problems of similar complexity.*

### 5.1.2. Relaxing the rate of decrease

The Lyapunov-based trigger rules in Section 3.2 required a rate of decrease in the form $V_\gamma(x_{t+1}) \leq \lambda V_\gamma(x_t)$. Lemma 5.1 immediately (if the upper bound can be established) yields such an inequality with $\lambda = 1 - c_3/c_2$. However, we expect this rate to be a conservative overestimate in general. For this reason, and for the sake of consistency with the subsequent sections in which $V_\gamma^s$ is not directly employed, we pursue a slightly different approach, exploiting another property of the augmented value function:

**Lemma 5.2.** *Assume that there exists an $\epsilon > 0$ such that $\gamma(A + BK)^i \mathcal{F}_\infty \oplus \epsilon \mathcal{B}_{n_x} \subseteq \bar{\mathcal{X}}_i$, $i \in \{0, \dots, N - 1\}$. Let $x \in \mathcal{X}_\mathrm{f}$, $((x[i])_{i \in \mathbb{N}}, (u[i])_{i \in \mathbb{N}}) \in \mathcal{T}(x)$, and $y \in \mathcal{Y}^\gamma(x)$. Then it* *holds that*

$$\bar{V}^{\rm s}(x - y) - \bar{V}^{\rm s}(x[i] - (A + BK)^i y) + \sum_{j=0}^{i-1} \ell(u[j] - Kx[j]) \geq (1 - \lambda^i)V_\gamma(x) \qquad (5.5)$$

*for all $i \in \mathbb{N}$ and some $\lambda \in [0, 1)$ which is independent of $x$ and $y$.*

The inequality in (5.5) allows us to check for an exponential decrease of the augmented value function without knowing the rate $\lambda$. Further, the following technical relation holds, which allows a relaxation of the decrease condition.

**Proposition 5.1.** *Let $\lambda, \nu \in [0, 1)$. There exists a $\mu \in [0, 1)$ such that*

$$1 - (1 - \nu^\tau)(1 - \lambda^\tau) \leq \mu^\tau \qquad (5.6)$$

*for all $\tau \in \mathbb{N}$.*

Using $\nu \in [0, 1)$ as a tuning parameter, we now construct trigger conditions based on another result involving the augmented value function.

**Theorem 5.2.** *Let a dynamical system $\mathfrak{f} : \mathcal{S} \to (\mathbb{R}^n)^{\mathbb{N}}$ be given. Assume that for all $(x_t)_{t \in \mathbb{N}} = \mathfrak{f}(s)$ with $s \in \mathcal{S}$ and $x_0 \in \mathcal{X}_{\rm f}$, and all $t \in \mathbb{N}$, there exists a $\tau \in \{1, \ldots, \min\{\theta, t+ 1\}\}$ such that*

$$V_\gamma(x_{t+1}) \leq V_\gamma(x_{t+1-\tau}) - (1 - \nu^\tau)\Bigg( \bar{V}^{\rm s}(x_{t+1-\tau} - y_{t+1-\tau})$$

$$- \bar{V}^{\rm s}(x_{t+1-\tau}[\tau] - (A + BK)^\tau y_{t+1-\tau}) + \sum_{k=0}^{\tau-1} \ell(u_{t+1-\tau}[k] - Kx_{t+1-\tau}[k]) \Bigg), \quad (5.7)$$

*where $((x_{t+1-\tau}[i])_{i \in \mathbb{N}}, (u_{t+1-\tau}[i])_{i \in \mathbb{N}}) \in \mathcal{T}(x_{t+1-\tau})$ and $y_{t+1-\tau} \in \mathcal{Y}^\gamma(x_{t+1-\tau})$. Then, it holds that $x_t \in \mathcal{X}$ for all $t \in \mathbb{N}$. If, additionally, there exists an $\epsilon > 0$ such that $\gamma(A + BK)^i \mathcal{F}_\infty \oplus \epsilon \mathcal{B}_{n_x} \subseteq \bar{\mathcal{X}}_i$ for all $i \in \{0, \ldots, N-1\}$, then the set $\gamma \mathcal{F}_\infty$ is $\theta$-uniformly asymptotically stable for $\mathfrak{f}$ with $\mathcal{X}_{\rm f}$ belonging to its region of attraction.*

*Proof.* A simple inductive argument shows that $V_\gamma(x_t)$ is finite for all $t \in \mathbb{N}$, which ensures $x_t \in \mathcal{X}_{\rm f} \subseteq \mathcal{X}$. Further, if there exists an $\epsilon > 0$ such that $\gamma(A+BK)^i \mathcal{F}_\infty \oplus \epsilon \mathcal{B}_{n_x} \subseteq \bar{\mathcal{X}}_i$ for all $i \in \{0, \ldots, N-1\}$, Lemma 5.1, Lemma 5.2, and Proposition 5.1 together imply that the conditions of Theorem 2.1 are satisfied, establishing $\theta$-uniform asymptotic stability of $\gamma \mathcal{F}_\infty$. $\qquad \square$

Defining the decrease condition according to (5.7), in contrast to the approach in Section 3.2, has the advantage that for all $\nu > 0$ the requirements are relaxed compared with the guaranteed decrease for a feedback that is updated at every point in time. In Section 3.2, we had to choose $\nu$ greater than $\lambda$, which had to be known. Note that $\lambda$ does not appear in (5.7).

### 5.1.3. Aperiodic control algorithms

Based on the theoretical results obtained so far, we now propose event-triggered and self-triggered predictive control algorithms for system (5.1). The implementation of these algorithms is discussed in Section 5.1.4.

Similar to the aperiodic controllers described in Chapter 3, communication will only take place at the transmission instances $t_i$, $i \in \mathbb{N}$, which will be determined either in an event-triggered or a self-triggered fashion, as defined in Section 2.3. Further, we employ a function $\kappa : \mathbb{R}^{n_x} \times \mathbb{N} \to \mathbb{R}^{n_u}$ generating the input to the system, where the first argument is the state at the last transmission instant and the second argument the time since the last transmission instant. Again, various definitions of $\kappa$ are possible. Analogously to the unconstrained case, we require $\kappa(x, 0) \in \kappa_{\mathrm{MPC}}(x)$ for all $x \in \mathscr{X}_{\mathrm{f}}$. That is, at every transmission instance the input to the system is consistent with the MPC control law. This ensures, by the proof of Lemma 5.1, that (5.7) holds with $\tau = 1$ whenever $x_{t+1} = A x_t + B \kappa(x_t, 0) + w_t$, if $x_t \subset \mathscr{X}_{\mathrm{f}}$ and $w_t \in \mathscr{W}$. Additionally, we require that $\kappa(x, \tau) \in \mathscr{U}$ for all $x \in \mathscr{X}_{\mathrm{f}}$ and all $\tau \in \mathbb{N}$. Of particular interest is the choice of

$$\kappa(x, \tau) = u[\tau] \tag{5.8}$$

for some $((x[i])_{i \in \mathbb{N}}, (u[i])_{i \in \mathbb{N}}) \in \mathscr{T}(x)$. This is the analogous form of the "model-based" controller, and has been employed, for example, in [Varutti et al., 2009, Eqtami et al., 2010, Lehmann et al., 2013].

In order to simplify the exposition in the remainder of the section, we define, for the respective closed-loop systems, the quantities

$$W_t^{\mathrm{et}} := \max_{t^{-1} \in \mathbb{N} \cap \{t - \theta + 1, \ldots, t\}} V_\gamma(x_{t^{-1}}) - (1 - \nu^\tau) \bigg( \bar{V}^{\mathrm{s}}(x_{t^{-1}} - y_{t^{-1}}) -$$

$$\bar{V}^{\mathrm{s}}(x_{t^{-1}}[\tau] - (A + BK)^\tau y_{t^{-1}}) + \sum_{j=0}^{\tau-1} \ell(u_{t^{-1}}[j] - K x_{t^{-1}}[j]) \bigg) \tag{5.9}$$

for all $t \in \mathbb{N}$, where $\tau = t + 1 - t^{-1}$, and

$$W_{i,k}^{\mathrm{st}} := \max_{t^{-1} \in \{t_0, \ldots, t_i\} \cap \{k - \theta, \ldots, t_i\}} V_\gamma(x_{t^{-1}}) - (1 - \nu^\tau) \bigg( \bar{V}^{\mathrm{s}}(x_{t^{-1}} - y_{t^{-1}})$$

$$- \bar{V}^{\mathrm{s}}(x_{t^{-1}}[\tau] - (A + BK)^\tau y_{t^{-1}}) + \sum_{j=0}^{\tau-1} \ell(u_{t^{-1}}[j] - K x_{t^{-1}}[j]) \bigg) \tag{5.10}$$

for all $i \in \mathbb{N}$ and all $k \in \{t_i + 1, \ldots, t_i + \theta\}$, where $\tau = k - t^{-1}$. In both cases, it holds that $(x_{t^{-1}}[j])_{j \in \mathbb{N}}, (u_{t^{-1}}[j])_{j \in \mathbb{N}}) \in \mathscr{T}(x_{t^{-1}})$ and $y_{t^{-1}} \in \mathscr{Y}^\gamma(x_{t^{-1}})$. The expressions in (5.9) and (5.10) provide the largest values for the right-hand side of (5.7), under certain restrictions on $\tau$.

**Event-triggered control**

Algorithm 3 defines the closed-loop control of system (5.1) with an event-triggered control law. The idea is to check, at every time point $t \in \{1, \ldots\}$ whether the inequality (5.7) is satisfied without transmitting the current state to the actuator.

---

**Algorithm 3** Lyapunov-based event-triggered predictive control

---

1: $i \leftarrow 0$
2: $t \leftarrow 0$
3: $t_0 \leftarrow 0$
4: Solve the optimization problem in (2.19) for $x = x_0$ and obtain $(\mathbf{x}_0, \mathbf{u}_0) \in \mathscr{T}(x_0)$ and $y_0 \in \mathscr{Y}^\gamma(x_0)$.
5: Apply $u_0 = \kappa(x_0, 0)$ to the system.
6: **for** $t \in \{1, \ldots\}$ **do**
7:    Solve the optimization problem in (2.19) for $x = x_t$ and obtain $(\mathbf{x}_t, \mathbf{u}_t) \in \mathscr{T}(x_t)$, and $y_t \in \mathscr{Y}^\gamma(x_t)$.
8:    Compute $W_t^{\text{et}}$.
9:    **if** $\max_{w \in \mathscr{W}} V_\gamma(Ax_t + B\kappa(x_{t_i}, t - t_i) + w) > W_t^{\text{et}}$ **then**
10:       $i \leftarrow i + 1$
11:       $t_i \leftarrow t$
12:    **end if**
13:    Apply $u_t = \kappa(x_{t_i}, t - t_i)$ to the system.
14: **end for**

---

**Theorem 5.3.** *Let $x_0 \in \mathcal{X}_\mathrm{f}$ and let $u_t$ be determined according to Algorithm 3 for all $t \in \mathbb{N}$. Then, it holds that $(x_t, u_t) \in \mathcal{X} \times \mathcal{U}$ for all $t \in \mathbb{N}$. Further, if there exists an $\epsilon > 0$ such that $\gamma(A + BK)^i \mathscr{F}_\infty \oplus \epsilon \mathscr{B}_{n_x} \subseteq \bar{\mathcal{X}}_i$ for all $i \in \{0, \ldots, N - 1\}$, then the set $\gamma \mathscr{F}_\infty$ is $\theta$-uniformly asymptotically stable for any dynamical system generating $(x_t)_{t \in \mathbb{N}}$, with $\mathcal{X}_\mathrm{f}$ belonging to its region of attraction.*

**Self-triggered control**

For simplicity, we restrict $\tau$ in (5.7), such that only system states at *past* time-steps have to be evaluated. Additionally, we enforce $t_{i+1} - t_i \leq \theta$, compare the discussion in Section 3.2.3. Finally, we assume the system state only to be available as a measurement at the transmission times $t_i$ and, for simplicity, only use the value of the augmented value function at these points in time as a reference for the decrease conditions.

We distinguish between two strategies. The first strategy is analogous to the self-triggered schemes in Section 3.2 and is based on checking whether the input generated by $\kappa$ guarantees (5.7) under the worst-case assumed future disturbances over the scheduled time span until the next sampling instant. Algorithm 4 summarizes this approach.

---

**Algorithm 4** Lyapunov-based self-triggered predictive control with pre-defined inputs

1: $i \leftarrow -1$
2: $t \leftarrow 0$
3: $t_0 \leftarrow 0$
4: **for** $t \in \{0, \ldots\}$ **do**
5:    **if** $t = t_{i+1}$ **then**
6:       $i \leftarrow i + 1$
7:       Solve the optimization problem in (2.19) for $x = x_t$ and obtain $(\mathbf{x}_t, \mathbf{u}_t) \in \mathcal{T}(x_t)$
        and $y_t \in \mathcal{Y}^\gamma(x_0)$.
8:       **for** $k \in \{t_i + 1, \ldots, t_i + \theta\}$ **do**
9:          Compute $W_{i,k}^{\text{st}}$.
10:      **end for**
11:      $t_{i+1} \leftarrow \sup \left\{ t' \in \{t_i + 1, \ldots, t_i + \theta\} \; \middle| \; \forall k \in \{t_i + 1, \ldots, t'\}, \right.$

$$\left. \max_{w[l] \in \mathcal{W}, \, l \in \{t_i, \ldots, k-1\}} V_\gamma \left( A^{k - t_i} x_{t_i} + \sum_{l = t_i}^{k-1} A^{k - 1 - l} (B \kappa(x_{t_i}, l - t_i) + w[l]) \right) \leq W_{i,k}^{\text{st}} \right\}$$

12:    **end if**
13:    Apply $u_t = \kappa(x_{t_i}, t - t_i)$ to the system.
14: **end for**

---

**Theorem 5.4.** *Let $x_0 \in \mathcal{X}_{\mathrm{f}}$ and let $u_t$ be determined according to Algorithm 4 for all $t \in \mathbb{N}$. Then, it holds that $(x_t, u_t) \in \mathcal{X} \times \mathcal{U}$ for all $t \in \mathbb{N}$. Further, if there exists an $\epsilon > 0$ such that $\gamma(A + BK)^i \mathcal{F}_\infty \oplus \epsilon \mathcal{B}_{n_x} \subseteq \bar{\mathcal{X}}_i$ for all $i \in \{0, \ldots, N - 1\}$, then the set $\gamma \mathcal{F}_\infty$ is $\theta$-uniformly asymptotically stable for any dynamical system generating $(x_t)_{t \in \mathbb{N}}$, with $\mathcal{X}_{\mathrm{f}}$ belonging to its region of attraction.*

*Proof.* The statement follows by the same reasoning employed to prove Theorem 5.3. $\square$

In the second strategy, we define an MPC problem where the time until the next sampling instant is an optimization variable and (5.7) is included as a constraint, in the spirit of [Gommans and Heemels, 2015]. Needless to say, the computational effort involved with the second strategy can be expected to be much higher than with the first one (in [Gommans and Heemels, 2015], the computational complexity was only moderately higher than for a standard MPC problem, which was achieved by a careful rearrangement of the optimization problem). However, the additional degrees of freedom offered may decrease the overall average sampling rate. Several possible relaxations will be discussed in Section 5.1.4. We now define the MPC optimization problem for this second strategy. Define first the set of admissible nominal input sequences for a given point $x \in \mathbb{R}^{n_x}$, an assumed time-span of $M \in \{1, \ldots\} \cup \{\infty\}$ until the next sampling instant, and given $W_k \in \mathbb{R}$, $k \in \{1, \ldots, M\}$, that is,

$$\mathscr{D}_M^{\mathrm{L}}(x, W_1, \ldots, W_M) := \Bigg\{ (u[i])_{i \in \{0,\ldots,M-1\}} \in \mathbb{R}^{Mn_{\mathrm{u}}} \Bigg|$$

$$(u[0], \ldots, u[M-1]) \in C_M, \tag{5.11a}$$

$$x[0] = x, \tag{5.11b}$$

$$x[i+1] = Ax[i] + Bu[i], \tag{5.11c}$$

$$i \in \{0, \ldots, M-1\},$$

$$\forall w[j] \in \mathscr{W}, \; j \in \{0, \ldots, M-1\}, \; \forall k \in \{1, \ldots, M\},$$

$$V_\gamma \left( x[k] + \sum_{j=0}^{k-1} A^{k-1-j} w[j] \right) \le W_k \tag{5.11d}$$

$$\Bigg\}.$$

For the constraint (5.11a), we make the assumption that $C_M \subseteq \mathscr{U}^M$ for all $M \in \{1, \ldots\} \cup \{\infty\}$ and $C_1 = \mathscr{U}$. This constraint can be employed to enforce additional desired properties of the input between transmission instants, such as $u[i] = 0$ or $u[i] = u[0]$ for $i \in \{1, \ldots, M-1\}$, in accordance with the "to-zero" or "to-hold" paradigms in the controller types listed in (3.6), compare [Gommans and Heemels, 2015, Section 3.2].

**Remark 5.3.** *Note that explicit constraints on the predicted state are absent in (5.11). However, (5.11d) enforces robust constraint satisfaction, due to the definition of $V_\gamma$.*

Algorithm 5 describes the self-triggered closed-loop predictive control with optimized inputs.

**Theorem 5.5.** *Let $x_0 \in \mathscr{X}_{\mathrm{f}}$ and let $u_t$ be determined according to Algorithm 5 for all $t \in \mathbb{N}$. Then, it holds that $(x_t, u_t) \in \mathscr{X} \times \mathscr{U}$ for all $t \in \mathbb{N}$. Further, if there exists an $\epsilon > 0$ such that $\gamma(A + BK)^i \mathscr{F}_\infty \oplus \epsilon \mathscr{B}_{n_x} \subseteq \mathscr{X}_i$ for all $i \in \{0, \ldots, N-1\}$, then the set $\gamma \mathscr{F}_\infty$ is $\theta$-uniformly asymptotically stable for any dynamical system generating $(x_t)_{t \in \mathbb{N}}$, with $\mathscr{X}_{\mathrm{f}}$ belonging to its region of attraction.*

**Remark 5.4.** *The choice in step 13 of Algorithm 5 may be performed by any optimization over the set $\mathscr{D}_M^{\mathrm{L}}(x_{t_i}, W_{i,t_i+1}^{\mathrm{st}}, \ldots, W_{i,t_i+M}^{\mathrm{st}})$.*

## 5.1.4. Implementation

We assume that the sets $\mathscr{X}$, $\mathscr{U}$, $\mathscr{W}$, and $\mathscr{X}_{\Gamma}$ are polytopes and that the functions $\ell$ and $\bar{V}^{\mathrm{s}}$ are convex. Our goal is to implement the algorithms by solving convex programs,

---

**Algorithm 5** Lyapunov-based self-triggered predictive control with optimized inputs

---

1: $i \leftarrow -1$
2: $t \leftarrow 0$
3: $t_0 \leftarrow 0$
4: **for** $t \in \{0, \ldots\}$ **do**
5:   **if** $t = t_{i+1}$ **then**
6:     $i \leftarrow i + 1$
7:     Solve the optimization problem in (2.19) for $x = x_t$ and obtain $(\mathbf{x}_t, \mathbf{u}_t) \in \mathscr{T}(x_t)$
    and $y_t \in \mathscr{Y}^\gamma(x_t)$.
8:     **for** $k \in \{t_i + 1, \ldots, t_i + \theta\}$ **do**
9:       Compute $W_{i,k}^{\mathrm{st}}$.
10:     **end for**
11:     $M^\star \leftarrow \sup\{M \in \{1, \ldots, \theta\} \mid \mathscr{D}_M^{\mathrm{L}}(x_{t_i}, W_{i,t_i+1}^{\mathrm{st}}, \ldots, W_{i,t_i+M}^{\mathrm{st}}) \neq \emptyset\}$
12:     $t_{i+1} \leftarrow t_i + M^\star$
13:     $(u[i])_{i \in \{0, \ldots, M^\star - 1\}} \leftarrow (u^\star[i])_{i \in \{0, \ldots, M^\star - 1\}}$ for some
    $(u^\star[i])_{i \in \{0, \ldots, M^\star - 1\}} \in \mathscr{D}_{M^\star}^{\mathrm{L}}(x_{t_i}, W_{i,t_i+1}^{\mathrm{st}}, \ldots, W_{i,t_i+M^\star}^{\mathrm{st}})$.
14:   **end if**
15:   Apply $u_t = u[t - t_i]$ to the system.
16: **end for**

---

preferably with only linear constraints. In order to implement the presented algorithms, it is necessary to evaluate the function $V_\gamma^{\mathrm{s}}$, in particular in order to compute the quantities $W_t^{\mathrm{et}}$ and $W_{i,k}^{\mathrm{st}}$ in (5.9) and (5.10). As discussed in Section 3.2.3, it is usually only possible to compute upper and lower approximations of the values of $V_\gamma^{\mathrm{s}}$, as the set $\gamma \mathscr{F}_\infty$ can, in general, not be represented by a finite number of inequalities. In the following, we assume that upper and lower bounds on $V_\gamma^{\mathrm{s}}$ and approximations $\Omega_1 \supseteq \gamma \mathscr{F}_\infty \supseteq \Omega_2$ are available, if required. For details, please refer to Chapter 3. In order to retain the guarantees of the algorithms, it is necessary to obtain lower approximations of $W_t^{\mathrm{et}}$ and $W_{i,k}^{\mathrm{st}}$. Such approximations are, for example, achieved by replacing

$$V_\gamma(x_{t-1}) - (1 - \nu^\tau)\left(\bar{V}^{\mathrm{s}}(x_{t-1} - y_{t-1}) - \bar{V}^{\mathrm{s}}(x_{t-1}[\tau] - (A + BK)^\tau y_{t-1})\right.$$
$$\left. + \sum_{j=0}^{\tau-1} \ell(u_{t-1}[j] - Kx_{t-1}[j])\right) \quad (5.12)$$

in the definitions by

$$J(x_{t-1}) + \min_{y \in \Omega_1} \left( \bar{V}^{\mathrm{s}}(x_{t-1} - y) + (1 - \nu^\tau)\bar{V}^{\mathrm{s}}(x_{t-1}[\tau] - (A + BK)^\tau y) \right)$$

$$- (1 - \nu^\tau) \left( \min_{y \in \Omega_2} \bar{V}^{\mathrm{s}}(x_{t-1} - y) + \sum_{j=0}^{\tau-1} \ell(u_{t-1}[j] - Kx_{t-1}[j]) \right). \quad (5.13)$$

Note that the minimization in the second addend introduces conservatism as we do not employ the (unknown) quantity $y_{t-1}$, which appears multiple times the original expression. With the approximations above, the trigger conditions in Algorithm 3 (step 8) and Algorithm 4 (step 11) can be implemented if upper bounds on expressions of the form

$$\max_{w \in \mathcal{G}_i} V_\gamma(x + w), \quad (5.14)$$

can be obtained, where $\mathcal{G}_i$ are arbitrary polytopes satisfying $\bigoplus_{j=0}^{i-1} A^j \mathcal{W} \subseteq \mathcal{G}_i \subseteq \mathbb{R}^{n_x}$ for $i \in \mathbb{N}$. Considering the convexity of the optimization problems involved, an upper approximation of (5.14) is achieved by evaluating

$$\min_{y \in \Omega_2} \bar{V}^{\mathrm{s}}(x + v - y) + J(x + v) \quad (5.15)$$

for $v$ taking the value of every vertex of $\mathcal{G}_i$. The computational effort involved with evaluating the second addend is expected to be much higher than that involved with evaluating the first; in the following, we describe how an upper bound on $J(x + v)$, for every $v \in \bigoplus_{j=0}^{i-1} A^j \mathcal{W}$, can be obtained by solving a single optimization problem. The idea is to let a single predicted nominal trajectory start at $x$ and parameterize a whole family of trajectories by $v$ which are generated from the predicted trajectory starting at $x$ by including a feedback term of the form $Kv$. In other words, if $(u[0], u[1] \ldots, u[N-1])$ is the input computed for $x$, then $(u[0] + K(v), u[1] + K(A + BK)v, \ldots, u[N-1] + K(A + BK)^{N-1}v)$ is the input sequence for $x + v$. This approach is based on the "advanced-step" technique, proposed in [Zavala and Biegler, 2009] to cope with computational delay in MPC. In order to guarantee that the input sequence including the feedback $Kv$ is feasible at $x + v$, the constraints in the MPC problem have to be tightened further: we define the modified constraint sets

$$\bar{\mathscr{D}}^k(x) := \left\{ (u[i])_{i \in \{0,\ldots,N-1\}} \in \mathbb{R}^{Nn_u} \; \middle| \right.$$

$$x[0] = x, \tag{5.16a}$$

$$x[i+1] = Ax[i] + Bu[i], \tag{5.16b}$$

$$u[i] \in \mathscr{U} \ominus K\mathscr{F}_i \ominus K(A+BK)^i \bigoplus_{l=0}^{k-1} A^l \mathscr{W}, \tag{5.16c}$$

$$x[i] \in \mathscr{X} \ominus \mathscr{F}_i \ominus (A+BK)^i \bigoplus_{l=0}^{k-1} A^l \mathscr{W}, \tag{5.16d}$$

$$i \in \{0,\ldots,N-1\},$$

$$x[N] \in \mathscr{X}_{\mathrm{T}} \ominus (A+BK)^N \bigoplus_{l=0}^{k-1} A^l \mathscr{W} \tag{5.16e}$$

$$\left. \vphantom{\bigoplus_{l=0}^{k-1}} \right\}$$

for $k \in \mathbb{N}$. It readily follows that if $\mathbf{u} := (u[0], u[1], \ldots, u[N-1]) \in \bar{\mathscr{D}}^k(x)$, then $\tilde{\mathbf{u}} := (u[0] + K\bar{w}, u[1] + K(A+BK)\bar{w} \ldots, u[N-1] + K(A+BK)^{N-1}\bar{w}) \in \mathscr{D}(x+\bar{w})$ and $\bar{J}(x, \mathbf{u}) = \bar{J}(x+\bar{w}, \tilde{\mathbf{u}})$ for all $\bar{w} \in \bigoplus_{j=0}^{k-1} A^j \mathscr{W}$. Hence, we have

$$\max_{\bar{w} \in \bigoplus_{j=0}^{k-1} A^j \mathscr{W}} J(x+\bar{w}) \leq \inf \left\{ \bar{J}(x, \mathbf{u}) \;\middle|\; \mathbf{u} \in \bar{\mathscr{D}}^k(x) \right\}. \tag{5.17}$$

Determining the right hand side of this inequality requires the solution of a single optimization problem.

Consider now Algorithm 5. With the approximations described so far, the algorithm can be implemented using convex programming by introducing additional decision variables, namely, predicted trajectories for every combination of $w_j \in \mathscr{W}$, $j \in \{0,\ldots,M-1\}$, for all $k \in \{1,\ldots,M\}$ in order to define the value of $V_\gamma$ in (5.11d). Of course, if the number of constraints and variables are to be finite, the same must hold for $M$. The number of decision variables can be reduced considerably if the modified constraint sets (5.16) are employed in order to define an upper bound on the values of $V_\gamma$ in (5.11d). In particular, we let a single nominal trajectory $(x[j|k])_{j \in \{0,\ldots,N-k\}}$ satisfying constraints similar to those in (5.16) start at each nominal predicted state $x[k]$ for $k \in \{1,\ldots,M\}$ in order to obtain an upper bound on $J$ for the future states in the time span $\{t_i + 1, \ldots, t_{i+1}\}$. Let $M$ be restricted to the set $\{1,\ldots,N\}$ and let

$\mathscr{D}_M^{\mathrm{L}}(x, W_1, \ldots, W_M)$ be replaced with

$$\bar{\mathscr{D}}_M^{\mathrm{L}}(x, W_1, \ldots, W_M) :=$$

$$\left\{ (u[i])_{i \in \{0, \ldots, M-1\}} \in \mathbb{R}^{Mn_{\mathrm{u}}} \;\middle|\; \right.$$

$$(u[0], \ldots, u[M-1]) \in C_M, \tag{5.18a}$$

$$x[0] = x, \tag{5.18b}$$

$$x[i+1] = Ax[i] + Bu[i], \tag{5.18c}$$

$$i \in \{0, \ldots, M-1\},$$

$$x[0|k] = x[k], \tag{5.18d}$$

$$x[j+1|k] = Ax[j|k] + Bu[j|k], \tag{5.18e}$$

$$u[j|k] \in \mathscr{U} \ominus K\mathscr{F}_j \ominus K(A+BK)^j \bigoplus_{l=0}^{k-1} A^l \mathscr{W}, \tag{5.18f}$$

$$x[j|k] \in \mathscr{X} \ominus \mathscr{F}_j \ominus (A+BK)^j \bigoplus_{l=0}^{k-1} A^l \mathscr{W}, \tag{5.18g}$$

$$j \in \{0, \ldots, N-k-1\},$$

$$x[N-k|k] \in \mathscr{X}_{\mathrm{T}}^k \ominus (A+BK)^{N-k} \bigoplus_{l=0}^{k-1} A^l \mathscr{W}, \tag{5.18h}$$

$$\max_{w \in \mathscr{G}_k} \min_{y \in \Omega_2} \bar{V}^{\mathrm{s}}(x[k] + w - y) + \sum_{j=0}^{N-k-1} \ell(u[j|k] - Kx[j|k]) \leq W_k, \tag{5.18i}$$

$$\left. k \in \{1, \ldots, M\} \right\},$$

where

$$\mathscr{X}_{\mathrm{T}}^k := \bigcap_{i \in \{N-k, \ldots, N-1\}} \left( (A+BK)^{i-(N-k)} \right)^{-1} \left( (\mathscr{X} \cap K^{-1}\mathscr{U}) \ominus \mathscr{F}_i \right) \cap \left( (A+BK)^k \right)^{-1} \mathscr{X}_{\mathrm{T}} \tag{5.19}$$

for all $k \in \{1, \ldots, M\}$.

The following result justifies the approach.

**Lemma 5.3.** *Let* $(u[0], u[1], \ldots, u[M-1]) \in \bar{\mathscr{D}}_M^{\mathrm{L}}(x, W_1, \ldots, W_M)$ *for some arbitrary* $M \in \{1, \ldots, N\}$, $x \in \mathbb{R}^{n_{\mathrm{x}}}$, *and* $(W_1, \ldots, W_M) \in \mathbb{R}^M$. *Let further* $\bar{x} = A^k x + \sum_{j=0}^{k-1} A^{k-1-j} (Bu[j] + w_j)$ *for some* $w_j \in \mathscr{W}$, $j \in \{0, \ldots, k-1\}$ *and some* $k \in \{1, \ldots, M\}$. *Then,* $V_\gamma(\bar{x}) \leq W_k$.

Note that the constraints defined by (5.18i) are linear constraints if $\ell$ is a convex piecewise affine function and are second-order cone constraints if $\ell$ is convex and quadratic.

In the actual implementation, we do not evaluate the set $\mathscr{D}_M^L$ for $M = 1$, but instead replace $u^\star[0]$ with $u_t[0]$ where $(u_t[i])_{i\in\mathbb{N}} = \mathbf{u}_t$ if we can infer $M^\star \leq 1$ in step 11 of Algorithm 5 and, furthermore, enforce $M^\star \geq 1$ whenever $x_{t_i} \in \mathcal{X}_f$.

## 5.2. Mixed set–Lyapunov approach

The Lyapunov-based approaches proposed in the previous section suffer from the same problems as their unconstrained counterparts in Section 3.2. In particular, the set $\gamma\mathscr{F}_\infty$ is usually not available explicitly such that outer and inner approximations have to be employed; furthermore, the effort involved with computing the worst-case future value of $V_\gamma^s$ may be undesirably high.

### 5.2.1. Feasibility by value function decrease, stability by set-membership condition

In this section, we propose a slightly modified approach where only the MPC value function $J$ is employed in the trigger conditions and stability is ensured by adapting the results of Section 3.3. As already pointed out in the previous section, the value function $J$ of the robust MPC scheme is zero on a certain set $\mathcal{X}_0$, which is potentially much larger set than $\gamma\mathscr{F}_\infty$; ensuring that $J$ decreases by a certain amount (which is necessarily also zero on $\mathcal{X}_0$) over time does not necessarily imply attractivity or stability[1] of $\gamma\mathscr{F}_\infty$. On the other hand, one might be tempted to adapt the results of Section 3.3 directly by (only) ensuring that for all $t \in \mathbb{N}$ it holds that $x_t \in \{x[\tau]\} \oplus \gamma\mathscr{F}_\tau$, where $\tau \in \{1, \ldots, \min\{t, \theta\}\}$ and $(x[0], \ldots, x[N-1])$ is a trajectory resulting from an optimal solution to the MPC problem at the state $x_{t-\tau}$. However, if $\gamma > 1$, it is not necessarily true that $\{x[\tau]\} \oplus \gamma\mathscr{F}_\tau \subseteq \mathcal{X}_f$ or even $\{x[\tau]\} \oplus \gamma\mathscr{F}_\tau \subseteq \mathcal{X}$, that is, by only ensuring the simple set membership $x_t \in \{x[\tau]\} \oplus \gamma\mathscr{F}_\tau$, it might very well happen that the MPC problem is no longer feasible or that the state constraints are violated at time $t$. As it turns out, a *combination* of a decrease condition on $J$ with the set-membership condition described above does guarantee the desired closed-loop properties.

First we strengthen (5.4b) in Assumption 5.1 to

$$\bar{V}^s(x) - \bar{V}^s((A + BK)x + Bv) + (1 - \nu)\ell(v) \geq \bar{c}_3|x|^a, \tag{5.20}$$

where $\nu \in [0, 1)$ is the chosen relaxation parameter for the decrease condition.

**Remark 5.5.** *If Assumption 5.1 holds in its original form, then the strengthened condition can be satisfied by replacing $\bar{V}^s$ with $(1-\nu)\bar{V}^s$ and $\bar{c}_i$ with $(1-\nu)\bar{c}_i$ for $i \in \{1, 2, 3\}$.*

---

[1]This reasoning does work for the original scheme proposed in [Chisci et al., 2001], because therein the input applied to the system at a given point in time is always optimal with respect to the value function evaluated at the current system state, implying $u_t = Kx_t$ if $J(x_t) = 0$. In our aperiodic framework we do not want the input to be explicitly dependent on the current system state at every point in time, however.

The following result allows us to obtain suitable trigger conditions that guarantee that the requirements of Theorem 5.2 hold in the closed-loop system, without taking $V_\gamma^s$ into account explicitly.

**Lemma 5.4.** *Let*

$$J(x_{t+1}) \leq J(x_{t+1-\tau}) - (1 - \nu^{\tau+1}) \sum_{k=0}^{\tau-1} \ell(u_{t+1-\tau}[k] - Kx_{t+1-\tau}[k]) \tag{5.21a}$$

*and*

$$x_{t+1} \in \{x_{t+1-\tau}[\tau]\} \oplus \gamma \mathscr{F}_\tau, \tag{5.21b}$$

*for some* $\tau \in \{1, \ldots, t+1\}$, *where* $((x_{t+1-\tau}[i])_{i \in \mathbb{N}}, (u_{t+1-\tau}[i])_{i \in \mathbb{N}}) \in \mathscr{T}(x_{t+1-\tau})$. *Let further* $y_{t+1-\tau} \in \mathscr{Y}^\gamma(x_{t+1-\tau})$. *Then,* (5.7) *holds.*

## 5.2.2. Aperiodic control algorithms

Mirroring the presentation in Section 5.1.3, we describe an event-triggered and two self-triggered schemes based on Lemma 5.4.

### Event-triggered control

Algorithm 6 defines the closed-loop control of system (5.1) with an event-triggered control law.

**Theorem 5.6.** *Let* $x_0 \in \mathscr{X}_f$ *and let* $u_t$ *be determined according to Algorithm 6 for all* $t \in \mathbb{N}$. *Then, it holds that* $(x_t, u_t) \in \mathscr{X} \times \mathscr{U}$ *for all* $t \in \mathbb{N}$. *Further, if there exists an* $\epsilon > 0$ *such that* $\gamma(A + BK)^i \mathscr{F}_\infty \oplus \epsilon \mathscr{B}_{n_x} \subseteq \bar{\mathscr{X}}_i$ *for all* $i \in \{0, \ldots, N-1\}$, *then the set* $\gamma \mathscr{F}_\infty$ *is* $\theta$-uniformly asymptotically stable for any dynamical system generating $(x_t)_{t \in \mathbb{N}}$, *with* $\mathscr{X}_f$ *belonging to its region of attraction.*

### Self-triggered control

Algorithm 7 defines a self-triggered controller based on pre-defined inputs, which is analogous to the event-triggered controller defined in Algorithm 6. As before, we enforce $t_{i+1} - t_i \leq \theta$.

**Theorem 5.7.** *Let* $x_0 \in \mathscr{X}_f$ *and let* $u_t$ *be determined according to Algorithm 7 for all* $t \in \mathbb{N}$. *Then, it holds that* $(x_t, u_t) \in \mathscr{X} \times \mathscr{U}$ *for all* $t \in \mathbb{N}$. *Further, if there exists an* $\epsilon > 0$ *such that* $\gamma(A + BK)^i \mathscr{F}_\infty \oplus \epsilon \mathscr{B}_{n_x} \subseteq \bar{\mathscr{X}}_i$ *for all* $i \in \{0, \ldots, N-1\}$, *then the set* $\gamma \mathscr{F}_\infty$ *is* $\theta$-uniformly asymptotically stable for any dynamical system generating $(x_t)_{t \in \mathbb{N}}$, *with* $\mathscr{X}_f$ *belonging to its region of attraction.*

*Proof.* The statement follows by the same reasoning employed to prove Theorem 5.6. $\quad\square$

---

**Algorithm 6** Mixed set–Lyapunov-based event-triggered predictive control

---

1: $i \leftarrow 0$
2: $t \leftarrow 0$
3: $t_0 \leftarrow 0$
4: Solve the optimization problem in (2.19) for $x = x_0$ and obtain $(\mathbf{x}_0, \mathbf{u}_0) \in \mathcal{T}(x_0)$.
5: Apply $u_0 = \kappa(x_0, 0)$ to the system.
6: **for** $t \in \{1, \ldots\}$ **do**
7:    Solve the optimization problem in (2.19) for $x = x_t$ and obtain $(\mathbf{x}_t, \mathbf{u}_t) \in \mathcal{T}(x_t)$.
8:    **if** $\forall \tau \in \{1, \ldots, \min\{t+1, \theta\}\}$,
       $\max_{w \in \mathscr{W}} J(Ax_t + B\kappa(x_{t_i}, t - t_i) + w) > J(x_{t+1-\tau}) - (1 - \nu^{\tau+1}) \sum_{k=0}^{\tau-1} \ell(u[k] - Kx[k])$
       or
       $\exists w \in \mathscr{W}, Ax_t + B\kappa(x_{t_i}, t - t_i) + w \notin \{x[\tau]\} \oplus \gamma \mathscr{F}_\tau$,
       where $((x[j])_{j \in \mathbb{N}}, (u[j])_{j \in \mathbb{N}}) = (\mathbf{x}_{t+1-\tau}, \mathbf{u}_{t+1-\tau})$
       **then**
9:       $i \leftarrow i + 1$
10:      $t_i \leftarrow t$
11:   **end if**
12:   Apply $u_t = \kappa(x_{t_i}, t - t_i)$ to the system.
13: **end for**

---

---

**Algorithm 7**
Mixed set–Lyapunov-based self-triggered predictive control with pre-defined inputs

---

1: $i \leftarrow -1$
2: $t \leftarrow 0$
3: $t_0 \leftarrow 0$
4: **for** $t \in \{0, \ldots\}$ **do**
5:    **if** $t = t_{i+1}$ **then**
6:       $i \leftarrow i + 1$
7:       Solve the optimization problem in (2.19) for $x = x_t$ and obtain $(\mathbf{x}_t, \mathbf{u}_t) \in \mathscr{T}(x_t)$.

8:       $t_{i+1} \leftarrow \sup \left\{ t' \in \{t_i + 1, \ldots, t_i + \theta\} \;\middle|\; \forall k \in \{t_i + 1, \ldots, t'\}, \right.$

$$\exists t_k^{-1} \in \{t_0, \ldots, t_i\} \cap \{k - \theta, \ldots, k - 1\},$$

$$\max_{w[l] \in \mathscr{W},\ l \in \{t_i, \ldots, k-1\}} J\left( A^{k-t_i} x_{t_i} + \sum_{l=t_i}^{k-1} A^{k-1-l}(B\kappa(x_{t_i}, l - t_i) + w[l]) \right)$$

$$\leq J(x_{t_k^{-1}}) - (1 - \nu^{k - t_k^{-1} + 1}) \sum_{j=0}^{k - t_k^{-1} - 1} \ell(x[j] - Ku[j])$$

$$\text{and } \forall w[l] \in \mathscr{W},\ l \in \{t_i, \ldots, k-1\},$$

$$A^{k-t_i} x_{t_i} + \sum_{l=t_i}^{k-1} A^{k-1-l}(B\kappa(x_{t_i}, l - t_i) + w[l])$$

$$\in \{x[k - t_k^{-1}]\} \oplus \gamma \mathscr{F}_{k - t_k^{-1}},$$

$$\left. \text{where } (x[j])_{j \in \mathbb{N}} = \mathbf{x}_{t_k^{-1}} \right\}$$

9:    **end if**
10:    Apply $u_t = \kappa(x_{t_i}, t - t_i)$ to the system.
11: **end for**

---

Attempting to adapt Algorithm 5 to the framework proposed in this section, we face the problem that the set-membership condition (5.21b) cannot be as easily simplified as the cost-decrease condition (5.7) (where the least restrictive $\tau$ in the decrease condition could be computed *outside* the definition of $\mathscr{D}_M^{\mathrm{L}}$, in the form of the quantities $W_{i,j}^{\mathrm{st}}$). The main difference between the cost-decrease condition and the set-inclusion is that the values of the cost function for different $t^{-1}$ can be ordered (compare (5.9) and (5.9)), which is, in general, not the case for the sets defining the set-membership conditions. For the moment, we therefore include $t^{-1}$ as an integer variable in the following constraint set. Define for all $M \in \{1, \ldots\} \cup \{\infty\}$, all $t \in \mathbb{N}$, all $x \in \mathbb{R}^{n_x}$, and all $(\mathbf{x}_{t^{-1}}, \mathbf{u}_{t^{-1}}))_{t^{-1} \in \mathscr{R}}$ with $(\mathbf{x}_{t^{-1}}, \mathbf{u}_{t^{-1}}) \in (\mathbb{R}^{n_x})^{\mathbb{N}} \times (\mathbb{R}^{n_u})^{\mathbb{N}}$ for all $t^{-1} \in \mathscr{R}$ and $\mathscr{R} \subseteq \mathbb{N}$,

$$\mathscr{D}_M^{\mathrm{m}}(t, x, (\mathbf{x}_{t^{-1}}, \mathbf{u}_{t^{-1}})_{t^{-1} \in \mathscr{R}}, \mathscr{R}) := \left\{ (u[i])_{i \in \{0, \ldots, M-1\}} \in \mathbb{R}^{M n_u} \, \middle| \right.$$

$$(u[0], \ldots, u[M-1]) \in C_M, \tag{5.22a}$$

$$x[0] = x, \tag{5.22b}$$

$$x[i+1] = Ax[i] + Bu[i], \tag{5.22c}$$

$$i \in \{0, \ldots, M-1\},$$

$$\forall w[j] \in \mathscr{W}, \ j \in \{0, \ldots, M-1\}, \ \forall k \in \{1, \ldots, M\},$$

$$\exists t_k^{-1} \in \bar{\mathscr{R}}_k, \ ((x_k[j])_{j \in \mathbb{N}}, (u_k[j])_{j \in \mathbb{N}}) = (\mathbf{x}_{t_k^{-1}}, \mathbf{u}_{t_k^{-1}}),$$

$$J\left( x[k] + \sum_{j=0}^{k-1} A^{k-1-j} w[j] \right),$$

$$\leq J(x_k[0]) - (1 - \nu^{t+k-t_k^{-1}+1}) \sum_{j=0}^{t+k-t_k^{-1}-1} \ell(u_k[j] - Kx_k[j]) \tag{5.22d}$$

$$x[k] + \sum_{j=0}^{k-1} A^{k-1-j} w[j] \in \{x_k[t+k-t_k^{-1}]\} \oplus \gamma \mathscr{F}_{t+k-t_k^{-1}} \tag{5.22e}$$

$$\left. \vphantom{\sum} \right\},$$

where $\bar{\mathscr{R}}_k := \mathscr{R} \cap \{t + k - \theta, \ldots, t + k - 1\}$, $k \in \{1, \ldots, M\}$.

As in Section 5.1.3, we assume that $C_M \subseteq \mathscr{U}^M$, $M \in \{1, \ldots\} \cup \{\infty\}$, and $C_1 = \mathscr{U}$. Algorithm 8 describes the self-triggered closed-loop predictive control with optimized inputs.

---

**Algorithm 8**

Mixed set–Lyapunov-based self-triggered predictive control with optimized inputs

---

1: $i \leftarrow -1$
2: $t \leftarrow 0$
3: $t_0 \leftarrow 0$
4: **for** $t \in \{0, \dots\}$ **do**
5:    **if** $t = t_{i+1}$ **then**
6:       $i \leftarrow i + 1$
7:       Solve the optimization problem in (2.19) for $x = x_t$ and obtain $(\mathbf{x}_t, \mathbf{u}_t) \in \mathscr{T}(x_t)$.
8:       $\mathscr{R}_i \leftarrow \{t_i + 1 - \min\{t_i + 1, \theta\}, \dots, t_i\} \cap \{t_0, \dots, t_i\}$
9:       $M^\star \leftarrow \sup\{M \in \{1, \dots, \theta\} \mid \mathscr{D}_M^{\mathrm{m}}(t_i, x_{t_i}, (\mathbf{x}_{t^{-1}}, \mathbf{u}_{t^{-1}})_{t^{-1} \in \mathscr{R}_i}, \mathscr{R}_i) \neq \emptyset\}$
10:      $t_{i+1} \leftarrow t_i + M^\star$
11:      $(u[i])_{i \in \{0, \dots, M^\star - 1\}} \leftarrow (u^\star[i])_{i \in \{0, \dots, M^\star - 1\}}$ for some
        $(u^\star[i])_{i \in \{0, \dots, M^\star - 1\}} \in \mathscr{D}_{M^\star}^{\mathrm{m}}(t_i, x_{t_i}, (\mathbf{x}_{t^{-1}}, \mathbf{u}_{t^{-1}})_{t^{-1} \in \mathscr{R}_i}, \mathscr{R}_i)$.
12:    **end if**
13:    Apply $u_t = u[t - t_i]$ to the system.
14: **end for**

---

**Theorem 5.8.** *Let $x_0 \in \mathscr{X}_{\mathrm{f}}$ and let $u_t$ be determined according to Algorithm 8 for all $t \in \mathbb{N}$. Then, it holds that $(x_t, u_t) \in \mathscr{X} \times \mathscr{U}$ for all $t \in \mathbb{N}$. Further, if there exists an $\epsilon > 0$ such that $\gamma(A + BK)^i \mathscr{F}_\infty \oplus \epsilon \mathscr{B}_{n_x} \subseteq \bar{\mathscr{X}}_i$ for all $i \in \{0, \dots, N-1\}$, then the set $\gamma \mathscr{F}_\infty$ is $\theta$-uniformly asymptotically stable for any dynamical system generating $(x_t)_{t \in \mathbb{N}}$, with $\mathscr{X}_{\mathrm{f}}$ belonging to its region of attraction.*

## 5.2.3. Implementation

Algorithm 6 and Algorithm 7 can be implemented by a straight-forward application of the results in Section 5.1.4 concerning upper bounds on expressions of the form $\max_{\bar{w} \in \bigoplus_{j=0}^{i-1} A^j \mathscr{W}} J(x + \bar{w})$. The major difference when compared to the Lyapunov-based approaches in Section 5.1 is that no approximations of $V_\gamma^{\mathrm{s}}$ are necessary and instead set-membership conditions equivalent to those in Section 3.3 have to be checked.

Considering the implementation of Algorithm 8, we would like to avoid solving the mixed-integer problems arising from (5.22d) and (5.22e). In the following, we propose a reformulation of these constraints leading to a convex optimization problem. This reformulation is based on two ideas. First, we use the results from the previous section, namely in the definition of (5.18), to obtain an upper bound on the expression $J\left(x[k] + \sum_{j=0}^{k-1} A^{k-1-j} w_j\right)$ by including additional predicted nominal states $(x[j|k])$ as decision variables. Second, we replace the optimization over the integer variables $t_k^{-1}$ by an optimization over real variables that are constrained to a simplex. More precisely, instead of searching for an integer $t^{-1} \in \mathscr{R}$ such that $x[k] + \sum_{j=0}^{k-1} A^{k-1-j} w_j \in$

$\{x_k[t + k - t^{-1}]\} \oplus \gamma \mathscr{F}_{t+k-t^{-1}}$, we will instead search for real numbers $\lambda_{k,t^{-1}} \in [0, \infty)$ with $\sum_{t^{-1} \in \mathscr{R}} \lambda_{k,t^{-1}} = 1$ such that

$$x[k] + \sum_{j=0}^{k-1} A^{k-1-j} w_j \in \bigoplus_{t^{-1} \in \mathscr{R}} \lambda_{k,t^{-1}} \left( \{x_k[t + k - t^{-1}]\} \oplus \gamma \mathscr{F}_{t+k-t^{-1}} \right). \tag{5.23}$$

Note that this second part of the reformulation is lossless: choosing $\lambda_{k,t^{-1}} = 1 \Leftrightarrow t^{-1} = t_k^{-1}$ for a fixed $t_k^{-1}$ returns the original constraint $x[k] + \sum_{j=0}^{k-1} A^{k-1-j} w_j \in \{x_k[t + k - t^{-1}]\} \oplus \gamma \mathscr{F}_{t+k-t^{-1}}$.

We implement the reformulation by replacing every instance of $\mathscr{D}_M^{\mathrm{m}}$ in Algorithm 8 with

$$\bar{\mathscr{D}}_M^{\mathrm{m}}(t, x, (\mathbf{x}_{t^{-1}}, \mathbf{u}_{t^{-1}})_{t^{-1} \in \mathscr{R}}, \mathscr{R}) := \left\{ (u[i])_{i \in \{0, \dots, M-1\}} \in \mathbb{R}^{Mn_{\mathrm{u}}} \; \middle| \right.$$

$$(u[0], \dots, u[M-1]) \in C_M \tag{5.24a}$$

$$x[0] = x \tag{5.24b}$$

$$x[i+1] = Ax[i] + Bu[i] \tag{5.24c}$$

$$i \in \{0, \dots, M-1\},$$

$$x[0|k] = x[k], \tag{5.24d}$$

$$x[j+1|k] = Ax[j|k] + Bu[j|k], \tag{5.24e}$$

$$u[j|k] \in \mathcal{U} \ominus K\mathscr{F}_i \ominus K(A+BK)^j \bigoplus_{l=0}^{k-1} A^l \mathscr{W} \tag{5.24f}$$

$$, x[j|k] \in \mathcal{X} \ominus \mathscr{F}_i \ominus (A+BK)^j \bigoplus_{l=0}^{k-1} A^l \mathscr{W}, \tag{5.24g}$$

$$j \in \{0, \dots, N-k-1\},$$

$$x[N-k|k] \in \mathcal{X}_{\mathrm{T}}^k \ominus (A+BK)^{N-k} \bigoplus_{l=0}^{k-1} A^l \mathscr{W}, \tag{5.24h}$$

$$\exists (\lambda_{k,t^{-1}})_{t^{-1} \in \bar{\mathscr{R}}_k}, \; \lambda_{k,t^{-1}} \in [0, \infty), \; t^{-1} \in \bar{\mathscr{R}}_k, \; \sum_{t^{-1} \in \bar{\mathscr{R}}_k} \lambda_{k,t^{-1}} = 1,$$

$$\sum_{j=0}^{N-k-1} \ell(u[j|k] - Kx[j|k]) \le \sum_{t^{-1} \in \mathscr{R}_k} \lambda_{k,t^{-1}} \Bigg( J(x_{t^{-1}})$$

$$- (1 - \nu^{t+k-t^{-1}+1}) \sum_{l=0}^{t+k-t^{-1}-1} \ell(u_{t^{-1}}[l] - Kx_{t^{-1}}[l]) \Bigg) \quad (5.24\text{i})$$

$$x[k] \in \left\{ \sum_{t^{-1} \in \bar{\mathscr{R}}_k} \lambda_{k,t^{-1}} x_{t^{-1}}[t+k-t^{-1}] \right\} \oplus \bigoplus_{t^{-1} \in \bar{\mathscr{R}}_k} \lambda_{k,t^{-1}} \bar{\mathscr{F}}_{k,t+k-t^{-1}} \quad (5.24\text{j})$$

$$k \in \{1, \dots, M\}$$
$$\Bigg\},$$

where $((x_{t^{-1}}[j])_{j \in \mathbb{N}}, (u_{t^{-1}}[j])_{j \in \mathbb{N}}) = (\mathbf{x}_{t^{-1}}, \mathbf{u}_{t^{-1}})$, $t^{-1} \in \mathscr{R}$,

$$\bar{\mathscr{F}}_{k,t+k-t^{-1}} := \left( \gamma \mathscr{F}_{t+k-t^{-1}} \ominus \bigoplus_{j=0}^{k-1} A^j \mathscr{W} \right), \quad (5.25)$$

$\bar{\mathscr{R}}_k := \mathscr{R} \cap \{t+k-\theta, \dots, t+k-1\} \cap \{t^{-1} \in \mathbb{N} \mid \bar{\mathscr{F}}_{k,t+k-t^{-1}} \ne \emptyset\}$, $k \in \{1, \dots, M\}$, and $\mathscr{X}_{\mathrm{T}}^k$, $k \in \{1, \dots, M\}$ are defined as in (5.19). Further, we assume that $\mathscr{X}$, $\mathscr{U}$, $\mathscr{W}$, $C_M$, and $\mathscr{X}_{\mathrm{T}}$ are polytopic sets and that $\ell$ and $\bar{V}^{\mathrm{s}}$ are convex functions. Note that if $\mathcal{S}_k$ are polyhedrons of the form $\mathcal{S}_k = \{x \in \mathbb{R}^n \mid H_k x_k \le h_k\}$, then the set membership

$$x \in \left\{ \sum_k \lambda_k x_k \right\} \oplus \bigoplus_k \lambda_k \mathcal{S}_k \quad (5.26)$$

is equivalent to

$$x = \sum_k \lambda_k x_k + \sum_k \bar{x}_k,$$
$$\text{and } \forall k, \ H_k \bar{x}_k \le \lambda_k h_k. \quad (5.27\text{a})$$

Hence, all constraints in the definition of $\bar{\mathscr{D}}_M^{\mathrm{m}}$ are convex.

It remains to show that the application of the modified algorithm retains the guarantees of Theorem 5.12:

**Lemma 5.5.** *Let $x_0 \in \mathscr{X}_{\mathrm{f}}$ and let $u_t$ be determined according to Algorithm 8 for all $t \in \mathbb{N}$, where $\mathscr{D}_M^{\mathrm{m}}$ is replaced with $\bar{\mathscr{D}}_M^{\mathrm{m}}$. Then, it holds that $(x_t, u_t) \in \mathscr{X} \times \mathscr{U}$ for all $t \in \mathbb{N}$. Further, if there exists an $\epsilon > 0$ such that $\gamma(A + BK)^i \mathscr{F}_\infty \oplus \epsilon \mathscr{B}_{n_x} \subseteq \bar{\mathscr{X}}_i$ for all $i \in \{0, \dots, N-1\}$, then the set $\gamma \mathscr{F}_\infty$ is $\theta$-uniformly asymptotically stable for any dynamical system generating $(x_t)_{t \in \mathbb{N}}$, with $\mathscr{X}_{\mathrm{f}}$ belonging to its region of attraction.*

**Remark 5.6.** *The relaxation approach presented here is closely related to interpolation-based control, see for example [Gutman and Cwikel, 1986, Blanchini, 1990], where a control law at a given point in the state space is defined as the convex interpolation of fixed control inputs at a set of vertices, of which in turn the given state is an interpolation. The technique was later extended to the interpolation between* sets, *see for example [Lee and Kouvaritakis, 2002, Bacic et al., 2003]; a contribution by the author can be found in [Brunner and Allgöwer, 2014].*

## 5.3. Purely set-based approach

In both Section 5.1 Section 5.2, an explicit decrease of the MPC value function (augmented, in the former case) in the trigger conditions was employed to ensure constraint satisfaction and stability in the closed-loop system. Here, we show how to get rid of the decrease conditions entirely, which greatly reduces the complexity of both the trigger conditions and, in the case of self-triggered control, the optimization problems involved.

### 5.3.1. Feasibility from set-membership conditions

In order to allow feasibility and constraint satisfaction to be concluded from the set-membership conditions in Section 5.2 alone, we employ modified MPC problems as defined in the following. Let $\gamma \in [1, \infty)$ and define the set of admissible nominal input sequences for a given point $x \in \mathbb{R}^{n_x}$

$$\mathscr{D}^\gamma(x) := \left\{ (u[i])_{i \in \{0,\dots,N-1\}} \in \mathbb{R}^{Nn_u} \; \middle| \right.$$

$$x[0] = x, \tag{5.28a}$$

$$x[i+1] = Ax[i] + Bu[i], \tag{5.28b}$$

$$u[i] \in \mathscr{U} \ominus K\gamma\mathscr{F}_i, \tag{5.28c}$$

$$x[i] \in \mathscr{X} \ominus \gamma\mathscr{F}_i, \tag{5.28d}$$

$$i \in \{0,\dots,N-1\},$$

$$\left. x[N] \in \mathscr{X}_{\mathrm{T}}^\gamma \phantom{\Big\}} \right\} \tag{5.28e}$$

and the subset

$$\mathscr{X}_{\mathrm{f}}^\gamma := \{x \in \mathbb{R}^{n_x} \mid \mathscr{D}^\gamma(x) \neq \emptyset\} \tag{5.29}$$

of the state space where the constraints are feasible. The modified terminal set $\mathscr{X}_{\mathrm{T}}^{\gamma}$ is assumed to be compact and to satisfy $\mathscr{X}_{\mathrm{T}}^{\gamma} \subseteq \mathscr{X} \ominus \gamma \mathscr{F}_N$, $K\mathscr{X}_{\mathrm{T}}^{\gamma} \subseteq \mathscr{U} \ominus K\gamma \mathscr{F}_N$, and $(A+BK)\mathscr{X}_{\mathrm{T}}^{\gamma} \subseteq \mathscr{X}_{\mathrm{T}}^{\gamma} \ominus (A+BK)^N \gamma \mathscr{W}$. Define further the cost function $\bar{J}^{\gamma} : \mathbb{R}^{n_x} \times \mathbb{R}^{Nn_u} \to \mathbb{R}$ by

$$\bar{J}^{\gamma}(x,(u[0],\ldots,u[N-1])) = \sum_{k=0}^{N-1} \ell(u[i] - Kx[i]), \tag{5.30}$$

where $x[0] = x$ and $x[i+1] = Ax[i] + Bu[i]$ for $i \in \{0,\ldots,N-1\}$ and $\ell$ is a continuous positive definite function. Define the MPC value function

$$J^{\gamma}(x) = \inf \left\{ \bar{J}^{\gamma}(x,\mathbf{u}) \mid \mathbf{u} \in \mathscr{D}^{\gamma}(x) \right\}, \tag{5.31}$$

the set of optimal trajectories

$$\begin{aligned}
\mathscr{T}^{\gamma}(x) = \Big\{ &(\mathbf{x},\mathbf{u}) \in (\mathbb{R}^{n_x})^{\mathbb{N}} \times (\mathbb{R}^{n_u})^{\mathbb{N}} \mid \mathbf{x} = (x[i])_{i\in\mathbb{N}}, \ \mathbf{u} = (u[i])_{i\in\mathbb{N}}, \\
&x[0] = x, \ x[i+1] = Ax[i] + Bu[i], \ i \in \{0,\ldots,N-1\}, \\
&u[i] = Kx[i], i \in \{N,\ldots\}, \\
&(u[0],\ldots,u[N-1]) \in \mathscr{D}^{\gamma}(x), \ J^{\gamma}(x) = \bar{J}^{\gamma}(x,(u[0],\ldots,u[N-1])) \Big\}, \tag{5.32}
\end{aligned}$$

and the set-valued controller

$$\kappa_{\mathrm{MPC}}^{\gamma}(x) = \left\{ u \in \mathbb{R}^{n_u} \mid \exists \mathbf{x} \in (\mathbb{R}^{n_x})^{\mathbb{N}}, \ \exists \tilde{\mathbf{u}} \in (\mathbb{R}^{n_u})^{\mathbb{N}}, \ (\mathbf{x},(u,\tilde{\mathbf{u}})) \in \mathscr{T}^{\gamma}(x) \right\}. \tag{5.33}$$

Finally, define the sets

$$\begin{aligned}
\bar{\mathscr{X}}_j^{\gamma} := \{ x \in \mathbb{R}^{n_x} \mid &(A+BK)^{i-j}x \in \mathscr{X} \ominus \gamma \mathscr{F}_i, \\
&K(A+BK)^{i-j}x \in \mathscr{U} \ominus K\gamma \mathscr{F}_i, \ i \in \{j,\ldots,N-1\}, \ (A+BK)^{N-j}x \in \mathscr{X}_{\mathrm{T}}^{\gamma} \} \tag{5.34}
\end{aligned}$$

for $j \in \{0,\ldots,N\}$. We have the following result.

**Theorem 5.9.** *Let Assumption 5.1 hold. Let further a dynamical system* $\mathfrak{f} : \mathcal{S} \to (\mathbb{R}^n)^{\mathbb{N}}$ *be given. Assume that for all* $(x_t)_{t\in\mathbb{N}} = \mathfrak{f}(s)$ *with* $s \in \mathcal{S}$ *and* $x_0 \in \mathscr{X}_{\mathrm{f}}^{\gamma}$, *and all* $t \in \mathbb{N}$, *there exists a* $\tau \in \{1,\ldots,\min\{\theta,t+1\}\}$ *such that*

$$x_{t+1} \in \{x[\tau]\} \oplus \gamma \mathscr{F}_{\tau}, \tag{5.35}$$

*where* $((x[i])_{i\in\mathbb{N}},(u[i])_{i\in\mathbb{N}}) \in \mathscr{T}^{\gamma}(x_{t+1-\tau})$. *Then, for all* $(x_t)_{t\in\mathbb{N}} = \mathfrak{f}(s)$ *with* $s \in \mathcal{S}$ *and* $x_0 \in \mathscr{X}_{\mathrm{f}}^{\gamma}$ *it holds that* $x_t \in \mathscr{X}$ *for all* $t \in \mathbb{N}$. *If, additionally, there exists an* $\epsilon > 0$ *such that* $\gamma(A+BK)^i \mathscr{F}_{\infty} \oplus \epsilon \mathscr{B}_{n_x} \subseteq \bar{\mathscr{X}}_i^{\gamma}$ *for all* $i \in \{0,\ldots,N-1\}$, *then the set* $\gamma \mathscr{F}_{\infty}$ *is* $\theta$-*uniformly asymptotically stable for* $\mathfrak{f}$ *with* $\mathscr{X}_{\mathrm{f}}^{\gamma}$ *belonging to its region of attraction.*

---

**Algorithm 9** Purely set-based event-triggered predictive control

1: $i \leftarrow 0$
2: $t \leftarrow 0$
3: $t_0 \leftarrow 0$
4: Solve the optimization problem in (5.31) for $x = x_0$ and obtain $(\mathbf{x}_0, \mathbf{u}_0) \in \mathcal{T}^\gamma(x_0)$.
5: Apply $u_0 = \kappa(x_0, 0)$ to the system.
6: **for** $t \in \{1, \ldots\}$ **do**
7:     Solve the optimization problem in (5.31) for $x = x_t$ and obtain $(\mathbf{x}_t, \mathbf{u}_t) \in \mathcal{T}^\gamma(x_t)$.
8:     **if** $\forall \tau \in \{1, \ldots, \min\{t + 1, \theta\}\}$,
        $\exists w \in \mathcal{W}, Ax + B\kappa(x_{t_i}, t - t_i) + w \notin \{x[\tau]\} \oplus \gamma \mathcal{F}_\tau$,
        where $((x[j])_{j \in \mathbb{N}}, (u[j])_{j \in \mathbb{N}}) = (\mathbf{x}_{t+1-\tau}, \mathbf{u}_{t+1-\tau})$
    **then**
9:         $i \leftarrow i + 1$
10:        $t_i \leftarrow t$
11:     **end if**
12:     Apply $u_t = \kappa(x_{t_i}, t - t_i)$ to the system.
13: **end for**

---

### 5.3.2. Aperiodic control algorithms

As in the previous sections, we describe an event-triggered and two self-triggered schemes based on Theorem 5.9. Here, the previous assumption on the function $\kappa$ is modified to $\kappa(x, 0) \in \kappa_{\mathrm{MPC}}^\gamma(x)$ for all $x \in \mathcal{X}_{\mathrm{f}}^\gamma$.

**Event-triggered control**

Algorithm 9 defines the closed-loop control of system (5.1) with an event-triggered control law.

**Theorem 5.10.** *Let $x_0 \in \mathcal{X}_{\mathrm{f}}^\gamma$ and let $u_t$ be determined according to Algorithm 9 for all $t \in \mathbb{N}$. Then, it holds that $(x_t, u_t) \in \mathcal{X} \times \mathcal{U}$ for all $t \in \mathbb{N}$. Further, if there exists an $\epsilon > 0$ such that $\gamma(A + BK)^i \mathcal{F}_\infty \oplus \epsilon \mathcal{B}_{n_x} \subseteq \tilde{\mathcal{X}}_i^\gamma$ for all $i \in \{0, \ldots, N - 1\}$, then the set $\gamma \mathcal{F}_\infty$ is $\theta$-uniformly asymptotically stable for any dynamical system generating $(x_t)_{t \in \mathbb{N}}$, with $\mathcal{X}_{\mathrm{f}}^\gamma$ belonging to its region of attraction.*

*Proof.* With the help of Theorem 5.9, the statement follows analogously to the results in the previous section. □

**Self-triggered control**

Algorithm 10 defines a self-triggered controller based on pre-defined inputs.

**Theorem 5.11.** *Let $x_0 \in \mathcal{X}_{\mathrm{f}}^\gamma$ and let $u_t$ be determined according to Algorithm 10 for all $t \in \mathbb{N}$. Then, it holds that $(x_t, u_t) \in \mathcal{X} \times \mathcal{U}$ for all $t \in \mathbb{N}$. Further, if there exists an*

---

**Algorithm 10** Purely set-based self-triggered predictive control with pre-defined inputs

1: $i \leftarrow -1$
2: $t \leftarrow 0$
3: $t_0 \leftarrow 0$
4: **for** $t \in \{0, \ldots\}$ **do**
5:     **if** $t = t_{i+1}$ **then**
6:       $i \leftarrow i + 1$
7:       Solve the optimization problem in (5.31) for $x = x_t$ and obtain $(\mathbf{x}_t, \mathbf{u}_t) \in \mathscr{T}^\gamma(x_t)$.

8:       $t_{i+1} \leftarrow \sup \Bigg\{ t' \in \{t_i + 1, \ldots, t_i + \theta\} \,\Big|\, \forall k \in \{t_i + 1, \ldots, t'\},$

$$\exists t_k^{-1} \in \{t_0, \ldots, t_i\} \cap \{k - \theta, \ldots, k - 1\},$$

$$\forall w[l] \in \mathscr{W}, \; l \in \{t_i, \ldots, k - 1\},$$

$$A^{k-t_i} x_{t_i} + \sum_{l=t_i}^{k-1} A^{k-1-l}(B\kappa(x_{t_i}, l - t_i) + w[l]) \in \{x[k - t_k^{-1}]\} \oplus \gamma \mathscr{F}_{k - t_k^{-1}},$$

$$\text{where } (x[j])_{j \in \mathbb{N}} = \mathbf{x}_{t_k^{-1}} \Bigg\}$$

9:     **end if**
10:     Apply $u_t = \kappa(x_{t_i}, t - t_i)$ to the system.
11: **end for**

---

$\epsilon > 0$ *such that* $\gamma(A + BK)^i \mathscr{F}_\infty \oplus \epsilon \mathscr{B}_{n_x} \subseteq \bar{\mathscr{X}}_i^\gamma$ *for all* $i \in \{0, \ldots, N - 1\}$, *then the set* $\gamma \mathscr{F}_\infty$ *is* $\theta$*-uniformly asymptotically stable for any dynamical system generating* $(x_t)_{t \in \mathbb{N}}$, *with* $\mathscr{X}_f$ *belonging to its region of attraction.*

*Proof.* With the help of Theorem 5.9, the statement follows analogously to the results in the previous section. $\qquad\square$

For the self-triggered scheme with optimized inputs, define for all $M \in \{1, \ldots\} \cup \{\infty\}$, all $t \in \mathbb{N}$, all $x \in \mathbb{R}^{n_x}$, and all $((\mathbf{x}_{t^{-1}}, \mathbf{u}_{t^{-1}}))_{t^{-1} \in \mathscr{R}}$ with $(\mathbf{x}_{t^{-1}}, \mathbf{u}_{t^{-1}}) \in (\mathbb{R}^{n_x})^\mathbb{N} \times (\mathbb{R}^{n_u})^\mathbb{N}$ for all $t^{-1} \in \mathscr{R}$ and $\mathscr{R} \subseteq \mathbb{N}$,

$$\mathscr{D}_M^{\mathrm{p}}(t, x, (\mathbf{x}_{t^{-1}}, \mathbf{u}_{t^{-1}})_{t^{-1} \in \mathscr{R}}, \mathscr{R}) := \Bigg\{ (u[i])_{i \in \{0, \ldots, M-1\}} \in \mathbb{R}^{M n_u} \,\Bigg|$$

$$(u[0], \ldots, u[M - 1]) \in C_M, \qquad (5.36a)$$
$$x[0] = x, \qquad (5.36b)$$
$$x[i + 1] = Ax[i] + Bu[i], \quad (5.36c)$$
$$i \in \{0, \ldots, M - 1\},$$

$$\forall w[j] \in \mathscr{W}, \; j \in \{0, \ldots, M-1\}, \; k \in \{1, \ldots, M\},$$

$$\exists t_k^{-1} \in \bar{\mathscr{R}}_k, \; ((x_k[j])_{j \in \mathbb{N}}, (u_k[j])_{j \in \mathbb{N}}) = (\mathbf{x}_{t_k^{-1}}, \mathbf{u}_{t_k^{-1}}),$$

$$x[k] + \sum_{j=0}^{k-1} A^{k-1-j} w[j] \in \{x_k[t + k - t_k^{-1}]\} \oplus \gamma \mathscr{F}_{t+k-t_k^{-1}} \tag{5.36d}$$

$$\Bigg\},$$

where $\bar{\mathscr{R}}_k := \mathscr{R} \cap \{t + k - \theta, \ldots, t + k - 1\}$, $k \in \{1, \ldots, M\}$. The only difference to (5.22) is the omission of (5.22d).

As in the previous sections, we assume that $C_M \subseteq \mathscr{U}^M$, $M \in \{1, \ldots\} \cup \{\infty\}$, and $C_1 = \mathscr{U}$. Algorithm 11 describes the self-triggered closed-loop predictive control with optimized inputs.

**Theorem 5.12.** *Let $x_0 \in \mathscr{X}_\mathrm{f}^\gamma$ and let $u_t$ be determined according to Algorithm 11 for all $t \in \mathbb{N}$. Then, it holds that $(x_t, u_t) \in \mathscr{X} \times \mathscr{U}$ for all $t \in \mathbb{N}$. Further, if there exists an $\epsilon > 0$ such that $\gamma(A + BK)^i \mathscr{F}_\infty \oplus \epsilon \mathscr{B}_{n_x} \subseteq \bar{\mathscr{X}}_i^\gamma$ for all $i \in \{0, \ldots, N-1\}$, then the set $\gamma \mathscr{F}_\infty$ is $\theta$-uniformly asymptotically stable for any dynamical system generating $(x_t)_{t \in \mathbb{N}}$, with $\mathscr{X}_\mathrm{f}$ belonging to its region of attraction.*

*Proof.* With the help of Theorem 5.9, the statement follows analogously to the results in the previous section. □

---

**Algorithm 11** Purely set-based self-triggered predictive control with optimized inputs

---
1: $i \leftarrow -1$
2: $t \leftarrow 0$
3: $t_0 \leftarrow 0$
4: **for** $t \in \{0, \ldots\}$ **do**
5:    **if** $t = t_{i+1}$ **then**
6:      $i \leftarrow i + 1$
7:      Solve the optimization problem in (5.31) for $x = x_t$ and obtain $(\mathbf{x}_t, \mathbf{u}_t) \in \mathscr{T}^\gamma(x_t)$.
8:      $\mathscr{R}_i \leftarrow \{t_i + 1 - \min\{t_i + 1, \theta\}, \ldots, t_i\} \cap \{t_0, \ldots, t_i\}$
9:      $M^\star \leftarrow \sup\{M \in \{1, \ldots, \theta\} \mid \mathscr{D}_M^\mathrm{p}(t_i, x_{t_i}, (\mathbf{x}_{t^{-1}}, \mathbf{u}_{t^{-1}})_{t^{-1} \in \mathscr{R}_i}, \mathscr{R}_i) \neq \emptyset\}$
10:     $t_{i+1} \leftarrow t_i + M^\star$
11:     $(u[i])_{i \in \{0, \ldots, M^\star - 1\}} \leftarrow (u^\star[i])_{i \in \{0, \ldots, M^\star - 1\}}$ for some $(u^\star[i])_{i \in \{0, \ldots, M^\star - 1\}} \in \mathscr{D}_{M^\star}^\mathrm{p}(t_i, x_{t_i}, (\mathbf{x}_{t^{-1}}, \mathbf{u}_{t^{-1}})_{t^{-1} \in \mathscr{R}_i}, \mathscr{R}_i)$.
12:    **end if**
13:    Apply $u_t = u[t - t_i]$ to the system.
14: **end for**

---

### 5.3.3. Implementation

The implementation of Algorithm 9 and Algorithm 10 is straightforward; it is merely required to solve the MPC problem in (5.31) and to check certain set memberships, which are similar to those employed in Section 3.3.

The implementation of Algorithm 11, however, requires the solution of mixed-integer optimization problems, due to constraint (5.36d). In the remainder of this subsection, we present a modified version of the algorithm based on a reformulation of the set $\mathscr{D}_M^{\mathrm{p}}$ in (5.36) along the same lines as Section 5.2.3. In particular, we assume in the following that every instance of $\mathscr{D}_M^{\mathrm{p}}$ in Algorithm 11 is replaced with

$$\bar{\mathscr{D}}_M^{\mathrm{p}}(t, x, (\mathbf{x}_{t^{-1}}, \mathbf{u}_{t^{-1}})_{t^{-1} \in \mathscr{R}}, \mathscr{R}) := \Bigg\{ (u[i])_{i \in \{0,\dots,M-1\}} \in \mathbb{R}^{Mn_u} \;\Bigg|$$

$$(u[0], \dots, u[M-1]) \in C_M, \tag{5.37a}$$

$$x[0] = x, \tag{5.37b}$$

$$x[i+1] = Ax[i] + Bu[i], \tag{5.37c}$$

$$i \in \{0, \dots, M-1\},$$

$$\exists (\lambda_{k,t^{-1}})_{t^{-1} \in \bar{\mathscr{R}}_k}, \; \lambda_{k,t^{-1}} \in [0, \infty), \; t^{-1} \in \bar{\mathscr{R}}_k, \; \sum_{t^{-1} \in \bar{\mathscr{R}}_k} \lambda_{k,t^{-1}} = 1,$$

$$x[k] \in \Bigg\{ \sum_{t^{-1} \in \bar{\mathscr{R}}_k} \lambda_{k,t^{-1}} x_{t^{-1}}[t+k-t^{-1}] \Bigg\} \oplus \bigoplus_{t^{-1} \in \bar{\mathscr{R}}_k} \lambda_{k,t^{-1}} \bar{\mathscr{F}}_{k,t+k-t^{-1}}, \tag{5.37d}$$

$$k \in \{1, \dots, M\}, \Bigg\},$$

where $((x_{t^{-1}}[j])_{j \in \mathbb{N}}, (u_{t^{-1}}[j])_{j \in \mathbb{N}}) = (\mathbf{x}_{t^{-1}}, \mathbf{u}_{t^{-1}})$, $t^{-1} \in \mathscr{R}$,

$$\bar{\mathscr{F}}_{k,t+k-t^{-1}} := \left( \gamma \mathscr{F}_{t+k-t^{-1}} \ominus \bigoplus_{j=0}^{k-1} A^j \mathscr{W} \right), \tag{5.38}$$

and $\bar{\mathscr{R}}_k := \mathscr{R} \cap \{t+k-\theta, \dots, t+k-1\} \cap \{t^{-1} \in \mathbb{N} \mid \bar{\mathscr{F}}_{k,t+k-t^{-1}} \neq \emptyset\}$, $k \in \{1, \dots, M\}$. Further, we assume that $\mathscr{X}$, $\mathscr{U}$, $\mathscr{W}$, $C_M$, and $\mathscr{X}_{\mathrm{T}}^{\gamma}$ are polytopic sets and that $\ell$ is a convex function. Hence, all constraints in (5.37) are linear and step 9 to step 11 of Algorithm 11 can be implemented with linear programming. Furthermore, considering the remarks on this point in Section 5.2.3, the reformulation is lossless.

It remains to show that the application of the modified algorithm retains the guarantees of Theorem 5.12:

**Lemma 5.6.** *Let $x_0 \in \mathcal{X}_f^\gamma$ and let $u_t$ be determined according to Algorithm 11 for all $t \in \mathbb{N}$, where $\mathscr{D}_M^p$ is replaced with $\tilde{\mathscr{D}}_M^p$. Then, it holds that $(x_t, u_t) \in \mathcal{X} \times \mathcal{U}$ for all $t \in \mathbb{N}$. Further, if there exists an $\epsilon > 0$ such that $\gamma(A + BK)^i \mathscr{F}_\infty \oplus \epsilon \mathscr{B}_{n_x} \subseteq \tilde{\mathcal{X}}_i^\gamma$ for all $i \in \{0, \ldots, N-1\}$, then the set $\gamma \mathscr{F}_\infty$ is $\theta$-uniformly asymptotically stable for any dynamical system generating $(x_t)_{t \in \mathbb{N}}$, with $\mathcal{X}_f^\gamma$ belonging to its region of attraction.*

# 5.4. Threshold-based event-triggered MPC: analysis and stochastic design

In this section, we consider a special case of the event-triggered algorithm presented in Section 5.3. In particular, we restrict $\tau$ in step 8 of Algorithm 9 to $t + 1 - t_i$ for $t \in \{t_i, \ldots, t_{i+1} - 1\}$, that is, we always use the state at the last transmission instance as a reference. The resulting trigger condition then amounts to checking whether the set membership

$$Ax_t + B\kappa(x_{t_i}, t - t_i) - x[t + 1 - t_i] \in \gamma \mathscr{F}_{t+1-t_i} \ominus \mathscr{W} \tag{5.39}$$

holds, where $((x[i])_{i \in \mathbb{N}}, (u[i])_{i \in \mathbb{N}}) \in \mathscr{T}^\gamma(x_{t_i})$. If, additionally, $\kappa(x_i, t - t_i) = u[t - t_i]$ for all $t \in \{t_i, \ldots, t_{i+1} - 1\}$, which we assume in the following, then this condition further simplifies to

$$x_t - x[t - t_i] \in A^{-1}\left(\gamma \mathscr{F}_{t+1-t-i} \ominus \mathscr{W}\right), \tag{5.40}$$

where we used the fact that $x[t + 1 - t_i] = Ax[t - t_i] + Bu[t - t_i]$.

The condition in (5.40) is just the threshold condition that was previously proposed in [Brunner et al., 2015b] and [Brunner et al., 2017c], which is hereby revealed to be a special case in a more general framework. Additionally, it is also a special case of threshold-based event-triggered MPC algorithms as, for example, proposed in [Eqtami et al., 2010, Lehmann et al., 2013], where the trigger conditions read

$$x_t - x[t - t_i] \notin \bar{\mathscr{T}}_t \tag{5.41}$$

for certain threshold sets $\bar{\mathscr{T}}_t \subseteq \mathbb{R}^{n_x}$, $t \in \mathbb{N}$.

Consider the event-triggered controller defined by Algorithm 12, where we make the simplifying assumption that the threshold sets depend only on the time elapsed since the last event. Event-triggered control algorithms of this form have the advantage that, in addition to saving communication, the average number of optimization problems that have to be solved online is also reduced (due to the fact that Algorithm 12 only requires optimization problems to be solved at the transmission instants $t_i$). In fact, the reduction of the overall (meaning average in time, not worst-case at a single instant) computational effort involved with MPC has been another motivation for employing aperiodic schemes, next to the reduction of communication, compare [Varutti et al.,

---

**Algorithm 12** Threshold-based event-triggered predictive control

---

1: $i \leftarrow 0$
2: $t \leftarrow 0$
3: $t_0 \leftarrow 0$
4: Solve the optimization problem in (5.31) for $x = x_0$ and obtain $(\mathbf{x}_0, \mathbf{u}_0) \in \mathscr{T}^\gamma(x_0)$.
5: $((x[j])_{j \in \mathbb{N}}, (u[j])_{j \in \mathbb{N}}) \leftarrow (\mathbf{x}_0, \mathbf{u}_0)$
6: Apply $u_0 = u[0]$ to the system.
7: **for** $t \in \{1, \ldots\}$ **do**
8:     **if** $(x_t - x[t - t_i]) \notin \mathscr{T}_{t-t_i}$
      **then**
9:       $i \leftarrow i + 1$
10:       $t_i \leftarrow t$
11:       Solve the optimization problem in (5.31) for $x = x_t$ and obtain $(\mathbf{x}_t, \mathbf{u}_t) \in \mathscr{T}^\gamma(x_t)$.
12:       $((x[j])_{j \in \mathbb{N}}, (u[j])_{j \in \mathbb{N}}) \leftarrow (\mathbf{x}_t, \mathbf{u}_t)$
13:     **end if**
14:     Apply $u_t = u[t - t_i]$ to the system.
15: **end for**

---

2009, Sijs et al., 2010, Eqtami et al., 2010, Lehmann et al., 2013, Gommans and Heemels, 2015].

We have the following result, with which constraint satisfaction and stability properties for the closed-loop scheme follow from Theorem 5.9.

**Lemma 5.7.** *Let $\gamma \in [1, \infty)$ satisfy the requirements in Theorem 3.5 with regard to the sets $\mathscr{T}_j$, $j \in \mathbb{N}$. Then, if $x_0 \in \mathscr{X}_f^\gamma$, it holds that $x_t \in \{x[t - t_i]\} \oplus \gamma \mathscr{F}_{t-t_i}$ with $((x[j])_{j \in \mathbb{N}}, (u[j])_{j \in \mathbb{N}}) \in \mathscr{T}(x_{t_i})$ for all $t \in \{t_i + 1, \ldots, t_{i+1}\}$ and $i \in \mathbb{N}$ as resulting from Algorithm 12.*

Lemma 5.7 allows to design an event-triggered MPC scheme which is least conservative for the given thresholds in terms of the region of attraction $\mathscr{X}_f^\gamma$ and the guaranteed worst-case asymptotic bound $\gamma \mathscr{F}_\infty$. Least conservative here means that set membership in (5.35) could not be guaranteed for any smaller value of $\gamma$; compare the proof of Lemma 5.7 and the discussion in Section 3.3.2.

**Stochastic threshold design**

It holds that $x_{t_i + j + 1} - x[j + 1] = A(x_{t_i + j} - x[j]) + w_{t_i + j}$ for all $j \in \{0, \ldots, t_{i+1} - t_i\}$, $i \in \mathbb{N}$, in the closed-loop system defined by Algorithm 12. The inter-event times $t_{i+1} - t_i$ therefore only depend on the realization of the disturbances $w_{t_i + j}$ and the threshold sets $\mathscr{T}_j$. Hence, if $w_t$ is an identically and independently distributed random variable, the same discussion as in Section 4.1 applies regarding the design of the sets $\mathscr{T}_j$ based on the probability distribution governing the disturbances. In fact, exactly the same techniques can be applied in order to assign the probability distribution of $t_{i+1} - t_i$. In the context of MPC it is, however, important to note that the value of $\gamma$, which in turn depends on the size of the threshold sets, influences the region of attraction $\mathcal{X}_f^\gamma$. If $\gamma$ is too large, $\mathcal{X}_f^\gamma$ might even become empty, in which case no valid controller has been obtained.

## 5.5. Numerical example

We investigated the average transmission rates resulting from the application of Algorithm 3 to Algorithm 11 for a particular example. Consider a system as in (3.1a) with system matrices

$$A = \begin{bmatrix} 1 & 0.3 \\ 0 & 1 \end{bmatrix}, \ B = \begin{bmatrix} 0.045 \\ 0.3 \end{bmatrix}, \tag{5.42}$$

which correspond to a continuous-time double integrator system discretized with a sample time of 0.3. The disturbance bound was chosen to $\mathcal{W} = [-1, 1] \times [-1, 1]$. We computed the linear feedback gain LQ-optimal, using the weighting matrices $Q = \begin{bmatrix} 1 & 0 \\ 0 & 1 \end{bmatrix}$ and $R = 1$ to obtain $K = [-0.7719 \ -1.4628]$. These are the same parameters as considered in the numerical examples in Chapter 3. Additionally, we chose the state and input constraint sets $\mathcal{X} = [-100, 100] \times [-100, 100]$ and $\mathcal{U} = [-40, 40]$. The cost function $\ell$ was defined by $\ell(v) = |v|$ in all examples. We defined $\bar{V}^s(x) := |Px|$ for all $x \in \mathbb{R}^{n_x}$, where

$$P = \begin{bmatrix}
1.0842 & 0 \\
0 & 1.0842 \\
1.3168 & 0.3195 \\
1.5062 & 0.6136 \\
1.6506 & 0.8770 \\
1.7492 & 1.1056 \\
1.8023 & 1.2960 \\
1.8113 & 1.4462 \\
1.7785 & 1.5548 \\
1.7070 & 1.6219 \\
1.6007 & 1.6482 \\
1.4639 & 1.6354 \\
1.3014 & 1.5860
\end{bmatrix} \tag{5.43}$$

was chosen such that Assumption 5.1 is satisfied. In particular, we computed $S$ such that $\{z \in \mathbb{R}^{n_x} \mid |Sz| \leq 1\}$ is the *maximal output admissible set* [Gilbert and Tan, 1991] for the system $z_{t+1} = \frac{1}{\lambda}(A + BK)z_t$ with $\lambda = 0.7947$ (equal to $0.1 + 0.9\rho(A + BK)$ with $\rho(A + BK)$ being the spectral radius of $A + BK$) and the constraints $|z_t| \leq 1$, $t \in \mathbb{N}$, (implying $|S(A+BK)z| \leq \lambda|Sz|$ for all $z \in \mathbb{R}^{n_x}$, compare [Blanchini et al., 1995]) and defined $P := \eta S$ for a suitable $\eta$ with the properties prescribed in the proof of Lemma A.7. Hence, all optimization problems to be solved online are linear programs. All terminal sets $\mathcal{X}_\mathrm{T}$ and $\mathcal{X}_\mathrm{T}^\gamma$ were computed as the maximal sets in the family of sets $\mathcal{X}$ satisfying $(A+BK)\mathcal{X} \oplus \gamma(A+BK)^N\mathcal{W} \subseteq \mathcal{X}$, $\mathcal{X} \oplus \gamma\mathcal{F}_N \subseteq \mathcal{X}$, and $\mathcal{X} \oplus K\gamma\mathcal{F}_N \subseteq \mathcal{U}$ for appropriate $\gamma \in [1, \infty)$, using the algorithm in [Kolmanovsky and Gilbert, 1998, Section 6].

As in Chapter 3, we considered three different ways of constraining the input $u_t$ for $t \in \{t_i + 1, \ldots, t_{i+1} - 1\}$, termed "to-zero" (if $u_t = 0$ for $t \in \{t_i + 1, \ldots, t_{i+1} - 1\}$), "to-hold" (if $u_t = u_{t_i}$ for $t \in \{t_i + 1, \ldots, t_{i+1} - 1\}$), and "model-based" (if $u_t = u_{t_i}[t - t_i]$ for $t \in \{t_i + 1, \ldots, t_{i+1} - 1\}$). In the implementation, these constraints were either realized by a suitable choice of the function $\kappa$ or, for the self-triggered algorithms with optimized inputs, by defining the sets $C_M$ appropriately (for the model-based approaches, we did not impose additional constraints here). We chose the prediction horizon to $N = 20$ and enforced $t_{i+1} - t_i \leq 10$ in the self-triggered case. Inner and outer approximations of $\gamma\mathcal{F}_\infty$ were computed as described in Section 3.2.4. In the self-triggered approaches with optimized inputs, we selected $(u^\star[i])_{i \in \{0, \ldots, M^\star - 1\}} \in \mathcal{D}_{M^\star}^T$ (for $T \in \{\mathrm{L}, \mathrm{m}, \mathrm{p}\}$ by choosing an input sequence minimizing the cost function $\sum_{i=0}^{M^\star - 1} \ell(u[i] - Kx[i])$. In the Lyapunov-based and mixed approaches we relaxed the respective cost function inequalities in the event-triggered controllers and self-triggered controllers with predefined inputs by 0.001, similar to Section 3.2.4.

We investigated the transient behavior of the closed-loop systems in terms of the average transmission rates by initializing the system at 30 random points in $(([-32, 32] \times [-32, 32]) \setminus ([-20, 20] \times [-20, 20]))$ and simulating for 30 time points. This set was chosen such that the algorithms are guaranteed to be initially feasible for all ranges of parameters. With each initial point we associated a disturbance sequence where $w_t$ was independently uniformly sampled on $\mathcal{W}$ for all $t \in \mathbb{N}$. For each combination of control parameters we chose these 30 initial conditions and disturbance realizations identically. The asymptotic behavior was investigated by initializing the system at the origin and simulating for 100 time points.

The results are reported in Appendix C, where we also depict the initial conditions chosen, the terminal set $\mathcal{X}_\mathrm{T}$, and an inner approximation of the set $\mathcal{F}_\infty$. From the data, the same dependence of the transmission rate on the parameters $\nu$, $\theta$, and $\gamma$ can be observed, namely that the rate decreases if the parameters are increased. For the mixed set–Lyapunov-based approaches, the influence of $\nu$ is almost non-existent, suggesting that the set-membership conditions are the deciding trigger conditions here[2]. Expectedly, $\nu$ has no influence on the asymptotic transmission rates, as the cost functions are zero for all time if the initial state is contained in $\gamma\mathcal{F}_\infty$.

When comparing the transmission rates for the different types of algorithms employed, we can see that the Lyapunov-based algorithms achieve generally lower rates than the

---

[2]In the numerical example in Section 6.2.3, we observed a stronger influence of $\nu$ on the transmission rate, suggesting a dependence on the system parameters and constraints.

mixed set–Lyapunov-based and purely set-based algorithms. The transient rates of the mixed set–Lyapunov-based approaches on the other hand are almost identical to those of the purely set-based approaches, the latter sometimes even achieving a slightly lower average rate. Finally, we conclude from the data that optimizing over the inputs in self-triggered control allows, in general, a great reduction in the average transmission rate when compared to simply checking whether the trigger conditions hold for a sequence of given inputs.

Given the comparable performance of the mixed set–Lyapunov-based and purely set-based algorithms, one might tend to favor the latter groups of control schemes, given their comparatively simpler trigger conditions. However, the additional required constraint tightening in the underlying MPC schemes—which increases with $\gamma$—leads to a reduced region in the state space where the algorithms are feasible. In Figure, 5.1, approximations of these sets[3] are depicted, demonstrating the conservatism resulting from additional constraint tightening involved with the purely set-based approaches.

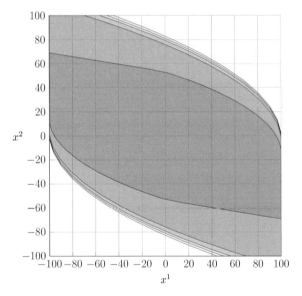

Figure 5.1.: Feasible regions $\mathcal{X}_f^\gamma$ for $\gamma \in \{1, 1.25, 1.5, 2, 5\}$; the largest region is associated with $\gamma = 1$ and is equal to $\mathcal{X}_f$, the smallest region is associated with $\gamma = 5$.

---

[3]We used the plot function implemented in YALMIP [Löfberg, 2004], based on shooting 100 rays in different directions, to obtain the feasible regions of the respective MPC problems.

# 5.6. Summary

In this chapter, we have proposed multiple algorithms dealing with a robust constrained stabilization problem with the additional objective of reducing the communication between the sensor and the actuator. Three types of schemes were considered: the first being based on guaranteeing that a certain Lyapunov function decreases over time; the second being based on guaranteeing that merely the value function of a certain MPC scheme decreases, but requiring to check additional set-membership conditions, and the third one being based on checking set memberships alone. We have related the proposed event-triggered algorithm of this third type to previously proposed threshold-based trigger rules and have shown how stability and constraint satisfaction can be ensured with minimal (in a certain sense) amount of constraint tightening. The stability properties of the second type were established by showing that the conditions of the first type were satisfied, which leads us to the conclusion that the presumed reduction in computational complexity comes with a certain increase in conservatism, leading to an increase of the average transmission rate, as could be verified in the examples. A similar relation holds between the third type and the second, although this was not reflected in a discernible difference in the transmission rates. However the implementation of the algorithms of the third type are much simpler than those of the second (and the first), but the increased conservatism entails a reduction of the subset of the state space where the control algorithms are feasible as a result of the required additional constraint tightening. A thorough comparison of the computational complexity of the different algorithms—which depends on the specific low-level implementation of the solvers and on how much the structure of the optimization problems is exploited—is, however, beyond the scope of this thesis.

# 6. Output-feedback Event-triggered Model Predictive Control

In this chapter, we consider the output-feedback analogue of the control problem in Chapter 5. While MPC also offers a solution in this setting, the constrained robust output-feedback case is much harder that the state-feedback case, as we will discuss below. In particular, the combination of MPC with set-valued estimation is still a challenging problem for which we offer our own contribution before applying the results to aperiodic control. Specifically, we consider the control of a system defined by

$$x_{t+1} = Ax_t + Bu_t + w_t \tag{6.1a}$$

$$y_t = Cx_t + v_t \tag{6.1b}$$

as described in Section 2.2, where bounds on the disturbance $w_t$ and the noise $v_t$ are given. In particular, we assume $w_t \in \mathscr{W}$ and $v_t \in \mathscr{V}$ for all $t \in \mathbb{N}$, where $\mathscr{W} \subseteq \mathbb{R}^{n_x}$ and $\mathscr{V} \subseteq \mathbb{R}^{n_y}$ are known compact sets that contain the origin. As in Section 3.2.2, we assume a linear output feedback controller of the form

$$\hat{x}_{t+1} = A\hat{x}_t + Bu_t + L(C(A\hat{x}_t + Bu_t) - y_{t+1}) \tag{6.2a}$$

$$u_t^{\text{lin}} = K\hat{x}_t, \tag{6.2b}$$

to be given, where the matrices $A + BK$ and $A + LCA$ are stable. We assume that $\hat{x}_t$ and $y_t$ are available to the controller at any time point $t$, but that $x_t$ cannot be measured directly. The resulting closed-loop control structure is depicted in Figure 6.1. Our goal is to stabilize a compact set in the joint state space composed of the state estimate $\hat{x}_t$ and the state estimation error $x_t - \hat{x}_t$ while satisfying constraints on the input and state in the form $(x_t, u_t) \in \mathscr{X} \times \mathscr{U}$, $t \in \mathbb{N}$, for given compact sets $\mathscr{X} \subseteq \mathbb{R}^{n_x}$ and $\mathscr{U} \subseteq \mathbb{R}^{n_u}$.

As mentioned above, the MPC problem in the output-feedback setting is considerably harder than for the state-feedback case. A particular issue arises from the sources of uncertainty present: while in the state-feedback setting only the future disturbances are unknown, in the output-feedback case the same holds true for the current system state. Hence, if satisfaction of hard state constraints is a control objective, it is necessary to obtain set-valued state estimates. In contrast to the bounds on the disturbances (which are assumed time-invariant and known in our setting), the bounds on the system state are time-varying and unknown *a priori*. This complicates the derivation of bounds on the future system state in the prediction and requires more elaborate steps to ensure recursive feasibility of the MPC problem.

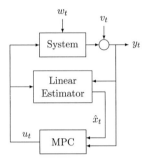

Figure 6.1.: Structure of the output-feedback control scheme.

Several approaches to output-feedback MPC have been proposed in the literature: In [Bemporad and Garulli, 2000], a robust output feedback MPC scheme based on set-valued estimates was proposed, where the estimates overapproximate the set of possible system states. Recursive feasibility of the MPC problem is not required in that work (and was not shown). Instead, it was proposed to apply the predicted input from the last time instant at which the MPC problem was feasible in the case of infeasibility at the current time instant. In [Chisci and Zappa, 2002], a recursively feasible output feedback MPC scheme based on ideal set-valued estimation was proposed, where the whole information gained from each measurement since initialization was taken into account. As already mentioned in [Chisci and Zappa, 2002], it is generally impossible to implement such a control scheme, as the amount of stored information and the complexity of the resulting optimization problems grow unbounded with time. This problem, which is inherent in set-valued estimation, was already observed in [Schweppe, 1968]. In previous output feedback MPC approaches for linear systems, this problem was usually overcome by using set-valued estimates of finite complexity based on linear estimators. Examples for such approaches can for example be found in [Chisci and Zappa, 2002] itself, and in [Mayne et al., 2006, Mayne et al., 2009, Goulart and Kerrigan, 2007, Sui et al., 2008, Kögel and Findeisen, 2015a]. These approaches have one important drawback which we want to alleviate with the method proposed in this chapter: the set-valued estimates that can be obtained from linear estimators are effectively just *a priori* computable bounds on the estimation error based on the disturbance bounds and the error dynamics. These error bounds usually disregard the dynamics between the estimation error and the estimator state, as otherwise the complexity of the bounds might grow unbounded with time. Hence, information is usually lost between the state estimates at two different time instants. Assume now that at time $t$ we perform a prediction of the set of possible states at time $t+i$ based on a certain set-valued state estimate obtained at time $t$. At time $t+1$ we also perform a prediction of the set of possible states at time $t+i$, but now based on a set-valued state estimate obtained at time $t+1$. With the possible loss of information in the set-valued estimates, it might happen that the prediction at time $t + 1$ of the set of states at $t + i$ is *not* a subset of the prediction at time $t$. Considering MPC and hard state constraints, this phenomenon might lead to the loss of feasibility in the optimization problem. In other words, for recursive feasibility to hold it is necessary that a set-valued estimate at a future time instant is always a subset of any set-valued prediction from an earlier time step, as observed in [Findeisen and Allgöwer, 2004]. In [Chisci and Zappa, 2002] and in (most of) the subsequent approaches, this problem was treated by approximating the worst-case loss of information already in the prediction at time $t$, thereby purposely not making use of the full information available. Beyond the possible loss of information—inherent to all set-valued estimators with finite complexity—set-valued estimates based on *a priori* computable error bounds will never be tighter than these very bounds. On the other hand, in set valued estimation schemes along the lines of [Schweppe, 1968], intersection of certain sets (in particular, pre-images of the output functions of the plant) is performed in order to obtain the estimates, possibly leading to much tighter error bounds if the realizations of the disturbance and noise are favorable. Other MPC approaches based on set-valued estimation (other than using only *a priori* error bounds for linear estimators) can for example be found in [Löfberg, 2002, Jia and Krogh, 2005, Sijs et al., 2010, Le et al., 2011, Subramanian et al., 2016], with various means of parameterizing and overapproximating the estimates. In [Chisci and Zappa, 2002], one of the various approaches proposed there is based on using the inequalities in the MPC problem to define the set-valued estimates. This approach guarantees recursive feasibility but entails a high, albeit bounded, computational complexity. For linear systems with uncertainties in the system matrices, predictive control approaches based on linear matrix inequalities were proposed, for example in [Kim et al., 2006]. For unbounded disturbances, where state constraint satisfaction cannot be guaranteed with certainty in general, stochastic predictive control methods were proposed, see [Farina et al., 2015] for one example.

Previous results on aperiodic output-feedback MPC are also available. In [Sijs et al., 2010] and in [Kögel and Findeisen, 2016], the amount of communication between the sensor and the controller is reduced by employing event-based transmission of measurements. At the controller side, bounds on the estimation error are obtained based on the knowledge of how the events at the sensor side are generated, thereby inferring knowledge about the system state even when no measurements arrive. In [Bernardini and Bemporad, 2012], the scheduling system at the sensor side receives information about the predicted states (computed at the controller side), and decides to transmit if the actual state deviates (under worst-case assumptions) too far from the predicted state. An event-triggered output-feedback MPC scheme is proposed in [Zhang et al., 2014], where the trigger conditions are defined in terms of the deviation of the predicted output from the actually measured output; a similar scheme is proposed in [Lu et al., 2017]. Considering self-triggered output-feedback control, in [Kögel and Findeisen, 2015b], at periodic time instants, a schedule for the times at which measurements are taken until the next such periodic instant is computed. In particular, these schedules are decision variables in the MPC problem at these time instants and influence the constraint tightening therein.

In this thesis, we propose an output feedback MPC scheme where the set-valued estimate of the current state is obtained based on the $M$ most recent measurements obtained from the plant, where $M$ is a fixed integer. This way, the complexity associated with the set-valued estimation is bounded, without the need for explicitly computing an outer approximation of the state estimate at each point in time. Using a fixed number of recent measurements in order to obtain a state estimate is known as moving horizon estimation (MHE), see for example [Rawlings and Mayne, 2009] and the references therein. Output feedback MPC incorporating MHE was recently proposed in [Copp and Hespanha, 2014] for nonlinear systems, although without considering robust state constraint satisfaction. In [Kögel and Findeisen, 2015b], output-feedback MPC for constrained linear systems was considered, where, similarly to the approach considered in this chapter, set-valued moving horizon estimation is employed. Consistency was achieved by intersecting the uncertainty bounds obtained at the current time step with the uncertainty bounds from the previous time steps. With the estimation structure based on $M$ recent measurements employed here, there still is the possible loss of information when advancing in time mentioned above, as we discard the oldest measurement when a new measurement arrives. We treat this problem by using only the $M-i$ most recent measurements when performing, at time $t$, the prediction of the state and inputs at time $t+i$. For predictions of $M$ and more steps ahead and predictions at the end of the horizon, we resort to methods similar to those proposed in [Chisci and Zappa, 2002, Mayne et al., 2006, Mayne et al., 2009] for linear estimators. Thereby we can prove recursive feasibility of the resulting MPC scheme, as for the first $M$ predicted steps *no* loss of information occurs, and for the remaining steps established methods of overapproximating the loss of information are in place. Considering stability of the closed-loop system, the same asymptotic bound on the system state as [Chisci and Zappa, 2002] holds. We will demonstrate the efficacy of the proposed approach in a numerical example, where we will show that our approach allows the enlargement of the feasible region of the MPC scheme beyond the capabilities of the linear-estimator-based approach in [Chisci and Zappa, 2002]. As the set-valued estimates used in the MPC scheme are in general far less conservative than set-valued estimates based on *a priori* error bounds for linear estimators, we also expect a general improvement in closed-loop performance.

In Section 6.1 we derive a (periodically updated) output-feedback MPC scheme that demonstrably improves upon earlier controllers. In Section 6.2, we apply the ideas of Section 5.2 in order to develop an event-triggered version of our scheme: the proposed aperiodic controller is based on guaranteeing a decrease of the MPC value function and the inclusion of the joint state being contained in a certain set, analogous to the output feedback schemes proposed in Section 3.3.3.

Parts of the introduction and most of Section 6.1 are also contained in both [Brunner et al., 2016e] and [Brunner et al., 2017d].

# 6.1. Set-valued moving horizon estimation in model predictive control

As in Section 3.2.2, we employ the estimation error $\tilde{x}_t := x_t - \hat{x}_t$ and the joint state $(\tilde{x}_t, \hat{x}_t)$ with dynamics

$$\begin{bmatrix} \tilde{x}_{t+1} \\ \hat{x}_{t+1} \end{bmatrix} = \hat{A} \begin{bmatrix} \tilde{x}_t \\ \hat{x}_t \end{bmatrix} + \begin{bmatrix} 0 \\ B \end{bmatrix} (u_t - K\hat{x}_t) + \hat{w}_t, \tag{6.3}$$

with $\hat{w}_t \in \hat{\mathscr{W}}$, where $\hat{A}$, $\hat{w}_t$, $\hat{\mathscr{W}}$, and $\hat{\mathscr{F}}_i$ for $i \in \mathbb{N} \cup \{\infty\}$ are defined as in Section 3.2.2, that is,

$$\hat{A} := \begin{bmatrix} A + LCA & 0 \\ -LCA & A + BK \end{bmatrix}, \quad \hat{B} := \begin{bmatrix} 0 \\ B \end{bmatrix}, \quad \hat{w}_t := \begin{bmatrix} I + LC & L \\ -LC & -L \end{bmatrix} \begin{bmatrix} w_t \\ v_{t+1} \end{bmatrix}, \tag{6.4}$$

for $t \in \mathbb{N}$,

$$\hat{\mathscr{W}} := \begin{bmatrix} I + LC & L \\ -LC & -L \end{bmatrix} (\mathscr{W} \times \mathscr{V}), \quad \text{and} \quad \hat{\mathscr{F}}_i = \bigoplus_{j=0}^{i-1} \hat{A}^j \hat{\mathscr{W}} \tag{6.5}$$

for $i \in \mathbb{N} \cup \{\infty\}$. Finally, we denote by

$$\begin{aligned} \tilde{y}_{t+1} &:= -CA\tilde{x}_t - Cw_t - v_{t+1} \\ &= C(A\hat{x}_t + Bu_t) - y_{t+1} \end{aligned} \tag{6.6}$$

for $t \in \mathbb{N}$ the *innovations* in the estimator dynamics (6.2a). Note that this quantity can be computed based on information known to the controller. Further, we assume that sets $\mathscr{E}_t \subseteq \mathbb{R}^{n_x}$ satisfying $\tilde{x}_t \in \mathscr{E}_t$ for all $t \in \mathbb{N}$ are known *a priori*. No further restrictions are placed on these sets for the moment.

## 6.1.1. General results

The main idea in tube MPC schemes as, for example, proposed in [Chisci et al., 2001, Chisci and Zappa, 2002] is to predict a nominal trajectory of the state (or the state estimate), that is, a trajectory of the system where the disturbances (or the innovations) are assumed zero. Additionally, sets containing the worst case future deviation of the real trajectory from the predicted nominal trajectory are computed and the constraints on the nominal trajectory are appropriately tightened in order to accommodate this uncertainty in the prediction. Here, it is important to take into account that feedback will be present at future points in time, which reduces the growth of the uncertainty in the prediction horizon. In the simplest tube MPC schemes, this feedback is assumed to be based on a fixed linear gain [Chisci et al., 2001, Mayne et al., 2005], while more elaborate schemes include the feedback as an optimization variable [Löfberg, 2003, Raković et al.,

2012a]. Here, we stick to the simple case of fixed linear feedback. In the output-feedback setting, the prediction uncertainty consists of the estimation error and the deviation of the state estimate from its prediction. We describe these bound by sets $\hat{\mathcal{X}}[i|t] \subseteq \mathbb{R}^{2n_x}$, where $t \in \mathbb{N}$ denotes the time of prediction and $i \in \mathbb{N}$ the prediction step.

In particular, let $((\hat{x}[i])_{i\in\mathbb{N}}, (u[i])_{i\in\mathbb{N}})$ denote a prediction of the nominal input and the associated nominal state estimate trajectory made at time $t$, that is, $\hat{x}[0] = \hat{x}_t$ and $\hat{x}[i+1] = A\hat{x}[i] + Bu[i]$ for $i \in \mathbb{N}$. Under the assumption of feedback of the form $u_{t+i} = u[i] + K(\hat{x}_{t+i} - \hat{x}[i])$ for $i \in \mathbb{N}$, the bounds on the uncertainty should satisfy

$$\begin{bmatrix} \tilde{x}_{t+i} \\ \hat{x}_{t+i} - \hat{x}[i] \end{bmatrix} \in \hat{\mathcal{X}}[i|t]. \tag{6.7}$$

for all $t \in \mathbb{N}$ and all $i \in \{0, \ldots, N\}$, with $N \in \{1, \ldots\}$ being the length of the prediction horizon. Note, however, that the actual input *applied* to the system at time $t + i$ might deviate from the input $u[i] + K(\hat{x}_{t+i} - \hat{x}[i])$ *predicted* at time $t$. The uncertainty description depends on the bounds on the estimation error known to the controller at time $t$ and, hence, is time-varying.

Following Section 2.5, we now define the structure of the MPC scheme. The set of admissible nominal input sequences at time $t$ is given by

$$\mathscr{D}_t(\hat{x}) := \left\{ (u[i])_{i\in\{0,\ldots,N-1\}} \in \mathbb{R}^{Nn_u} \right.$$

$$\hat{x}[0] = \hat{x} \tag{6.8a}$$

$$\hat{x}[i+1] = A\hat{x}[i] + Bu[i] \tag{6.8b}$$

$$\left\{ \begin{bmatrix} \hat{x}[i] \\ u[i] \end{bmatrix} \right\} \oplus \begin{bmatrix} I & I \\ 0 & K \end{bmatrix} \hat{\mathcal{X}}[i|t] \subseteq \mathcal{X} \times \mathcal{U}, \quad i \in \{0, \ldots, N-1\} \tag{6.8c}$$

$$\hat{x}[N] \in \hat{\mathcal{X}}_{\mathrm{T},t} \tag{6.8d}$$

$$\left. \vphantom{\begin{bmatrix} \hat{x}[i] \\ u[i] \end{bmatrix}} \right\}$$

and

$$\hat{\mathcal{X}}_{\mathrm{f},t} := \{\hat{x} \in \mathbb{R}^{n_x} \mid \mathscr{D}_t(\hat{x}) \neq \emptyset\} \tag{6.9}$$

denotes the subset of the estimator state space where the constraints are feasible at time $t$. The sets $\mathcal{X}_{\mathrm{T},t}$ constitute terminal constraints.

**Remark 6.1.** *To be precise, $\mathscr{D}_t$ is a function not of time, but of the information available at time t. The available information in this setting, however, is strictly a function of time and of the realization of the disturbance and noise, that is, it is independent of the decisions made by the controller. Hence, the chosen simpler notation is justified.*

Define further the cost function $\bar{J} : \mathbb{R}^{n_x} \times \mathbb{R}^{N n_u}$ by

$$\bar{J}(\hat{x}, (u[0], \ldots, u[N-1])) = \sum_{k=0}^{N-1} \ell(u[i] - K\hat{x}[i]), \tag{6.10}$$

where $\hat{x}[0] = \hat{x}$ and $\hat{x}[i+1] = A\hat{x}[i] + Bu[i]$ for $i \in \{0, \ldots, N-2\}$ and $\ell$ is a continuous positive definite function. Finally, define the MPC value function

$$J_t(\hat{x}) := \inf \left\{ \bar{J}(\hat{x}, \mathbf{u}) \mid \mathbf{u} \in \mathscr{D}_t(\hat{x}) \right\}, \tag{6.11}$$

the set of optimal trajectories

$$\begin{aligned}
\mathscr{T}_t(\hat{x}) = \Big\{ (\hat{\mathbf{x}}, \mathbf{u}) &\in (\mathbb{R}^{n_x})^{\mathbb{N}} \times (\mathbb{R}^{n_u})^{\mathbb{N}} \mid \hat{\mathbf{x}} = (x[i])_{i \in \mathbb{N}}, \ \mathbf{u} = (u[i])_{i \in \mathbb{N}}, \\
&\hat{x}[0] = \hat{x}, \ \hat{x}[i+1] = A\hat{x}[i] + Bu[i], \ i \in \{0, \ldots, N-1\}, \\
&(u[0], \ldots, u[N-1]) \in \mathscr{D}(\hat{x}) \\
&u[i] = K\hat{x}[i], i \in \{N, \ldots\}, \ \bar{J}(\hat{x}, (u[0], \ldots, u[N-1])) = J_t(\hat{x}) \Big\}, \tag{6.12}
\end{aligned}$$

and the set-valued controller

$$\kappa_{\mathrm{MPC},t}(\hat{x}) = \left\{ u \in \mathbb{R}^{n_u} \mid \exists \hat{\mathbf{x}} \in (\mathbb{R}^{n_x})^{\mathbb{N}}, \ \exists \tilde{\mathbf{u}} \in (\mathbb{R}^{n_u})^{\mathbb{N}}, \ (\hat{\mathbf{x}}, (u, \tilde{\mathbf{u}})) \in \mathscr{T}_t(\hat{x}) \right\}. \tag{6.13}$$

We make the following assumptions on the sets $\hat{\mathcal{X}}[i|t]$ and $\mathcal{X}_{\mathrm{T},t}$; realizations of these sets satisfying the assumptions will be discussed in Section 6.1.2.

**Assumption 6.1.** *For all $t \in \mathbb{N}$, it holds that*

*(i)* $(\tilde{x}_t, 0) \in \hat{\mathcal{X}}[0|t]$,

*(ii)* $\hat{\mathcal{X}}[i|t+1] \oplus \left\{ \begin{bmatrix} 0 \\ I \end{bmatrix} (A+BK)^i L\tilde{y}_{t+1} \right\} \subseteq \hat{\mathcal{X}}[i+1|t]$ *for $i \in \{0, \ldots, N-1\}$,*

*(iii)* $\hat{\mathcal{X}}_{\mathrm{T},t}$ *is compact,*

*(iv)* $\begin{bmatrix} I \\ K \end{bmatrix} \hat{\mathcal{X}}_{\mathrm{T},t} \oplus \begin{bmatrix} I & I \\ 0 & K \end{bmatrix} \hat{\mathcal{X}}[N|t] \subseteq \mathcal{X} \times \mathcal{U}$, *and*

*(v)* $(A+BK)\hat{\mathcal{X}}_{\mathrm{T},t} \oplus \left\{ (A+BK)^N L\tilde{y}_{t+1} \right\} \subseteq \hat{\mathcal{X}}_{\mathrm{T},t+1}$.

Here, Assumption 6.1(i) ensures that $\hat{\mathcal{X}}[0|t]$ constitutes a bound on the the true estimation error at time $t$, Assumption 6.1(ii) expresses the consistency between state predictions made at different points in time discussed in the chapter introduction, and Assumption 6.1(iii) to Assumption 6.1(v) fix the properties of the terminal sets used to prove recursive feasibility later in the section.

The closed-loop control of system (6.1) with the output-feedback MPC scheme is described in Algorithm 13.

---

**Algorithm 13** Output-feedback model predictive control

---

1: Initialize state estimate $\hat{x}_0$.
2: **for** $t \in \mathbb{N}$ **do**
3:     Obtain measurement $y_t$.
4:     **if** $t \geq 1$ **then**
5:         Update state estimate $\hat{x}_t$ by iterating (6.2a).
6:     **end if**
7:     Obtain uncertainty bounds $\hat{\mathcal{X}}[i|t]$ for $i \in \{0, \ldots, N-1\}$.
8:     Obtain terminal set $\hat{\mathcal{X}}_{\mathrm{T},t}$.
9:     Solve (6.11), obtain a $u \in \kappa_{\mathrm{MPC},t}(\hat{x}_t)$.
10:    Apply $u_t = u$ to the system.
11: **end for**

---

**Main properties of the output-feedback MPC scheme**

We first state a technical result concerning the existence of a minimizer of the optimization problem in (6.11).

**Lemma 6.1.** *If $\hat{x} \in \hat{\mathcal{X}}_{\mathrm{f},t}$, then $\kappa_{\mathrm{MPC},t}(\hat{x}) \neq \emptyset$.*

*Proof.* The statement follows from standard arguments; in particular, due to the compactness of $\mathcal{X}$ and $\mathcal{U}$ and the closedness of $\mathcal{X}_{\mathrm{T},t}$, $\mathcal{D}_t(\hat{x})$ is also compact for all $\hat{x} \in \mathbb{R}^{n_x}$. Together with the continuity of $\ell$, this yields the result, compare Proposition 2.4 in [Rawlings and Mayne, 2009].    □

The next statement establishes recursive feasibility of the MPC scheme in closed loop operation, that is, the nonemptiness of $\kappa_{\mathrm{MPC},t}(\hat{x}_t)$ in step 9 of Algorithm 13 under certain assumptions.

**Theorem 6.1.** *Let $t \in \mathbb{N}$, $\hat{x}_t \in \hat{\mathcal{X}}_{\mathrm{f},t}$, and $u_t \in \kappa_{\mathrm{MPC},t}(\hat{x}_t)$. Then it holds that $\hat{x}_{t+1} \in \hat{\mathcal{X}}_{\mathrm{f},t+1}$, where $\hat{x}_{t+1} = A\hat{x}_t + Bu_t + L\tilde{y}_{t+1}$.*

Constraint satisfaction and convergence then follow readily, using similar arguments as in [Chisci and Zappa, 2002]:

**Corollary 6.1.** *Let $\hat{x}_0 \in \hat{\mathcal{X}}_{\mathrm{f},0}$ and let $u_t \in \kappa_{\mathrm{MPC},t}(\hat{x}_t)$ in closed-loop with (6.1) and (6.2a) for all $t \in \mathbb{N}$. Then it holds that $(x_t, u_t) \in \mathcal{X} \times \mathcal{U}$ for all $t \in \mathbb{N}$. Further $\lim_{t \to \infty} |(\tilde{x}_t, \hat{x}_t)|_{\hat{\mathcal{F}}_\infty} = 0$.*

*Proof.* By Theorem 6.1, it holds that $\hat{x}_t \in \hat{\mathcal{X}}_{\mathrm{f},t}$ for all $t \in \mathbb{N}$. Considering constraint (6.8c) in the MPC scheme for $i = 0$ and Assumption 6.1(i) then yields $(x_t, u_t) \in \mathcal{X} \times \mathcal{U}$ for all $t \in \mathbb{N}$. Considering the candidate trajectory $((\hat{x}'[0], \ldots, \hat{x}'[N]), (u'[0], \ldots, u'[N-1]))$ employed in the proof of Theorem 6.1, we see that by optimality of $u_t$ and the positive definiteness of $\ell$ it holds that $0 \leq J_{t+1}(\hat{x}_{t+1}) \leq J_t(\hat{x}_t) - \ell(u_t - K\hat{x}_t)$ and, hence,

$\lim_{t\to\infty}(u_t - K\hat{x}_t) = 0$. Considering (6.3), it follows that $\lim_{t\to\infty} |(\tilde{x}_t, \hat{x}_t)|_{\hat{\mathscr{F}}_\infty} = 0$, thereby completing the proof. $\qquad\square$

In order to establish stability of $\hat{\mathscr{F}}_\infty$ in addition to convergence, we require further technical definitions and assumptions. Define

$$\hat{\mathscr{X}}[i] := \bigcup_{t\in\mathbb{N}} \hat{\mathscr{X}}[i|t], \tag{6.14}$$

$$\hat{\mathscr{X}}_{\mathrm{T}} := \bigcap_{t\in\mathbb{N}} \hat{\mathscr{X}}_{\mathrm{T},t}, \tag{6.15}$$

and

$$\bar{\mathscr{X}}_0 := \left\{ \hat{x} \in \mathbb{R}^{n_{\mathsf{x}}} \ \middle| \ \left\{ \begin{bmatrix} (A+BK)^i \hat{x} \\ K(A+BK)^i \hat{x} \end{bmatrix} \right\} \oplus \begin{bmatrix} I & I \\ 0 & K \end{bmatrix} \hat{\mathscr{X}}[i] \subseteq \mathscr{X} \times \mathscr{U}, \ i \in \{0, \ldots, N-1\}, \right.$$
$$\left. (A+BK)^N \hat{x} \in \hat{\mathscr{X}}_{\mathrm{T}} \right\}. \tag{6.16}$$

It holds that $\bar{\mathscr{X}}_0$ is compact and is a subset of $\hat{\mathscr{X}}_{\mathrm{f},t}$ for all $t \in \mathbb{N}$. In particular, when compared to $\hat{\mathscr{X}}_{\mathrm{f},t}$, $\bar{\mathscr{X}}_0$ contains the additional constraint that the input sequence is generated by the linear control law $u = K\hat{x}$.

We can now state a stability result for the closed-loop system:

**Lemma 6.2.** *Let there exist an $\epsilon > 0$ such that $\hat{\mathscr{F}}_\infty \oplus \epsilon\mathscr{B}_{2n_{\mathsf{x}}} \subseteq \mathbb{R}^{n_{\mathsf{x}}} \times \bar{\mathscr{X}}_0$. Let further $\hat{\mathscr{X}}[i|t] \subseteq \eta\mathscr{B}_{n_{\mathsf{x}}}$ for all $i \in \{0, \ldots, N-1\}$ and all $t \in \mathbb{N}$ and some $\eta \in [0,\infty)$. Finally, let there exist a continuous positive definite function $\bar{V}^{\mathsf{s}} : \mathbb{R}^{2n_{\mathsf{x}}} \to \mathbb{R}$ and constants $\bar{c}_1$, $\bar{c}_2$, $\bar{c}_3$, $a \in (0,\infty)$ such that for all $z \in \mathbb{R}^{2n_{\mathsf{x}}}$ and all $v \in \mathbb{R}^{n_{\mathsf{u}}}$ it holds that*

$$\bar{c}_1 |z|^a \leq \bar{V}^{\mathsf{s}}(z) \leq \bar{c}_2 |z|^a \tag{6.17a}$$

$$\bar{V}^{\mathsf{s}}(z) - \bar{V}^{\mathsf{s}}(\hat{A}z + \left[\begin{smallmatrix} 0 \\ B \end{smallmatrix}\right] v) + \ell(v) \geq \bar{c}_3 |z|^a. \tag{6.17b}$$

*Then, $\hat{\mathscr{F}}_\infty$ is 1-uniformly asymptotically stable for any dynamical system generating $(\tilde{x}_t, \hat{x}_t)_{t\in\mathbb{N}}$ for the closed-loop consisting of (6.1) and (6.2a) where $u_t \in \kappa_{\mathrm{MPC},t}(\hat{x}_t)$ for all $t \in \mathbb{N}$, with $[I\ 0]\hat{\mathscr{X}}[0|0] \times \hat{\mathscr{X}}_{\mathrm{f},0}$ belonging to its region of attraction.*

**Remark 6.2.** *The assumption that the sets $\hat{\mathscr{X}}[i|t]$ are uniformly bounded is satisfied under mild assumptions for the specific bounds proposed later on. As already remarked in Section 5.1, for quadratic or convex piecewise linear $\ell$, a function $\bar{V}^{\mathsf{s}}$ with the required properties always exists.*

## 6.1.2. Realization with set-valued moving horizon estimation

In this section, we propose a particular realization of the sets $\hat{\mathcal{X}}[i|t]$ in (6.8) based on set-valued moving horizon estimation. Further, we provide appropriate terminal sets $\hat{\mathcal{X}}_{\mathrm{T},t}$. Assume for the moment that at time $t$ a bound $\tilde{x}_t \in \mathcal{E}_t$ is known. Then, under the assumed feedback described in Section 6.1.1, the tightest choice for $\hat{\mathcal{X}}[i|t]$ satisfying (6.7) is

$$\hat{\mathcal{X}}[i|t] = \hat{A}^i(\mathcal{E}_t \times \{0\}) \oplus \hat{\mathcal{F}}_i. \tag{6.18}$$

However, with this choice, Assumption 6.1(ii) places restrictions on $\mathcal{E}_t$ which require this set to be updated at each time step in a manner which, in general, leads to an unbounded computational complexity, compare the discussion in [Chisci and Zappa, 2002, Section 5]. An alternative solution, described in the same work, is to separate the estimator dynamics from the estimation error dynamics and to assume *a priori* bounds $\tilde{x}_t \in \mathcal{E}_t$ on the estimation error to be known[1]. Considering that $\hat{x}_{t+1} = A\hat{x}_t + Bu_t + L\tilde{y}_{t+1} = A\hat{x}_t + Bu_t + (-LCA)\tilde{x}_t + (-LC)w_t + (-L)v_{t+1}$ and the assumption (in the prediction) that $u_{t+i} = u[i] + K(\hat{x}_{t+i} - \hat{x}[i])$, leads to

$$\hat{\mathcal{X}}[i|t] = \mathcal{E}_{t+i} \times \bar{\mathcal{G}}[i|t], \tag{6.19}$$

where

$$\bar{\mathcal{G}}[i|t] := \bigoplus_{j=0}^{i-1} (A + BK)^{i-1-j}((-LCA)\mathcal{E}_{t+j} \oplus (-LC)\mathcal{W} \oplus (-L)\mathcal{V}). \tag{6.20}$$

While satisfying Assumption 6.1(ii) and ensuring that the complexity remains bounded, this approach has a significant drawback: due to the separation of the dynamics, the uncertainty is overapproximated. In particular, one neglects that the estimation error and the estimate are affected by the *same* realization of the disturbance $w_t$ and the noise $v_{t+1}$. This conservatism vanishes in general only for the choice $L = 0$. If $A$ has positive eigenvalues, however, there exists a lower bound on $L$ necessary for the stability of $A + LCA$. Here, we propose an approach that allows the less conservative bounds in (6.18) to be employed at least for the beginning of the prediction horizon, while a bound similar to (6.19) will be used for the remaining prediction steps. As we will see, this leads to a great reduction in the overall conservatism, demonstrated by an increase in the feasible region $\hat{\mathcal{X}}_{\mathrm{f},0}$.

**Set-valued moving horizon estimation**

The first key to the novel approach is to employ a moving-horizon set-valued estimator, whose inputs at time $t$ are (arbitrary) *a priori* bounds $\tilde{x}_{t-j} \in \mathcal{E}_{t-j}$, for some $j \in \mathbb{N}$

---

[1]In [Chisci and Zappa, 2002] it was additionally assumed that $\mathcal{E}_t = \mathcal{E}$ for $t \in \mathbb{N}$. An extension to time-varying error bounds satisfying $\mathcal{E}_{t+1} \subseteq \mathcal{E}_t$ for all $t \in \mathbb{N}$ was proposed in [Mayne et al., 2009].

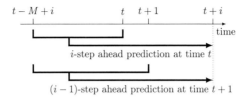

Figure 6.2.: Proposed prediction structure.

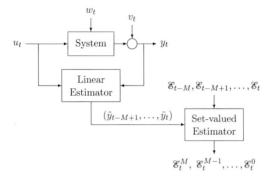

Figure 6.3.: Structure of the estimation scheme.

on the estimation error and the innovations $\tilde{y}_\tau$ for a certain range of $\tau$. In particular, the estimator uses (at most) the $M$, henceforth referred to as the "estimation horizon", most recent values of the innovation to compute a bound on the estimation error at a certain point in time. The second key to the approach is that the set of innovations used to compute the bounds on the estimation error at time $t$ depend on the prediction step for which these bounds are used; that is, for prediction steps *farther* into the future, *less* innovations are employed. To be precise, for the $i$ step ahead prediction at time $t$, the innovations $(\tilde{y}_{t+1-M+i}, \ldots, \tilde{y}_t)$ are employed. This concept is illustrated in Figure 6.2. Thus, the set-valued estimator yields multiple bounds on the estimation error at time $t$, parameterized by the number $j$ of recent innovations used in its computation and the time $t - j$ at which the *a priori* bound is required. We denote this estimate by $\mathscr{E}_t^j$. The estimation structure is depicted in Figure 6.3. Following the exposition in Section 2.4.2, we define the sets $\mathscr{E}_t^j$ for $t \in \mathbb{N}$ and $j \in \{0, \ldots, j\}$ by

$$
\mathscr{E}_t^j := \left\{ \tilde{x}_t' \in \mathbb{R}^{n_x} \left| \begin{array}{c} \tilde{x}_{t-j}' \in \mathscr{E}_{t-j}, \ (w_k', v_{k+1}') \in \mathscr{W} \times \mathscr{V}, \\ \tilde{x}_{k+1}' = (A + LCA)\tilde{x}_k' + (I + LC)w_k' + Lv_{k+1}' \\ \tilde{y}_{k+1} = -CA\tilde{x}_k' - Cw_k' - v_{k+1}', \\ k \in \{t - j, \ldots, t - 1\} \end{array} \right. \right\}. \tag{6.21}
$$

Clearly, it holds that $\tilde{x}_t \in \mathscr{E}_t^j$ and $\mathscr{E}_t^0 = \mathscr{E}_t$ for all $t \in \mathbb{N}$.

**Uncertainty bounds**

Let $\hat{M} := \min\{N, M\}$ and define for all $t \in \mathbb{N}$ and all $i \in \{0, \ldots, \hat{M}\}$

$$
\hat{\mathscr{X}}^{\mathrm{I}}[i|t] := \hat{A}^i \left( \mathscr{E}_t^{\min\{t, M-i\}} \times \{0\} \right) \oplus \hat{\mathscr{F}}_i. \tag{6.22}
$$

These sets describe the uncertainty in the beginning of the prediction horizon and are based on the less conservative joint dynamics as employed in (6.18). Note that the way $\hat{\mathscr{X}}^{\mathrm{I}}[i|t]$ is defined, no measurements at times before $t = 0$ are required to compute the set. Define further for all $t \in \mathbb{N}$ and all $i \in \{\hat{M}, \ldots, N\}$

$$
\hat{\mathscr{X}}^{\mathrm{II}}[i|t] := \hat{A}^{\hat{M}} \left( \hat{\mathscr{E}}_{t+i-\hat{M}} \times \{0\} \right) \oplus \hat{\mathscr{F}}_{\hat{M}} \oplus \left( \{0\} \times \mathscr{G}[i|t] \right), \tag{6.23}
$$

where

$$
\hat{\mathscr{E}}_j := (A + LCA)^{\min\{j, M-\hat{M}\}} \mathscr{E}_{\max\{0, j-M+\hat{M}\}} \oplus \bigoplus_{k=0}^{\min\{j, M-\hat{M}\}-1} (A + LCA)^k \left( (I + LC)\mathscr{W} \oplus L\mathscr{V} \right) \tag{6.24}
$$

and

$$
\mathscr{G}[i|t] := \bigoplus_{k=0}^{i-1-\hat{M}} (A + BK)^{i-1-k} \left( (-LCA)\hat{\mathscr{E}}_{t+k} \oplus (-LC)\mathscr{W} \oplus (-L)\mathscr{V} \right). \tag{6.25}
$$

The sets $\hat{\mathscr{X}}^{\mathrm{II}}[i|t]$ describe the uncertainty in the latter part of the prediction horizon and are based on the more conservative separate dynamics as employed in (6.19). Note, however, that the term $\hat{A}^{\hat{M}} \left( \hat{\mathscr{E}}_{t+i-\hat{M}} \times \{0\} \right) \oplus \hat{\mathscr{F}}_{\hat{M}}$ in the definition of $\hat{\mathscr{X}}^{\mathrm{II}}[i|t]$ allows to retain some of the low conservatism (determined by $\hat{M}$) even for the latter part, such that the sets $\hat{\mathscr{X}}^{\mathrm{II}}[i|t]$ are still of lower conservatism than the sets in (6.19) for $\hat{M} > 0$. For $\hat{M} = 0$ (implying $M = 0$), the expressions in (6.19) and (6.23) are identical. Further, considering the definition of the sets, it holds that $\mathscr{E}_t^{\min\{t, M-\hat{M}\}} \subseteq \hat{\mathscr{E}}_t$ for all $t \in \mathbb{N}$, and, hence, $\hat{\mathscr{X}}^{\mathrm{I}}[\hat{M}|t] \subseteq \hat{\mathscr{X}}^{\mathrm{II}}[\hat{M}|t]$ for all $t \in \mathbb{N}$ and $\hat{M} \in \mathbb{N}$.

We have the following result.

**Lemma 6.3.** *Let*

$$\hat{\mathcal{X}}[i|t] = \begin{cases} \hat{\mathcal{X}}^{\mathrm{I}}[i|t] & \text{if } i \in \{0, \ldots, \hat{M}-1\} \\ \hat{\mathcal{X}}^{\mathrm{II}}[i|t] & \text{if } i \in \{\hat{M}, \ldots, N\}. \end{cases} \tag{6.26}$$

*Then Assumption 6.1(i) and Assumption 6.1(ii) are satisfied.*

For $M = 0$ and $\mathcal{E}_t = \mathcal{E}$, $t \in \mathbb{N}$, the uncertainty description in (6.26) becomes identical to the low complexity description proposed in [Chisci and Zappa, 2002, Section 5.3].

### Terminal set

From a computational point of view, one would like to compute all terminal sets $\hat{\mathcal{X}}_{\mathrm{T},t}$ explicitly offline. This of course requires the number of different sets $\hat{\mathcal{X}}_{\mathrm{T},t}$ to be finite. In order to achieve this requirement, we additionally assume that there exists an $s \in \mathbb{N}$, such that the *a priori* error bounds satisfy $\mathcal{E}_t = \mathcal{E}_s$ for $t \in \{s, \ldots\}$. Considering that the set dynamics induced by the dynamics of the estimation error, (namely $\mathcal{E}^+ = (A + LCA)\mathcal{E} \oplus (I + LCA)\mathcal{W} \oplus L\mathcal{V})$, are convergent, compare [Mayne et al., 2009, Kögel and Findeisen, 2016, Subramanian et al., 2016], this assumption is justified. Then, with $\hat{\mathcal{X}}[N|t] = \hat{\mathcal{X}}^{\mathrm{II}}[N|t]$ as defined in (6.26), Assumption 6.1(iv) becomes

$$\begin{bmatrix} I \\ K \end{bmatrix} \hat{\mathcal{X}}_{\mathrm{T},t} \oplus \begin{bmatrix} I & I \\ 0 & K \end{bmatrix} \hat{\mathcal{X}}^{\mathrm{II}}[N|s'] \subseteq \mathcal{X} \times \mathcal{U} \tag{6.27}$$

for all $t \geq s'$ and $s' := s + \max\{0, M - N\}$. (This requirement on $t$ follows from the definition of $\hat{\mathcal{X}}^{\mathrm{II}}[N|t]$). Here it is important to note that $\hat{\mathcal{X}}^{\mathrm{II}}[N|t]$ does not depend on online set-valued estimation; that is, the set can be computed from *a priori* known quantities. Define now $\hat{\mathcal{X}}_{\mathrm{T},s'}$ as the maximal set in the family of sets given by

$$\left\{ \mathcal{X}' \subseteq \left( \begin{bmatrix} I \\ K \end{bmatrix} \right)^{-1} \left( (\mathcal{X} \times \mathcal{U}) \ominus \begin{bmatrix} I & I \\ 0 & K \end{bmatrix} \tilde{\mathcal{X}}^{\mathrm{II}}[N|s'] \right) \right|$$

$$(A + BK)\mathcal{X}' \oplus (A + BK)^N \left( (-LCA)\mathcal{E}_s \oplus (-LC)\mathcal{W} \oplus (-L)\mathcal{V} \right) \subseteq \mathcal{X}' \right\}. \tag{6.28}$$

Define further $\hat{\mathcal{X}}_{\mathrm{T},t} := \hat{\mathcal{X}}_{\mathrm{T},s'}$ for all $t \geq s'$, for which we immediately conclude that Assumption 6.1(iv) and Assumption 6.1(v) are satisfied. The satisfaction of Assumption 6.1(iii) follows from [Kolmanovsky and Gilbert, 1998, Theorem 5.2].

Finally, define for all $t \in \{0, \ldots, s' - 1\}$

$$
\hat{\mathcal{X}}_{\mathrm{T},t} := \left( \begin{bmatrix} I \\ K \end{bmatrix} \right)^{-1} \left( (\mathcal{X} \times \mathcal{U}) \ominus \begin{bmatrix} I & I \\ 0 & K \end{bmatrix} \tilde{\mathcal{X}}^{\mathrm{II}}[N|t] \right)
$$

$$
\cap (A + BK)^{-1} \left( \hat{\mathcal{X}}_{\mathrm{T},t+1} \ominus (A + BK)^N \left( (-LCA)\mathcal{E}_t \right. \right.
$$

$$
\left. \left. \oplus (-LC)\mathcal{W} \oplus (-L)\mathcal{V} \right) \right). \quad (6.29)
$$

By definition, these sets satisfy Assumption 6.1(iii) to Assumption 6.1(v).

### 6.1.3. Implementation

In the following, we assume that the sets $\mathcal{W}$, $\mathcal{V}$, $\mathcal{X}$, $\mathcal{U}$, and $\mathcal{E}_t$, $t \in \mathbb{N}$, are polytopes. Then, by employing the algorithm in [Kolmanovsky and Gilbert, 1998], polytopic sets $\hat{\mathcal{X}}_{\mathrm{T},t}$, $t \in \mathbb{N}$, may be obtained using the procedure outlined in the previous subsection. Hence, if the constraint (6.8c) can be shown to be linear and the cost function $\ell$ is convex, the MPC problem defined by (6.11) will be a linearly constrained convex problem.

We will establish linearity of (6.8c) along the same lines as the discussion at the end of Section 3.3.4. Consider first the case $\hat{\mathcal{X}}[i|t] = \hat{\mathcal{X}}^{\mathrm{I}}[i|t]$, in which the constraint is equivalent to

$$
\left\{ \begin{bmatrix} \hat{x}[i] \\ u[i] \end{bmatrix} \right\} \oplus \begin{bmatrix} I & I \\ 0 & K \end{bmatrix} \tilde{A}^i \left( \mathcal{E}_t^{\min\{t, M-i\}} \times \{0\} \right) \subseteq (\mathcal{X} \times \mathcal{U}) \ominus \begin{bmatrix} I & I \\ 0 & K \end{bmatrix} \hat{\mathcal{F}}_i. \quad (6.30)
$$

Here, $\hat{\mathcal{F}}_i$ is an (*a priori* known) polytope and so is the right-hand side of the set inclusion, which we assume admits a representation of the form[2] $\{z \in \mathbb{R}^{2n_\mathrm{x}} \mid Gz \leq g\}$ for a matrix $G \in \mathbb{R}^{q \times 2n_\mathrm{x}}$ and a vector $g \in \mathbb{R}^q$. Further, by its definition in (6.21), $\mathcal{E}_t^{\min\{t, M-i\}}$ is also a polytope, which depends on the innovations used in the set-valued estimation, but which does not depend on decision variables in the MPC problem. We assume that it admits a representation of the form $\{He \mid e \in \mathbb{R}^r, Fe \leq f\}$ for matrices $H \in \mathbb{R}^{n_\mathrm{x} \times r}$, $F \in \mathbb{R}^{p \times r}$, and a vector $f \in \mathbb{R}^p$. Hence, (6.30) is equivalent to the requirement that

$$
\forall \bar{e} \in \{e \in \mathbb{R}^r \mid Fe \leq f\}, \quad \left( \begin{bmatrix} \hat{x}[i] \\ u[i] \end{bmatrix} + \begin{bmatrix} I & I \\ 0 & K \end{bmatrix} \tilde{A}^i \begin{bmatrix} I \\ 0 \end{bmatrix} H\bar{e} \right) \in \{z \in \mathbb{R}^{2n_\mathrm{x}} \mid Gz \leq g\}.
$$

$$
(6.31)
$$

Using the main result in [Hennet, 1989], (6.31) is equivalent to the existence of a matrix

---

[2] For simplicity, we neglect here that the dimension of the matrices defining the polytopes under investigation in this section depend on $N$, $M$, $t$, and $i$.

$P \in \mathbb{R}^{q \times p}$ satisfying

$$P \geq 0, \tag{6.32a}$$

$$PF = G \begin{bmatrix} I & I \\ 0 & K \end{bmatrix} \tilde{A}^i \begin{bmatrix} I \\ 0 \end{bmatrix} H, \tag{6.32b}$$

$$\text{and } Pf \leq g - G \begin{bmatrix} \hat{x}[i] \\ u[i] \end{bmatrix}. \tag{6.32c}$$

Hence, by the introduction of the additional decision variable $P$ subject to (6.32), constraint (6.8c) becomes linear.

Consider now the case $\hat{\mathcal{X}}[i|t] = \hat{\mathcal{X}}^{\mathrm{II}}[i|t]$. As $\hat{\mathcal{X}}^{\mathrm{II}}[i|t]$ is defined as a Minkowski sum of polytopes which are all fixed *a priori*, the constraint (6.8c) is equivalent to

$$\begin{bmatrix} \hat{x}[i|t] \\ u[i|t] \end{bmatrix} \in (\mathcal{X} \times \mathcal{U}) \ominus \begin{bmatrix} I & I \\ 0 & K \end{bmatrix} \hat{\mathcal{X}}^{\mathrm{II}}[i|t], \tag{6.33}$$

where the right-hand side of the set-membership condition is a polytope that can be computed offline. Hence, the constraint is linear.

**Complexity**

In the following, we discuss the complexity of the MPC problem defined by (6.11) in terms of the dependence of the number of variables and constraints on the horizons $M$ and $N$ and on the system dimensions. For simplicity, we assume here that $n_{\mathrm{x}}$, $n_{\mathrm{u}}$, and $n_{\mathrm{y}}$ are of the same order, denoted by $n$, and that the number of hyperplanes used to describe the sets $\mathcal{X}$, $\mathcal{U}$, $\mathcal{W}$, $\mathcal{V}$, $\mathcal{E}_t$, and $\hat{\mathcal{X}}_{\mathrm{T},t}$ depends linearly on $n$. One important property of the Pontryagin difference is that if $\mathcal{X} \subseteq \mathbb{R}^{n_{\mathrm{x}}}$ is a polyhedron, so is $\mathcal{X} \ominus \mathcal{Y}$ for *any* nonempty set $\mathcal{Y} \subseteq \mathbb{R}^{n_{\mathrm{x}}}$ and, furthermore, the number of hyperplanes necessary to describe $\mathcal{X} \ominus \mathcal{Y}$ is not higher than the number necessary to describe $\mathcal{X}$ [Kolmanovsky and Gilbert, 1998, Theorem 2.3]. We readily conclude that the number of variables and constraints resulting from (6.8a), (6.8b), and (6.8d) is of order $O(Nn)$ and the number of variables and constraints resulting from (6.8c) for $i \geq \hat{M}$ is of order $O((N - \hat{M})n)$. Consider now the dimensions of the matrices in (6.32) for prediction step $i$. The number of variables is $qp$ and the number constraints is $qp + qr + q$, where $q = O(n)$, $p = O(Mn)$, and $r = O(Mn)$. Hence, the number of variables and constraints resulting from (6.8c) for $i \in \{0, \ldots, \hat{M} - 1\}$ is of order $O(\hat{M}Mn^2)$, resulting in an overall number of constraints and variables of order $O(\hat{M}Mn^2 + Nn)$.

### 6.1.4. Numerical example

We consider a system as in (6.1) with matrices

$$A = \begin{bmatrix} 1 & 0.5 \\ 0 & 1 \end{bmatrix}, B = \begin{bmatrix} 0.125 \\ 0.5 \end{bmatrix}, C = \begin{bmatrix} 1 & 0 \end{bmatrix}, \tag{6.34}$$

which correspond to a continuous-time double integrator system discretized with a sample time of 0.5. The disturbance bounds were chosen to $\mathscr{W} = [-0.1, 0.1] \times [-0.1, 0.1]$ and $\mathscr{V} = [-0.1, 0.1]$; the constraints on the state and input to $\mathscr{X} = [-40, 40] \times [-40, 40]$ and $\mathscr{U} = [-4, 4]$. We computed both the linear feedback and the estimator gain LQ-optimal, using the weighting matrices $Q = \begin{bmatrix} 1 & 0 \\ 0 & 1 \end{bmatrix}$ and $R = 1$ to obtain $K = [-0.6514 \ -1.3142]$. The gain $L$ was computed as the transpose of the LQ-optimal feedback gain for the matrix pair $(A^\mathsf{T}, A^\mathsf{T} C^\mathsf{T})$ and weighting matrices $Q = \begin{bmatrix} 10 & 0 \\ 0 & 100 \end{bmatrix}$ and $R = 1$, resulting in $L = [-0.9787 \ -1.4590]^\mathsf{T}$. Thus, the eigenvalues of $A + BK$ are at $0.6307 \pm 0.1628\mathrm{i}$ and the eigenvalues of $A + LCA$ (both) at $0.1459$, which is consistent with the common advice to choose the dynamics of the estimator four to five times as fast as the closed-loop dynamics under state feedback.

For simplicity, we chose the *a priori* bound on the estimation error to $\mathscr{E}_t = \mathscr{E}$ for all $t \in \mathbb{N}$, where $\mathscr{E}$ was computed as a 0.1-outer approximation of the minimal robust positively invariant set for the error dynamics $\tilde{x}_{t+1} = (A + LCA)\tilde{x}_t + (I + LC)w_t + v_{t+1}$, using the algorithm proposed in [Raković et al., 2005]. (This choice of $\mathscr{E}$ is consistent with a scenario where the estimation error has already converged to its worst-case asymptotic bound. A more realistic scenario could be simulated by multiplying this set $\mathscr{E}$ with some factor such that a set known to contain the initial estimation error is contained in the scaled-up error bound.)

In Figure 6.4, we depict the feasible regions[3] at initialization $\mathscr{R}_0$ for a prediction horizon of $N = 5$ and an estimation horizon $M$ between 0 and 5. Note that $M = 0$ corresponds to the output-feedback MPC scheme proposed in [Chisci and Zappa, 2002, Section 5.3], demonstrating the significant reduction in conservatism achieved even for short estimation horizons.

As is well known, the feasible region of an MPC algorithm generally increases with longer prediction horizons. For the system and controller parameters chosen here, we could not report an increase of the feasible region for $M = 0$ beyond a choice of $N = 50$. However, as shown in Figure 6.5, increasing $M$ allows a further enlargement of this set.

Finally, consider now a reduction of the input constraint set from $\mathscr{U} = [-4, 4]$ to $\mathscr{U} = [-3, 3]$. For the chosen parameters, it is no longer possible to obtain a non-empty terminal set (and hence, no non-empty feasible region) for $M \in \{0, 1\}$ and $N = 50$. However, increasing $M$ to 2 achieves the feasible region depicted in Figure 6.6. This shows that the proposed MPC algorithm is able to solve problems that cannot be handled by the scheme in [Chisci and Zappa, 2002, Section 5.3].

---

[3]We used the plot function implemented in YALMIP [Löfberg, 2004], based on shooting 100 rays in different directions, to obtain the feasible regions of the MPC problems.

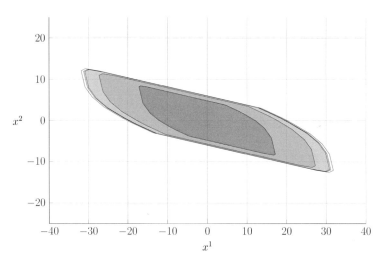

Figure 6.4.: Feasible regions $\mathcal{R}_0$ for $N = 5$ and $M \in \{0, \ldots, 5\}$, the innermost region being associated with $M \in \{0, 1\}$.

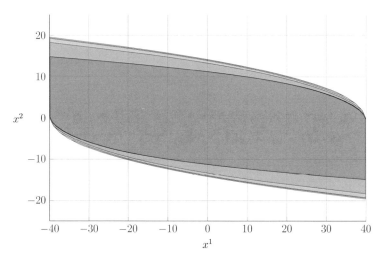

Figure 6.5.: Feasible regions $\mathcal{R}_0$ for $N = 50$ and $M \in \{0, \ldots, 5\}$. For $M = 0$, the feasible region could not be observed to increase with $N$ beyond $N = 50$.

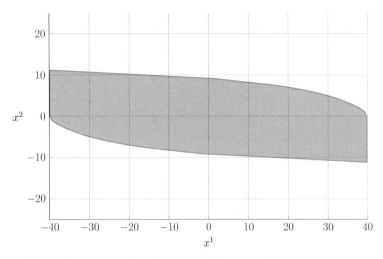

Figure 6.6.: Feasible region $\mathscr{R}_0$ for $N = 50$ and $M = 2$ with the input constraint set restricted to $\mathscr{U} = [-3, 3]$. For $M \in \{0, 1\}$, the feasible region is empty.

**Discussion**

The reason why we are able to enlarge the feasible region in our scheme is only indirectly a consequence of the set-valued estimator employed providing a tighter bound on the current state. In fact, as we only investigated the feasible region at initialization, no set-valued estimation was performed at all; we only used the *a priori* bound $\mathscr{E}$ on the estimation error. A more direct cause lies in the uncertainty bounds defined in (6.26): a larger estimation horizon $M$ implies that $\tilde{\mathcal{X}}^{\mathrm{I}}[i|t]$ appears more often than $\tilde{\mathcal{X}}^{\mathrm{II}}[i|t]$ in the constraints of the optimal control problem, in particular in the first steps of the prediction horizon. The fact that the sets $\tilde{\mathcal{X}}^{\mathrm{I}}[i|t]$ offer a tighter bound on the future uncertainty than $\tilde{\mathcal{X}}^{\mathrm{II}}[i|t]$ is the first reason for the increased size of the region of attraction. Further, the sets $\tilde{\mathcal{X}}^{\mathrm{II}}[i|t]$ themselves, with their definition in (6.23), generally decrease in size with larger $M$ (depending on $\mathscr{E}_t$). This also affects the maximal size of the terminal set, due to Assumption 6.1(iv). Note, however, that employing the uncertainty bounds $\tilde{\mathcal{X}}^{\mathrm{I}}[i|t]$ *requires* online set-valued estimation in order to ensure recursive feasibility.

For our comparison with [Chisci and Zappa, 2002] we want to point out that (i) alternative, less conservative but significantly more computation-heavy, schemes were proposed in [Chisci and Zappa, 2002], which we did not consider in the simulations here and (ii) a different choice of $K$ and $L$ might also lead to a non-empty feasible region for the restricted input constraint set, even for $M = 0$.

# 6.2. Event-triggered output-feedback control

In this section, we present an event-triggered output-feedback control scheme based on the output-feedback controller introduced in the last section and the trigger principle proposed in Section 5.2. Three tuning parameters are required: $\gamma \in [1, \infty)$ which defines the scaling of the stabilized set, $\nu \in [0, 1)$ which defines the relaxation of the decrease rate of the MPC cost function, and $\theta \in \{1, \dots\} \cup \{\infty\}$, which relaxes the stability requirements on the closed-loop system. Further, let $\kappa_t : \mathbb{R}^{n_x} \times \mathbb{N} \to \mathbb{R}^{n_u}$, $t \in \mathbb{N}$, be functions satisfying $\kappa_t(\hat{x}, 0) \in \kappa_{\mathrm{MPC},t}(\hat{x})$ for all $\hat{x} \in \hat{\mathcal{X}}_{\mathrm{f},0}$ and all $t \in \mathbb{N}$. Finally, let $\tilde{\mathcal{X}}_t \subseteq \mathbb{R}^{(t+1)n_x}$ be any sets satisfying $(\tilde{x}_0, \dots, \tilde{x}_t) \in \tilde{\mathcal{X}}_t$, $t \in \mathbb{N}$, and assume these sets to be known at time $t \in \mathbb{N}$.

Algorithm 14 describes the closed-loop control.

## 6.2.1. Closed-loop properties

Similar to the elaborations in Chapter 5, we introduce the sets

$$
\bar{\mathcal{X}}_j := \left\{ \hat{x} \in \mathbb{R}^{n_x} \middle| \left\{ \begin{bmatrix} (A+BK)^{i-j}\hat{x} \\ K(A+BK)^{i-j}\hat{x} \end{bmatrix} \right\} \oplus \begin{bmatrix} I & I \\ 0 & K \end{bmatrix} \hat{\mathcal{X}}[i] \subseteq \mathcal{X} \times \mathcal{U}, \right.
$$
$$
\left. i \in \{j, \dots, N-1\}, \ (A+BK)^{N-j}\hat{x} \in \hat{\mathcal{X}}_{\mathrm{T}} \right\}, \quad (6.35)
$$

for $j \in \{0, \dots, N-1\}$, consistent with (6.16). Further, let $\bar{V}^{\mathrm{s}} : \mathbb{R}^{2n_x} \to \mathbb{R}$ be any continuous positive definite function and define

$$
V_\gamma^{\mathrm{s}}(z) := \min_{z' \in \gamma \hat{\mathcal{F}}_\infty} \bar{V}^{\mathrm{s}}(z - z') \tag{6.36}
$$

$$
\mathcal{X}^\gamma(z) := \left\{ z' \in \gamma \hat{\mathcal{F}}_\infty \ \middle| \ \bar{V}^{\mathrm{s}}(z - z') = V_\gamma^{\mathrm{s}}(z) \right\} \tag{6.37}
$$

for all $z \in \mathbb{R}^{2n_x}$ and all $\gamma \in [1, \infty)$. Similar to the previous section, and analogous to Section 5.2, we assume that there exist constants $\bar{c}_1, \bar{c}_2, \bar{c}_3, a \in (0, \infty)$ such that for all $z \in \mathbb{R}^{2n_x}$ and all $v \in \mathbb{R}^{n_u}$ it holds that

$$
\bar{c}_1 |z|^a \leq \bar{V}^{\mathrm{s}}(z) \leq \bar{c}_2 |z|^a \tag{6.38a}
$$
$$
\bar{V}^{\mathrm{s}}(z) - \bar{V}^{\mathrm{s}}\left(\hat{A}z + \begin{bmatrix} 0 \\ B \end{bmatrix} v\right) + (1-\nu)\ell(v) \geq \bar{c}_3 |z|^a. \tag{6.38b}
$$

---

**Algorithm 14** Output-feedback event-triggered predictive control

---

1: $i \leftarrow 0$
2: $t \leftarrow 0$
3: $t_0 \leftarrow 0$
4: Solve the optimization problem in (6.11) for $\hat{x} = \hat{x}_0$ and obtain $(\hat{\mathbf{x}}_0, \mathbf{u}_0) \in \mathscr{T}_t(\hat{x}_0)$.
5: Apply $u_0 = \kappa_0(\hat{x}_0, 0)$ to the system.
6: **for** $t \in \{1, \ldots\}$ **do**
7:     Solve the optimization problem in (6.11) for $\hat{x} = \hat{x}_t$ and obtain $(\hat{\mathbf{x}}_t, \mathbf{u}_t) \in \mathscr{T}_t(\hat{x}_t)$.
8:     **if** $\forall \tau \in \{1, \ldots, \min\{t+1, \theta\}\}, \ \exists \hat{w} \in \hat{\mathscr{W}}, \exists (\tilde{x}_0', \ldots, \tilde{x}_t') \in \tilde{\hat{\mathcal{X}}}_t,$

$$J_{t+1}(A\hat{x}_t + B\kappa_{t_i}(\hat{x}_{t_i}, t - t_i) + \tilde{w}) > J_{t+1-\tau}(\hat{x}_{t+1-\tau}) - (1 - \nu^{\tau+1}) \sum_{k=0}^{\tau-1} \ell(u[k] - K\hat{x}[k])$$

    or

$$\hat{A}\begin{bmatrix} \tilde{x}_t' \\ \hat{x}_t \end{bmatrix} + \begin{bmatrix} 0 \\ B \end{bmatrix} (\kappa_{t_i}(\hat{x}_{t_i}, t - t_i) - K\hat{x}_t) + \hat{w} \notin \left\{ \begin{bmatrix} 0 \\ \hat{x}[\tau] \end{bmatrix} + \hat{A}^\tau \begin{bmatrix} \tilde{x}_{t+1-\tau}' \\ 0 \end{bmatrix} \right\} \oplus \gamma \hat{\mathscr{F}}_\tau,$$

    where $((\hat{x}[j])_{j \in \mathbb{N}}, (u[j])_{j \in \mathbb{N}}) = (\hat{\mathbf{x}}_{t+1-\tau}, \mathbf{u}_{t+1-\tau})$ and $\tilde{w} = -LCA\tilde{x}_t' + [0 \ I]\hat{w}$
    **then**
9:         $i \leftarrow i + 1$
10:        $t_i \leftarrow t$
11:     **end if**
12:     Apply $u_t = \kappa_{t_i}(\hat{x}_{t_i}, t - t_i)$ to the system.
13: **end for**

---

We state the following result, analogous to Lemma 5.4.

**Lemma 6.4.** *Let*

$$J_{t+1}(\hat{x}_{t+1}) \leq J_t(\hat{x}_t) - (1 - \nu^{\tau+1}) \sum_{k=0}^{\tau-1} \ell(u_{t+1-\tau}[k] - K\hat{x}_{t+1-\tau}[k]) \tag{6.39}$$

*and*

$$z_{t+1} \in \left\{ \begin{bmatrix} 0 \\ \hat{x}[\tau] \end{bmatrix} + \hat{A}^\tau \begin{bmatrix} \tilde{x}_{t+1-\tau} \\ 0 \end{bmatrix} \right\} \oplus \gamma \hat{\mathscr{F}}_\tau \tag{6.40}$$

*for some $\tau \in \mathbb{N}$, where $z_{t+1} := (\tilde{x}_{t+1}, \hat{x}_{t+1})$ and $((\hat{x}[i])_{i \in \mathbb{N}}, (u[i])_{i \in \mathbb{N}}) \in \mathscr{T}_{t+1-\tau}(\hat{x}_{t+1-\tau})$.*

*Let further $z'_{t+1-\tau} \in \mathcal{Z}^\gamma(z_{t+1-\tau})$, where $z_{t+1-\tau} := (\tilde{x}_{t+1-\tau}, \hat{x}_{t+1-\tau})$. Then,*

$$
J_{t+1}(\hat{x}_{t+1}) + V_\gamma^s(z_{t+1}) \le J_{t+1-\tau}(\hat{x}_{t+1-\tau}) + V_\gamma^s(z_{t+1-\tau}) - (1 - \nu^\tau)\Bigg( \bar{V}^s(z_{t+1-\tau} - z'_{t+1-\tau})
$$

$$
- \bar{V}^s \left( \hat{A}^\tau(z_{t+1-\tau} - z'_{t+1-\tau}) + \sum_{j=0}^{\tau-1} \hat{A}^{\tau-1-j} \begin{bmatrix} 0 \\ B \end{bmatrix} (u[j] - K\hat{x}[j]) \right) + \sum_{k=0}^{\tau-1} \ell(u[k] - K\hat{x}[k]) \Bigg).
$$

$$(6.41)$$

We summarize the properties of the closed-loop system in the following theorem:

**Theorem 6.2.** *Let $\mathfrak{f} : \mathcal{S} \to (\mathbb{R}^{2n_x})^{\mathbb{N}}$ be the dynamical system generating $((\tilde{x}_t, \hat{x}_t))_{t \in \mathbb{N}}$ for the closed-loop system consisting of (6.1) and (6.2a), where $u_t$ is determined according to Algorithm 14. If $\hat{x}_0 \in \hat{\mathcal{X}}_{\mathrm{f},0}$, then the algorithm is well-defined in the sense that the optimization problem in (6.11) has a solution when required. Furthermore, it holds that $(x_t, u_t) \in \mathcal{X} \times \mathcal{U}$ for $t \in \mathbb{N}$. If, additionally, there exists an $\epsilon \in (0, \infty)$ such that $\gamma \hat{A}^i \hat{\mathcal{F}}_\infty \oplus \hat{A}^i(-\mathcal{E}_t \times \{0\}) \oplus \epsilon \mathcal{B}_{2n_x} \subseteq \mathbb{R}^{n_x} \times \hat{\mathcal{X}}_i$ for all $i \in \{0, \ldots, N-1\}$ and all $t \in \mathbb{N}$, then the set $\gamma \hat{\mathcal{F}}_\infty$ is $\theta$-uniformly asymptotically stable for $\mathfrak{f}$ with $[I\ 0]\hat{\mathcal{X}}[0|0] \times \hat{\mathcal{X}}_{\mathrm{f},0}$ belonging to its region of attraction.*

## 6.2.2. Implementation

We assume that all involved sets are polytopes, such that the optimization problem (6.11) appearing in Algorithm 14 can be solved as described in Section 6.1.3. Hence, the main question regarding the implementation of Algorithm 14 is how to check the trigger condition in step 8. Here, the second condition is equivalent to the set-memberships discussed in 3.3.3 and can be evaluated analogously as discussed in Section 3.3.4.

For the first condition, one has to compute an upper bound on the worst case value of the MPC value function $J_{t+1}$ at the next time step. Here, it is important to note that even if all involved sets and functions are convex, the worst case realization of the disturbances does not necessarily occur at one of the vertices of the uncertainty set: due to the intersection of sets performed in the set-valued estimator, larger realized disturbances might actually lead to a smaller bound on the estimation error. We circumvent this problem by simply dropping the information gained from $y_{t+1}$ when computing the bound on $J_{t+1}$ at time $t$ (incurring a certain level of additional conservatism in the process).

## Vertex based bound

Let

$$
\mathscr{D}'_t(\hat{x}) := \left\{ (u[i])_{i \in \{0, \dots, N-1\}} \in \mathbb{R}^{Nn_u} \; \middle| \right.
$$

$$
\hat{x}[0] = \hat{x} \tag{6.42a}
$$

$$
\hat{x}[i+1] = A\hat{x}[i] + Bu[i] \tag{6.42b}
$$

$$
\left\{ \begin{bmatrix} \hat{x}[i] \\ u[i] \end{bmatrix} \right\} \oplus \begin{bmatrix} I & I \\ 0 & K \end{bmatrix} \hat{\mathscr{X}}'[i|t] \subseteq \mathscr{X} \times \mathscr{U}, \quad i \in \{0, \dots, N-1\} \tag{6.42c}
$$

$$
\hat{x}[N] \in \hat{\mathscr{X}}_{\mathrm{T},t} \tag{6.42d}
$$

$$
\left. \vphantom{\begin{bmatrix} \hat{x}[i] \\ u[i] \end{bmatrix}} \right\},
$$

which is identical to $\mathscr{D}_t$ in (6.8), with the difference that the sets $\hat{\mathscr{X}}'[i|t]$ are based on $\mathscr{E}_t'^j$ instead of $\mathscr{E}_t^j$ (compare (6.21) and (6.22)), where

$$
\mathscr{E}_t'^j := \left\{ \tilde{x}_t' \in \mathbb{R}^{n_{\mathrm{x}}} \; \middle| \; \begin{array}{l} \tilde{x}'_{t-j} \in \mathscr{E}_{t-j}, \; (w'_k, v'_{k+1}) \in \mathscr{W} \times \mathscr{V}, \\ \tilde{x}'_{k+1} = (A + LCA)\tilde{x}'_k + (I + LC)w'_k + Lv'_{k+1} \\ k \in \{t-j, \dots, t-1\} \\ \tilde{y}_{l+1} = -CA\tilde{x}'_l - Cw'_l - v'_{l+1}, \\ l \in \{t-j, \dots, t-2\} \end{array} \right\}.
$$

It holds that $\mathscr{E}_t'^j \supseteq \mathscr{E}_t^j$ for all $t \in \mathbb{N}$ and all $j \in \{0, \dots, t\}$, and, hence

$$
\inf \left\{ \bar{J}(\hat{x}, \mathbf{u}) \mid \mathbf{u} \in \mathscr{D}'_t(\hat{x}) \right\} \geq \inf \left\{ \bar{J}(\hat{x}, \mathbf{u}) \mid \mathbf{u} \in \mathscr{D}_t(\hat{x}) \right\} \tag{6.43}
$$

for all $t \in \mathbb{N}$ and all $\hat{x} \in \mathbb{R}^{n_{\mathrm{x}}}$. Most importantly, $\mathscr{D}'_t$ is determined by information available at time $t-1$.

Let furthermore $\tilde{\mathscr{W}}_t \subseteq \mathbb{R}^{n_u}$, $t \in \mathbb{N}$, be polytopes satisfying $-LCA\tilde{x}'_t + [0 \; I]\hat{w} \in \tilde{\mathscr{W}}_t$ for all $(\tilde{x}'_0, \dots, \tilde{x}'_t) \in \tilde{\mathscr{X}}_t$, and all $\hat{w} \in \tilde{\mathscr{W}}$. It holds that

$$
\max_{\tilde{w} \in \tilde{\mathscr{W}}_t} J_{t+1}(A\hat{x}_t + B\kappa_t(\hat{x}_{t_i}, t - t_i) + \tilde{w})
$$
$$
\leq \max_{\tilde{w} \in \tilde{\mathscr{W}}_t} \inf \left\{ \bar{J}(A\hat{x}_t + B\kappa_t(\hat{x}_{t_i}, t - t_i) + \tilde{w}, \mathbf{u}) \mid \mathbf{u} \in \mathscr{D}'_{t+1}(A\hat{x}_t + B\kappa_t(\hat{x}_{t_i}, t - t_i) + \tilde{w}) \right\}.
$$
$$
\tag{6.44}
$$

If $\ell$ is convex, the right-hand side of this inequality can be determined by solving the optimization problem for every vertex of the set $\tilde{\mathscr{W}}_t$.

**Bound based on further constraint tightening.**

Instead of solving one optimization problem for each of the vertices of the uncertainty set, one may instead solve a single optimization problem yielding an upper bound for every member of this set, similar to the approach presented in Section 5.1.4. Define

$$\mathscr{D}_t''(\hat{x}) := \left\{ (u[i])_{i \in \{0,\dots,N-1\}} \in \mathbb{R}^{Nn_u} \; \middle| \right.$$

$$\hat{x}[0] = \hat{x} \tag{6.45a}$$

$$\hat{x}[i+1] = A\hat{x}[i] + Bu[i] \tag{6.45b}$$

$$\left\{ \begin{bmatrix} \hat{x}[i] \\ u[i] \end{bmatrix} \right\} \oplus \begin{bmatrix} I & I \\ 0 & K \end{bmatrix} \hat{\mathcal{X}}'[i|t] \oplus \begin{bmatrix} I \\ K \end{bmatrix} (A+BK)^i \tilde{\mathscr{W}}_t \subseteq \mathcal{X} \times \mathcal{U}, \quad i \in \{0,\dots,N-1\} \tag{6.45c}$$

$$\left. \{\hat{x}[N]\} \oplus (A+BK)^N \tilde{\mathscr{W}}_t \subseteq \hat{\mathcal{X}}_{\mathrm{T},t} \right\}. \tag{6.45d}$$

It readily follows that if $\mathbf{u} := (u[0], u[1], \dots, u[N-1]) \in \mathscr{D}_t''(\hat{x})$, then $\tilde{\mathbf{u}} := (u[0] + K\tilde{w}, u[1] + K(A+BK)\tilde{w} \dots, u[N-1] + K(A+BK)^{N-1}\tilde{w}) \in \mathscr{D}_t(\hat{x}+\tilde{w})$ and $\bar{J}(\hat{x}, \mathbf{u}) = \bar{J}(\hat{x}+\tilde{w}, \tilde{\mathbf{u}})$ for all $\tilde{w} \in \mathscr{W}_t$. Hence, we have

$$\max_{\tilde{w} \in \mathscr{W}_t} J_{t+1}(A\hat{x}_t + B\kappa_t(\hat{x}_{t_i}, t - t_i) + \tilde{w})$$

$$\leq \inf \left\{ \bar{J}(A\hat{x}_t + B\kappa_t(\hat{x}_{t_i}, t - t_i), \mathbf{u}) \; \middle| \; \mathbf{u} \in \mathscr{D}_{t+1}''(A\hat{x}_t + B\kappa_t(\hat{x}_{t_i}, t - t_i)) \right\}. \tag{6.46}$$

Determining the right hand side of this inequality requires the solution of a single optimization problem. An additional advantage of this approach is that the vertices of the set $\mathscr{W}_t$ need not be known and it might be in fact be defined by a moving-horizon set-valued estimator itself.

## 6.2.3. Numerical Examples

Consider a system as in (6.1) with system matrices

$$A = \begin{bmatrix} 1 & 0.2 \\ 0 & 1 \end{bmatrix}, \; B = \begin{bmatrix} 0.02 \\ 0.2 \end{bmatrix}, \; C = \begin{bmatrix} 1 & 0 \end{bmatrix} \tag{6.47}$$

which correspond to a continuous-time double integrator system discretized with a sample time of 0.2. The disturbance bounds were chosen to $\mathscr{W} = [-0.1, 0.1] \times [-0.1, 0.1]$ and $\mathscr{V} = [-0.1, 0.1]$, the constraints on the state and input to $\mathcal{X} = [-100, 100] \times [-100, 100]$ and $\mathcal{U} = [-10, 10]$. We computed both the linear feedback and the estimator gain LQ-optimal, using the weighting matrices $Q = \begin{bmatrix} 1 & 0 \\ 0 & 1 \end{bmatrix}$ and $R = 1$ to obtain $K = [-0.8412 \; -$

1.5460]. The gain $L$ was computed as the transpose of the LQ-optimal feedback gain for the matrix pair $(A^\mathsf{T}, A^\mathsf{T}C^\mathsf{T})$ and weighting matrices $Q = \left[\begin{smallmatrix} 5 & 0 \\ 0 & 250 \end{smallmatrix}\right]$ and $R = 1$, resulting in $L = \left[\begin{smallmatrix} -0.9559 \\ -3.3201 \end{smallmatrix}\right]$. Thus, the absolute values of the eigenvalues of $A + BK$ are about 4 times as large as the absolute values of the eigenvalues of $A + LCA$.

We implemented Algorithm 14 based on the overapproximation of the one-step-ahead cost function via the constraint tightening in (6.45). The functions $\kappa_t$ were chosen as $\kappa_t : (\hat{x}, \tau) \mapsto u[\tau]$, where $(u[\tau])_{\tau \in \mathbb{N}} = \mathbf{u}_t$ and $\mathbf{u}_t$, $t \in \mathbb{N}$ is defined in step 4 and step 7 of Algorithm 14. For simplicity, we chose the *a priori* bound on the estimation error to $\mathscr{E}_t = \mathscr{E}$ for all $t \in \mathbb{N}$, where $\mathscr{E}$ was computed as a 0.1-outer approximation of the minimal robust positively invariant set for the error dynamics $\tilde{x}_{t+1} = (A + LCA)\tilde{x}_t + (I + LC)w_t + v_{t+1}$, using the algorithm proposed in [Raković et al., 2005]. We defined the set $\tilde{\mathscr{X}}_t$ analogously to (3.66) by

$$
\tilde{\mathscr{X}}_t := \left\{ (\tilde{x}'_0, \ldots, x'_t) \in \mathbb{R}^{(t+1)n_x} \;\middle|\; \begin{array}{c} \tilde{x}'_{\max\{0,t-M\}} \in \mathscr{E}_{\max\{0,t-M\}}, \\ w'_k \in \mathscr{W}, \; v'_{k+1} \in \mathscr{V}, \\ \tilde{x}'_{k+1} = (A + LCA)\tilde{x}'_k + (I + LC)w'_k + Lv'_{k+1}, \\ C(A\hat{x}_k + Bu_k) - y_{k+1} = -CA\tilde{x}'_k - Cw'_k - v'_{k+1}, \\ k \in \{\max\{0, t-M\}, \ldots, t-1\} \end{array} \right\}
$$

(6.48)

To simplify the computations involved with the tightened constraints in (6.45), we defined $\tilde{\mathscr{W}}_t := (-LCA)\mathscr{E} \oplus [0 \; I]\hat{\mathscr{W}}$ for all $t \in \mathbb{N}$. The prediction and estimation horizons were chosen to $N = 20$ and $M = 5$ in all simulations. We relaxed the trigger condition in step 8 of Algorithm 14 to

$$
J_{t+1}(A\hat{x}_t + B\kappa_{t_i}(\hat{x}_{t_i}, t - t_i) + \tilde{w})
$$

$$
> 0.001 + J_{t+1-\tau}(\hat{x}_{t+1-\tau}) - (1 - \nu^{\tau+1}) \sum_{k=0}^{\tau-1} \ell(u[k] - K\hat{x}[k]), \quad (6.49)
$$

in order to tolerate numerical inaccuracies in the solution of the optimization problems.

In Table 6.1, we report the average transmission rates (number of transmissions during the simulation divided by the number of time points in the simulation) for various values of $\theta$ and $\gamma$ and the system initialized at the origin ($x_0 = \hat{x}_0 = (0, 0)$). The parameter $\nu$—irrelevant in this example as $J_t(\hat{x}_t) = 0$ for all $t \in \mathbb{N}$ due to the system being initialized at the origin—was chosen to 0.9. The disturbance and noise sequences were generated by sampling $w_t$ and $v_t$ identically uniformly and independently on the set $\mathscr{W}$ and $\mathscr{V}$, respectively. For every combination of the parameters $\theta$ and $\gamma$, we computed the average transmission rate over 10 realizations of the disturbance and noise sequence and kept these 10 sequences identical for the simulation with the different parameter combinations.

In order to investigate the transient behavior in the closed-loop system, we initialized the system at $x_0 = \hat{x}_0 = (-40, 0)$ and computed the average transmission rate over the

Table 6.1.: Average transmission rates for the system being initialized at the origin and 50 time points in the simulation, depending on the parameters $\theta$ and $\gamma$ (average over 10 simulation runs each).

| | | $\theta$ | | |
|---|---|---|---|---|
| $\gamma$ | 1 | 2 | 3 | 4 |
| 1.0 | 1.00 | 1.00 | 1.00 | 1.00 |
| 1.2 | 1.00 | 0.75 | 0.63 | 0.47 |
| 1.4 | 1.00 | 0.55 | 0.32 | 0.24 |
| 1.6 | 1.00 | 0.39 | 0.22 | 0.18 |
| 1.8 | 1.00 | 0.29 | 0.18 | 0.12 |
| 2.0 | 1.00 | 0.20 | 0.12 | 0.09 |

first 10 time points, as, for this initial condition, the MPC value function $J_t$ becomes zero at approximately $t = 10$. In Table 6.2, we report the average transmission rates for various choices of the parameters $\theta$, $\gamma$, and $\nu$. During the simulations, the solver employed (cplexqp) reported numerical problems at some time instants.

In both Table 6.2 and Table 6.1 the same influence of the parameters on the transmission rate as in the state-feedback case considered in Chapter 5 becomes apparent. We observed a stronger influence of the parameter $\nu$, which presumably is a consequence of different system parameters.

Table 6.2.: Transmission rates for the system being initialized at $(-40, 0)$ and 10 time points in the simulation, depending on the parameters $\nu$, $\theta$, and $\gamma$ (average over 50 simulation runs each).

| | | | | | | | $\nu$ | | | | | | | | |
|---|---|---|---|---|---|---|---|---|---|---|---|---|---|---|
| | 0.00 | | | | 0.25 | | | | 0.50 | | | | 0.75 | | | | 0.99 | | |
| | | $\theta$ | | | | $\theta$ | | | | $\theta$ | | | | $\theta$ | | | | $\theta$ | |
| $\gamma$ | 1 | 2 | 4 | $\gamma$ | 1 | 2 | 4 | $\gamma$ | 1 | 2 | 4 | $\gamma$ | 1 | 2 | 4 | $\gamma$ | 1 | 2 | 4 |
| 1.00 | 1.00 | 1.00 | 1.00 | 1.00 | 1.00 | 1.00 | 1.00 | 1.00 | 1.00 | 1.00 | 1.00 | 1.00 | 1.00 | 1.00 | 1.00 | 1.00 | 1.00 | 1.00 | 1.00 |
| 1.25 | 1.00 | 0.96 | 0.92 | 1.25 | 1.00 | 0.96 | 0.92 | 1.25 | 1.00 | 0.94 | 0.88 | 1.25 | 1.00 | 0.83 | 0.68 | 1.25 | 1.00 | 0.83 | 0.68 |
| 1.50 | 1.00 | 0.92 | 0.81 | 1.50 | 1.00 | 0.92 | 0.81 | 1.50 | 1.00 | 0.88 | 0.75 | 1.50 | 1.00 | 0.72 | 0.49 | 1.50 | 1.00 | 0.70 | 0.46 |
| 2.00 | 1.00 | 0.86 | 0.74 | 2.00 | 1.00 | 0.86 | 0.74 | 2.00 | 1.00 | 0.79 | 0.63 | 2.00 | 1.00 | 0.61 | 0.34 | 2.00 | 1.00 | 0.60 | 0.33 |

## 6.2.4. Outlook: extension to self-triggered control

In all of this work, we made the assumption that in the self-triggered setting no measurements could be obtained between transmission times. For the output-feedback MPC scheme presented in this chapter, this implies that the assumption of estimation error bounds $\mathscr{E}_t$ being known *a priori* is no longer justified considering that the estimation error dynamics depend on the realization of the transmission times as described in Section 3.2.2. A particular difficulty arising from this issue is that the terminal sets can also not be computed *a priori*, due to their definition in terms of $\mathscr{E}_t$. Parameterizing both the bounds $\mathscr{E}_t$ and the terminal sets $\hat{\mathcal{X}}_{\mathrm{T},t}$ by some finite dimensional vector, say $\mathscr{E}_t = c_t \mathscr{E}$ for $c_t \in [0, \infty)$ and $\mathscr{E}$ fixed and $\hat{\mathcal{X}}_{\mathrm{T},t} = \hat{\mathcal{X}}_{\mathrm{T}}(c_t)$, could remedy this problem.

If, on the other hand, measurements *are* being taken at every point in time, then an extension of the output-feedback scheme to self-triggered control, along the lines of Algorithm 7 and Algorithm 8 is straight-forward.

## 6.3. Summary

We have presented a novel output-feedback MPC scheme for linear systems where information gained from online set-valued estimation is taken into account. The computational complexity is only moderately increased when compared to earlier schemes without online set-valued estimation. Moreover, the complexity of the resulting algorithm can be controlled explicitly through a tuning parameter of the scheme, that is, through the maximum number of past measurements that are taken into account by the set-valued estimator. We showed—via numerical examples—that our method allows a significant increase of the feasible region when compared to earlier, simple, output-feedback MPC schemes. Based on this novel controller, we derived an event-triggered output-feedback MPC scheme, using the theoretical foundations laid in the previous chapter.

# 7. Conclusions

We have presented multiple novel schemes for the aperiodic control of linear systems, taking various aspects and limitations of the given control task into account. Namely, we considered different assumptions on the disturbances affecting the system, treated both state-feedback and output-feedback scenarios, and considered constraints on the system variables. Set-theoretic properties of disturbed linear systems were utilized to a great extent, distinguishing the approaches in the thesis. The proposed controllers demonstrably achieved a reduction in the required amount of communication, in many cases guaranteeing comparable or even identical closed-loop properties as controllers updated at every point in time, where the trade-off between performance, network usage, and computational complexity was made accessible through tuning parameters.

The chief design goal considered throughout most of the thesis was the asymptotic stability of a certain given subset of the state space under worst-case disturbances. This goal was mainly achieved via two different means: by guaranteeing a decrease of a certain Lyapunov-type function in the next time step or by guaranteeing the inclusion of the state at the next time step in a certain set, both under the worst case disturbances affecting the system in the meantime. A reduction of the required communication was achieved by checking—at every point in time or at every communication instant—whether the required decrease or inclusion could be guaranteed without updating the control law at every point in time. As one central result of the thesis, it could be demonstrated that comparing the worst case state at the next point in time to states in the *past* instead of just the current state greatly relaxes the requirements on the controller, thereby allowing more communications to be saved. On the other hand, the definition of stability employed also had to be relaxed in the process, leading to a weaker system theoretic quality when compared to the standard definition. However, the difference in the definitions is quantifiable by the number of time steps one is allowed to go back in time for the comparison of system states and vanishes if this number is chosen to be zero. The main reason why a comparison with past system states promotes a decrease in overall communication is the fact that the stability guarantees hold for worst-case disturbances which are seldom realized: a comparison with past system states allows one to evaluate how far from the worst-case the effect of the past disturbances acting on the system actually was. In case of favorable disturbances the system can tolerate larger uncertainty—and, hence, less feedback— in the near future, allowing to forgo updating the controller momentarily.

Summarizing these concepts in a different way highlights that there is possible room for improvement in this approach: the derived controllers guarantee stability of a certain set under worst-case disturbances and save communications under less-than-worst-case disturbances. In particular, two questions warrant further investigation in this regard. First, it is of interest how much communications the controllers are able to save if the worst-case disturbances are actually realized and how the "worst-case" disturbances look for a particular closed-loop system. Second, if the actually realized disturbances are of a significantly smaller magnitude than assumed for the worst case, one often expects a sensibly designed closed-loop system to also react with a significantly lower state or output magnitude; the related stability concepts are known under the names input-to-state stability and input-output stability, respectively. As the controllers presented in this thesis are designed to exploit less-than-worst-case disturbances solely for the purpose of reducing communications, it seems unlikely that such a behavior could be expected of the related closed-loop systems. Discussions on this matter can, for example, be found in [Stöcker and Lunze, 2013] and [Brunner et al., 2016b], where the latter offers an ad hoc extension to a certain class of event-triggered controllers, related to those investigated in the thesis, which allows input-to-state stability to be recovered.

Most of the approaches presented in the thesis were derived for both the state-feedback and the output-feedback case, where the latter scenario was addressed by extending the state-space to include an observer state. While the results were relatively straight-forward for most of the event-triggered controllers under investigations, we placed an additional restriction on the design of all self-triggered controllers which severely increased the complexity of the output-feedback scenario: measurements of the plant output were only allowed at communication instants. In particular, this restriction leads to a general increase in the estimator uncertainty, as well as the necessity of taking the dependence of this uncertainty on the scheduled communication instants into account. This issue precluded us from extending the results of the last chapter to self-triggered control in a simple manner.

In most of our investigations, we did not consider a disturbance model of any greater detail than assuming a bound pointwise in time to be known. This simplicity is consequently reflected in a likewise simple guarantee on the closed-loop system behavior in the form of an asymptotic bound. In Chapter 4, we assumed the disturbance to be a stochastic variable with a known probability density function; this allowed conclusions to be drawn about the occurrence of transmission instants in the closed-loop system and was exploited in the design of controllers with a desired distribution of inter-transmission times. However, the disturbances were limited to be independent at different time points and were partly additionally assumed to be identically distributed.

Constraints on the input and state were treated by designing aperiodic robust model predictive controllers. Here, it became especially apparent that the self-triggered approaches in our framework are more computationally involved than their event-triggered counterparts, owing to the necessity of taking into account disturbances occurring at multiple future points in time. Careful steps had to be taken in order to avoid—particularly complex—mixed-integer optimizations problems. For the output-feedback case, we first offered our own contribution to the list of available MPC algorithms, establishing a novel way to incorporate set-valued estimation into a predictive control framework; the extension to an aperiodic—namely event-triggered—algorithm followed from applying the results of the previous chapters.

# Outlook

We believe that the methods for designing aperiodic controllers developed in this thesis are applicable to various scenarios and setups not considered here.

## Stability concept

The stability concept introduced in Chapter 2 relies on a backwards-in-time horizon in which the state of the system is allowed to have an influence on the distance of the current state to the stabilized set. The same horizon appeared as a tuning parameter in the aperiodic control schemes. It is straight forward to extend this notion of stability to different stability concepts—which have also been employed in the literature on aperiodic control of disturbed systems—such as input-to-state stability or input-output stability.

As mentioned above, these stability concepts provide a type of gain between the disturbance magnitude and the response of the closed-loop system in terms of the state or output magnitude. If the amount of communication is of interest, however, one would also expect a similar gain between the disturbance magnitude and the average transmission rate in a sensibly designed closed-loop system. As stability in one form or another is always an additional requirement, this would lead to a multiple way trade-off between different objectives—desirably directly accessible through tuning parameters of the controller.

## Computational considerations

While computational aspects were not the focus of the thesis, it is apparent that the presented aperiodic control schemes generally require significantly more involved computations than their periodic counterparts. Considering the ubiquitous use of polyhedrons in the thesis, the algorithms can be expected not to scale well with the system dimensions in general. Tailored approximations techniques that limit the complexity of the involved sets are of interest in this respect.

In a practical setting, the higher hardware and energy demand ensuing from an increased computational complexity has to be weighed against the reduction in communication. *A priori* bounds on the computational complexity would facilitate an effective design procedure.

## System classes

The systems to be controlled considered in the thesis were all additively perturbed linear time-invariant systems. An extension of the presented approaches to a wider class of systems, such as linear systems with multiplicative uncertainty, piecewise affine systems, or general nonlinear systems would be desirable; however, the proposed schemes made extensive use of the superposition principle of linear systems, allowing the uncertainty induced by the additive disturbances to be decoupled from the nominal dynamics. If this is no longer the case, any set-based law used for determining the transmission instant will depend, in a possibly very complex fashion, on the current system state. Further, linear systems have the advantage of preserving convexity of sets under their dynamics, which is essential for many methods to be computationally tractable.

For additively perturbed linear systems the complexity—and versatility—of the model can also be increased by considering different disturbance models, such as disturbances generated by (possibly disturbed) exosystems. This framework would also allow stochastic disturbances with a certain type of dependence between the disturbances at different points in time to be included.

## System Structure

Finally, another topic—well-studied in the literature—which may benefit from the application of set-theoretic methods is the aperiodic control of distributed systems with multiple communication channels between different agents instead of the single channel considered in the thesis. If the components have to communicate over a shared network, the mechanism for scheduling the communications is additionally assigned the task of allocating resources among different agents, adding another layer of complexity. For such distributed systems, the matter of scalability is usually important as the number of subsystems is generally very high. Considering that, as mentioned above, the set-based methods considered here are not expected to scale well if simply applied to high-dimensional systems, a sensible decomposition with a resulting limited size of the resulting sets will become important.

# A. Auxiliary Results

In this section of the appendix, we summarize some auxiliary results employed in the thesis.

**Lemma A.1.** *Let $\mathcal{X}, \mathcal{Y} \subseteq \mathbb{R}^n$ $c, d \in [0, \infty)$, $A \in \mathbb{R}^{m \times n}$. It holds that (i) if $\mathcal{X}, \mathcal{Y}$ are both compact (convex) [polyhedrons] then also $\mathcal{X} \oplus \mathcal{Y}$ is compact (convex) [a polyhedron]. (ii) $A\mathcal{X} \oplus A\mathcal{Y} = A(\mathcal{X} \oplus \mathcal{Y})$, (iii) if $\mathcal{X}$ is convex, then $c\mathcal{X} \oplus d\mathcal{X} = (c + d)\mathcal{X}$*

These results are standard in the literature, some of which can be found for example in [Schneider, 1993, Kolmanovsky and Gilbert, 1998, Blanchini and Miani, 2008, Raković and Kouramas, 2006a] or follow from straight-forward manipulations.

**Lemma A.2.** *Let $\mathcal{X}, \mathcal{Y}, \mathcal{Z} \subseteq \mathbb{R}^n$, $\mathcal{U} \in \mathbb{R}^m$, and $A \in \mathbb{R}^{m \times n}$. (i) if $\mathcal{X}$ is compact (convex) [a polyhedron] then also $\mathcal{X} \ominus \mathcal{Y}$ is compact (convex) [a polyhedron]. (ii) $(\mathcal{X} \oplus \mathcal{Y}) \subseteq \mathcal{Z} \Leftrightarrow \mathcal{X} \subseteq \mathcal{Z} \ominus \mathcal{Y}$ (iii) $(\mathcal{X} \ominus \mathcal{Y}) \oplus \mathcal{Y} \subseteq \mathcal{X}$ (iv) $\mathcal{X} \ominus (\mathcal{Y} \oplus \mathcal{Z}) = (\mathcal{X} \ominus \mathcal{Y}) \ominus \mathcal{Z}$ (v) $\mathcal{X} \ominus (\mathcal{Y} \cup \mathcal{Z}) = (\mathcal{X} \ominus \mathcal{Y}) \cap (\mathcal{X} \ominus \mathcal{Z})$ (vi) if $\mathcal{X}$ and $\mathcal{Y}$ are convex, compact, and nonempty, then $(\mathcal{X} \oplus \mathcal{Y}) \ominus \mathcal{Y} = \mathcal{X}$ (vii) $A^{-1}(\mathcal{X} \ominus A\mathcal{Y}) = A^{-1}\mathcal{X} \ominus \mathcal{Y}$.*

Proofs for most of these statements can be found in [Schneider, 1993] and [Kolmanovsky and Gilbert, 1998] or follow from straight-forward computations. Here, we only prove statement (vii).

*Proof.* It holds that

$$x \in A^{-1}(\mathcal{X} \ominus A\mathcal{Y})$$
$$\Leftrightarrow \quad Ax \in \mathcal{X} \ominus A\mathcal{Y}$$
$$\Leftrightarrow \quad \{Ax\} \oplus A\mathcal{Y} \subseteq \mathcal{X}$$
$$\overset{\text{Lemma A.1}}{\Leftrightarrow} A(\{x\} \oplus \mathcal{Y}) \subseteq \mathcal{X}$$
$$\Leftrightarrow \quad \{x\} \oplus \mathcal{Y} \subseteq A^{-1}\mathcal{X}$$
$$\Leftrightarrow \quad x \in A^{-1}\mathcal{X} \ominus \mathcal{Y}, \tag{A.1}$$

thereby completing the proof. $\square$

**Lemma A.3.** *For any sets $\mathcal{X}, \mathcal{Y} \subseteq \mathbb{R}^n$ and any $u, v \in \mathbb{R}^n$, it holds that $|u + v|_{\mathcal{X} \oplus \mathcal{Y}} \leq |u|_{\mathcal{X}} \oplus |v|_{\mathcal{Y}}$.*

*Proof.* Let $u, v \in \mathbb{R}^n$ be arbitrary. It holds that

$$
\begin{aligned}
|u + v|_{\mathcal{X} \oplus \mathcal{Y}} &= \inf_{z \in \mathcal{X} \oplus \mathcal{Y}} |u + v - z| \\
&= \inf_{x \in \mathcal{X}, y \in \mathcal{Y}} |u + v - (x + y)| \\
&\leq \inf_{x \in \mathcal{X}, y \in \mathcal{Y}} |u - x| + |v - y| \\
&= \inf_{x \in \mathcal{X}} |u - x| + \inf_{y \in \mathcal{Y}} |v - y| \\
&= |u|_{\mathcal{X}} + |v|_{\mathcal{Y}},
\end{aligned}
\tag{A.2}
$$

thereby completing the proof. $\qquad \square$

**Lemma A.4.** *Let a function $f : \mathbb{R} \to \mathbb{R}$ be given. If there exists a continuous function $g : \mathbb{R} \times \mathbb{R} \to \mathbb{R}$ where $g(w, w) = 0$ for all $w \in \mathbb{R}$ and $|f(y) - f(x)| \leq g(y, x)$ for all $x, y \in \mathbb{R}$ for which $y \geq x$, then $f$ is continuous.*

*Proof.* By assumption, for every $z \in \mathbb{R}$ and every $\epsilon > 0$ there exists a $\delta > 0$ such that $|w - z| < \delta$ implies $|g(z, w) - g(w, w)| = |g(z, w)| < \epsilon$ and $|g(w, z) - g(w, w)| = |g(w, z)| < \epsilon$. Hence, for this particular $z \in \mathbb{R}$ and this particular $\epsilon > 0$, $|w - z| < \delta$ implies $|f(w) - f(z)| \leq g(w, z) < \epsilon$ if $w > z$ and $|f(z) - f(w)| \leq g(z, w) < \epsilon$ if $z \geq w$, proving, as $z$ and $\epsilon$ were arbitrary, that $f$ is continuous. $\qquad \square$

**Lemma A.5.** *Let $\mathcal{T} : [0, \infty) \to 2^{\mathbb{R}^n}$ where $\mathcal{T}(r)$ is closed for all $r \in [0, \infty)$ and $r \leq r'$ implies $\mathcal{T}(r) \subseteq \mathcal{T}(r')$. Let further $\rho : \mathbb{R}^n \to [0, \infty)$ be a probability density function and let $\bar{\mathcal{T}} = \bigcup_{r \in [0, \infty)} \mathcal{T}(r)$. It holds that*

$$
\lim_{r \to \infty} \int_{\mathcal{T}(r)} \rho(x) \, \mathrm{d}x = \int_{\bar{\mathcal{T}}} \rho(x) \, \mathrm{d}x.
\tag{A.3}
$$

*Proof.* The statement follows as an immediate extension of Lemma 3.4 in [Bartle, 1995] by noting that $\bar{\mathcal{T}} = \bigcup_{k \in \mathbb{N}} \mathcal{T}(k)$ and $\lim_{\substack{k \in \mathbb{N} \\ k \to \infty}} \int_{\mathcal{T}(k)} \rho(x) \, \mathrm{d}x = \lim_{\substack{r \in \mathbb{R} \\ r \to \infty}} \int_{\mathcal{T}(r)} \rho(x) \, \mathrm{d}x$ due to the assumed monotonicity. $\qquad \square$

**Lemma A.6** (cf. [Brunner et al., 2017c]). *Let $\mathcal{F} \subseteq \mathbb{R}^n$ be a bounded polyhedron containing the origin defined by $\mathcal{F} := \{x \in \mathbb{R}^n | H_i x \leq h_i, i \in \{1, \ldots, p\}\}$ with $H_i \in \mathbb{R}^{1 \times n}$ and $h_i \in \mathbb{R}$ for $i \in \{1, \ldots, p\}$. Assume that the interior of $\mathcal{F}$ is non-empty. Let $\mathcal{W} \subseteq \mathbb{R}^n$ be a polyhedron defined by $\mathcal{W} := \mathrm{convh}\{w_1, \ldots, w_q\}$ with $w_j \in \mathbb{R}^n$, $j \in \{1, \ldots, q\}$. Define the set-valued function $\mathcal{T} : [0, \infty) \to 2^{\mathbb{R}^n}, r \mapsto r\mathcal{F} \ominus \mathcal{W}$ and the function $f : [0, \infty) \to [0, \infty), r \mapsto \mathrm{vol}(\mathcal{T}(r))$. It holds that $f$ is continuous.* $\qquad \square$

*Proof.* As $\mathscr{F}$ is convex and contains the origin, it holds that $r\mathscr{F} \subseteq r'\mathscr{F}$ for $r \leq r'$ and hence also $r\mathscr{F} \ominus \mathscr{W} \subseteq r'\mathscr{F} \ominus \mathscr{W}$, such that $f$ is monotonically nondecreasing . Using Theorem 2.2 in [Kolmanovsky and Gilbert, 1998] (with reference to [Schneider, 1993]), for all $r \in [0,\infty)$ it holds that

$$\mathscr{T}(r) = \left\{ x \in \mathbb{R}^n \;\middle|\; H_i x \leq rh_i - \max_{j \in \{1,\dots,q\}} H_i w_j, \; i \in \{1,\dots,p\} \right\}. \tag{A.4}$$

In the following, we employ the terminology and results of [Rockafellar and Wets, 2009] regarding set-valued functions.[1] Consider first the case that $\mathscr{T}(r)$ is empty for all $r \in [0,\infty)$. Then it holds that $f(r) = 0$ for all $r \in [0,\infty)$ and the proof is complete. Assume now that there exists an $\bar{r} \in [0,\infty)$ such that $\mathscr{T}(\bar{r})$ is non-empty. The graph of $\mathscr{T}$, denoted by $\mathrm{gph}(\mathscr{T})$, is given by

$$\mathrm{gph}(\mathscr{T}) = \{(x,r) \in \mathbb{R}^n \times [0,\infty) \mid x \in \mathscr{T}(r)\}$$
$$= \left\{ (x,r) \in \mathbb{R}^n \times [0,\infty) \;\middle|\; H_i x \leq rh_i - \max_{j \in \{1,\dots,q\}} H_i w_j, \; i \in \{1,\dots,p\} \right\}. \tag{A.5}$$

It holds that $\mathrm{gph}(\mathscr{T})$ is closed, such that by Theorem 5.7 in [Rockafellar and Wets, 2009] the function $\mathscr{T}$ is outer semi-continuous on the interval $[0,\infty)$. Define

$$r^\star = \inf\{r \in [0,\infty) \mid \exists x \in \mathscr{T}(r)\}$$
$$= \inf\{r \in [0,\bar{r}] \mid \exists x \in \mathscr{T}(r)\}. \tag{A.6}$$

As $\mathrm{gph}(\mathscr{T})$ is convex, by Theorem 5.9 in [Rockafellar and Wets, 2009] the function $\mathscr{T}$ is inner semi-continuous on $(r^\star,\infty)$. Hence, $\mathscr{T}$ is continuous on $(r^\star,\infty)$. As $\mathscr{T}(r) = \emptyset$ for $r \in [0,r^\star)$, it follows that $f(r) = 0$ for $r \in [0,r^\star)$. Assume now that $f(r^\star) > 0$. It follows that the interior of $\mathscr{T}(r^\star)$ is non-empty and there exist an $x \in \mathbb{R}^n$ and an $\eta > 0$ such that $\{x\} \oplus \eta\mathscr{B}_n \subseteq \mathscr{T}(r^\star)$. Let further $\mu > 0$ such that $\mu\mathscr{W} \subseteq \mathscr{B}_n$. Hence,

$$x \in (r^\star\mathscr{F} \ominus \mathscr{W}) \ominus \eta\mathscr{B}_n$$
$$= r^\star\mathscr{F} \ominus (\mathscr{W} \oplus \eta\mathscr{B}_n)$$
$$\subseteq r^\star\mathscr{F} \ominus (\mathscr{W} \oplus \eta\mu\mathscr{W})$$
$$= r^\star\mathscr{F} \ominus (1 + \eta\mu)\mathscr{W}. \tag{A.7}$$

It follows that

$$\frac{1}{1+\eta\mu}x \in \left( \frac{1}{1+\eta\mu}r^\star\mathscr{F} \ominus \mathscr{W} \right)$$
$$= \mathscr{T}\left( \frac{1}{1+\eta\mu}r^\star \right), \tag{A.8}$$

---

[1]We only use the term "set-valued function" here. In [Rockafellar and Wets, 2009], there is a distinction between the terms "set-valued mapping" and "set-valued function", which we do not take into account here.

which implies that $\mathcal{T}(r) \neq \emptyset$ for an $r < r^\star$, contradicting the definition of $r^\star$ in (A.6). Hence it holds that $f(r^\star) = 0$, and thus $f(r) = 0$ for all $r \in [0, r^\star]$.

To prove continuity of $f$, we will first establish right-continuity, then left-continuity. As $\mathcal{T}$ is outer semi-continuous for all $r \geq 0$, by Proposition 5.12 in [Rockafellar and Wets, 2009], it holds that for all $r' \geq 0$, all $\rho \in [0, \infty)$ and all $\epsilon > 0$ there exists a $\delta > 0$ such that for all $r \in [r', r' + \delta)$ it holds that $\mathcal{B}_\rho \cap \mathcal{T}(r) \subseteq \mathcal{T}(r') \oplus \epsilon \mathcal{B}_n$. As $\mathcal{T}$ is compact valued and satisfies $\mathcal{T}(r_1) \subseteq \mathcal{T}(r_2)$ for all $r_1 \leq r_2$, one can always choose $\rho$ such that $\mathcal{B}_\rho \cap \mathcal{T}(r) = \mathcal{T}(r)$ for all $r \in [r', r' + \delta']$ for some $\delta' \in (0, \infty)$. Hence, for all $r' \geq 0$ and all $\epsilon > 0$ there exists a $\delta > 0$ such that for all $r \in [r', r' + \delta)$ it holds that $\mathcal{T}(r) \subseteq \mathcal{T}(r') \oplus \epsilon \mathcal{B}_n$. By the Steiner-Minkowski formula (see, for example, [Schneider, 1993]), for all $r' \geq 0$ there exists a continuous function $p_{r'} : [0, \infty) \to [0, \infty)$ with $p_{r'}(0) = 0$ such that for all $\epsilon > 0$, $\mathrm{vol}(\mathcal{T}(r') \oplus \epsilon \mathcal{B}_n) = \mathrm{vol}(\mathcal{T}(r')) + p_{r'}(\epsilon) = f(r') + p_{r'}(\epsilon)$. Hence, for all $r' \geq 0$ and all $\epsilon > 0$ there exists a $\delta > 0$ such that for all $r \in [r', r' + \delta)$ it holds that $f(r') \leq f(r) \leq f(r') + p_{r'}(\epsilon)$. As $p_{r'}$ is continuous and $p_{r'}(0) = 0$ it follows that $f$ is right-continuous at $r'$. Next, we will show that $f$ is left-continuous. For all $r' > 0$ where $f(r') = 0$ it also holds that $f(r) = 0$ for $r \in (0, r']$, such that $f$ is left-continuous at $r'$. Assume in the following that $r' > 0$ and $f(r') > 0$. This implies that $r' > r^\star$ as $f(r^\star) = 0$. Hence, $\mathcal{T}$ is inner semi-continuous at $r'$ such that by Proposition 5.12 in [Rockafellar and Wets, 2009] for all $\epsilon > 0$ there exists a $\delta \in (0, r')$ such that for all $r \in (r' - \delta, r')$ it holds that $\mathcal{T}(r') \subseteq \mathcal{T}(r) \oplus \epsilon \mathcal{B}_n$. Further, as $f(r') > 0$, the interior of $\mathcal{T}(r')$ is non-empty such that there exist an $x' \in \mathbb{R}^n$ and an $\eta' > 0$ such that $x' \oplus \eta' \mathcal{B}_n \subseteq \mathcal{T}(r')$ or, equivalently, $\mathcal{B}_n \subseteq \frac{1}{\eta'}(\mathcal{T}(r') \oplus \{-x'\})$. Hence, $\mathcal{T}(r') \subseteq \mathcal{T}(r) \oplus \epsilon \mathcal{B}_n$ implies $\mathcal{T}(r') \subseteq \mathcal{T}(r) \oplus \frac{\epsilon}{\eta'}(\mathcal{T}(r') \oplus \{-x'\})$. If $\epsilon \in (0, \eta')$, it follows that

$$\left(1 - \frac{\epsilon}{\eta'}\right)\mathcal{T}(r') \subseteq \mathcal{T}(r) \oplus \frac{\epsilon}{\eta'}\{-x'\} \tag{A.9}$$

and, hence, $\left(1 - \frac{\epsilon}{\eta'}\right)^n f(r') \leq f(r) \leq f(r')$. As $\left(1 - \frac{\epsilon}{\eta'}\right)^n$ is continuous in $\epsilon$ and becomes 1 for $\epsilon = 0$, it follows that $f$ is left-continuous at $r'$. From the above we conclude that $f$ is continuous on $[0, \infty)$ and the proof is complete. $\qquad \square$

**Lemma A.7** (cf. [Brunner et al., 2017c]). *Let $A \in \mathbb{R}^{n_x \times n_x}$, $B \in \mathbb{R}^{n_x \times n_u}$, and $K \in \mathbb{R}^{n_x \times n_u}$. Let further $A + BK$ be stable. (i) If $L \in \mathbb{R}^{n_u \times n_u}$ with $L^\mathsf{T} = L \succ 0$, then there exists a matrix $S \in \mathbb{R}^{n_x \times n_x}$ with $S^\mathsf{T} = S \succ 0$ such that the function $\bar{V}^s : \mathbb{R}^{n_x} \to \mathbb{R}, x \mapsto x^\mathsf{T} S x$ satisfies Assumption 5.1 where $\ell : \mathbb{R}^{n_u} \to \mathbb{R}, v \mapsto v^\mathsf{T} L v$. (ii) If $L \in \mathbb{R}^{p \times n_u}$ with $\mathrm{rk}(L) = n_u$, then there exists a matrix $S \in \mathbb{R}^{r \times n_x}$ such that the function $\bar{V}^s : \mathbb{R}^{n_x} \to \mathbb{R}, x \mapsto |Sx|$ satisfies Assumption 5.1 where $\ell : \mathbb{R}^{n_u} \to \mathbb{R}, v \mapsto |Lv|$.*

*Proof.* (i) As $A + BK$ is stable, there exist matrices $Q \in \mathbb{R}^{n_x \times n_x}$ and $P \in \mathbb{R}^{n_x \times n_x}$ with $Q^\mathsf{T} = Q^\mathsf{T} \succ 0$, $P^\mathsf{T} = P^\mathsf{T} \succ 0$, and $(A + BK)^\mathsf{T} P(A + BK) = P - Q$. Let now $\eta \in (0, \infty)$ be chosen such that

$$L - \eta\left(2B^\mathsf{T} S(A + BK)Q^{-1}(A + BK)^\mathsf{T} SB + B^\mathsf{T} SB\right) \succeq 0. \tag{A.10}$$

It follows that

$$\eta x^\mathsf{T} Px - \eta\left((A+BK)x + Bv\right)^\mathsf{T} P\left((A+BK)x + Bv\right) + v^\mathsf{T} Lv$$
$$= \eta x^\mathsf{T} Px - \eta x^\mathsf{T}(A+BK)^\mathsf{T} P(A+BK)x - \eta 2x^\mathsf{T}(A+BK)^\mathsf{T} PBv$$
$$\quad - \eta v^\mathsf{T} B^\mathsf{T} PBv + v^\mathsf{T} Lv$$
$$= \eta x^\mathsf{T} Qx - \eta 2x^\mathsf{T}(A+BK)^\mathsf{T} PBv - \eta v^\mathsf{T} B^\mathsf{T} PBv + v^\mathsf{T} Lv$$
$$= \eta x^\mathsf{T} Qx$$
$$\quad + \underbrace{\eta\left(\frac{Q}{\sqrt{2}}x - \sqrt{2}(A+BK)^\mathsf{T} PBv\right)^\mathsf{T} Q^{-1}\left(\frac{Q}{\sqrt{2}}x - \sqrt{2}(A+BK)^\mathsf{T} PBv\right)}_{\geq 0}$$
$$\quad - \frac{\eta}{2}x^\mathsf{T} Qx + \underbrace{v^\mathsf{T} Lv - \eta v^\mathsf{T}\left(2B^\mathsf{T} P(A+BK)Q^{-1}(A+BK)^\mathsf{T} PB + B^\mathsf{T} PB\right)v}_{\geq 0 \text{ by (A.10)}}$$
$$\geq \frac{\eta}{2}x^\mathsf{T} Qx. \tag{A.11}$$

for all $x \in \mathbb{R}^{n_x}$ and all $v \in \mathbb{R}^{n_u}$. Hence, with the choice of $S := \eta P$ the requirements in Assumption 5.1 are satisfied.

(ii) As $A+BK$ is stable, there exists a matrix $P \in \mathbb{R}^{r \times n_x}$ with $\mathrm{rk}(P) = n_x$ and scalars $c_1, c_2 \in (0, \infty)$, $\lambda \in (0,1)$, such that $c_1|x| \leq |Px| \leq c_2|x|$ and $|P(A+BK)x| \leq \lambda|x|$ for all $x \in \mathbb{R}^{n_x}$; further, there exists a scalar $c_\ell \in (0, \infty)$ with $|Lv| \geq c_\ell$ for all $v \in \mathbb{R}^{n_u}$, see [Lazar, 2010, Theorem II.5, Corollary II.8]. Let now $\eta \in (0, \infty)$ be chosen such that $c_\ell - \eta|P||B| \geq 0$. It follows that

$$\eta|Px| - \eta|P((A+BK)x + B)v| + |Lv|$$
$$\geq \eta|Px| - \eta|P(A+BK)x| - \eta|PBv| + |Lv|$$
$$\geq \eta(1-\lambda)|Px| - \eta|PBv| + |Lv|$$
$$\geq \eta(1-\lambda)|Px| - \eta|P||B||v| + c_\ell|v|$$
$$\geq \eta(1-\lambda)c_1|x| \tag{A.12}$$

for all $x \in \mathbb{R}^{n_x}$ and all $v \in \mathbb{R}^{n_u}$. Hence, with the choice of $S := \eta P$ the requirements in Assumption 5.1 are satisfied. $\qquad\square$

**Lemma A.8.** *Let $\mathcal{X}_\mathrm{T} \subseteq \mathbb{R}^{n_x}$ such that $\mathcal{X}_\mathrm{T} \subseteq \mathcal{X} \ominus \mathcal{F}_N$, $K\mathcal{X}_\mathrm{T} \subseteq \mathcal{U} \ominus K\mathcal{F}_N$, and $(A+BK)\mathcal{X}_\mathrm{T} \subseteq \mathcal{X}_\mathrm{T} \ominus (A+BK)^N\mathcal{W}$, where $\mathcal{F}_i = \bigoplus_{j=0}^{i-1}(A+BK)^j\mathcal{W}$ for $i \in \mathbb{N}$. Then, it holds that $(A+BK)^i\mathcal{X}_\mathrm{T} \subseteq \mathcal{X} \ominus \mathcal{F}_{N+i}$, $K(A+BK)^i\mathcal{X}_\mathrm{T} \subseteq \mathcal{U} \ominus K\mathcal{F}_{N+i}$, and $(A+BK)^i\mathcal{X}_\mathrm{T} \subseteq \mathcal{X}_\mathrm{T} \ominus (A+BK)^N\mathcal{F}_i$, $i \in \mathbb{N}$.*

*Proof.* Consider first the last property, which we prove by induction. For $i = 0$, the inclusion holds trivially, providing the base case. Let the statement now hold true for any $i \in \mathbb{N}$. Then

$$
\begin{aligned}
(A + BK)^{i+1} \mathscr{X}_{\mathrm{T}} \oplus (A + BK)^N \mathscr{F}_{i+1} &= (A + BK) \left( (A + BK)^i \mathscr{X}_{\mathrm{T}} \oplus (A + BK)^N \mathscr{F}_i \right) \\
&\quad \oplus (A + BK)^N \mathscr{W} \\
&\subseteq (A + BK) \mathscr{X}_{\mathrm{T}} \oplus (A + BK)^N \mathscr{W} \\
&\subseteq \mathscr{X}_{\mathrm{T}},
\end{aligned}
\tag{A.13}
$$

where the first line follows from the definition of $\mathscr{F}_i$, the second-to-last line from the induction assumption, and the last line from the case $i = 0$. This completes the inductive step and the proof of $(A + BK)^i \mathscr{X}_{\mathrm{T}} \subseteq \mathscr{X}_{\mathrm{T}} \ominus (A + BK)^N \mathscr{F}_i$, $i \in \mathbb{N}$.

Then, it also holds that

$$
\begin{aligned}
(A + BK)^i \mathscr{X}_{\mathrm{T}} \oplus \mathscr{F}_{N+i} &= (A + BK)^i \mathscr{X}_{\mathrm{T}} \oplus (A + BK)^N \mathscr{F}_i \oplus \mathscr{F}_N \\
&\subseteq \mathscr{X}_{\mathrm{T}} \oplus \mathscr{F}_N \\
&\subseteq \mathscr{X}
\end{aligned}
\tag{A.14}
$$

where the first line follows from (2.4), the second line from (A.13), and the last line from the original assumption on $\mathscr{X}_{\mathrm{T}}$. The proof for the statement that $K(A + BK)^i \mathscr{X}_{\mathrm{T}} \subseteq \mathscr{U} \ominus K \mathscr{F}_{N+i}$ follows analogously. $\qquad\square$

# B. Proofs of Statements

## B.1. Proof of Theorem 2.1

The proof is a simple extension of standard Lyapunov proofs for exponential stability, compare [Hahn, 1967, proof of Theorem 26.5].

First, note that in order to establish the requirements in Definition 2.1, we may equivalently show that for all $t^1 \in \mathbb{N}$ and all $t^0 \in \{0, \ldots, t^1\}$ there exists a $t^{-1} \in \{t^0 - \theta + 1, \ldots, t^0\} \cap \mathbb{N}$ with $|x_{t^1}|_{\mathcal{X}} \leq \beta(|x_{t^{-1}}|_{\mathcal{X}}, t^1 - t^{-1})$. Second, we may assume that $x_t \in \mathcal{X}_{\mathrm{f}}$ for all $t \in \mathbb{N}$.

In the following, we employ strong induction on $t^1 \in \mathbb{N}$ in order to show that for all $t^1 \in \mathbb{N}$ and all $t^0 \in \{0, \ldots, t^1\}$ there exists a $t^{-1} \in \{t^0 - \theta + 1, \ldots, t^0\} \cap \mathbb{N}$ such that $V(x_{t^1}) \leq \eta^{t^1 - t^{-1}} V(x_{t^{-1}})$. The statement holds trivially for $t^1 = 0$, providing the base case. Assume now that the statement holds for all $t \in \{0, \ldots, t^1\}$ for some $t^1 \in \mathbb{N}$. By assumption, there exists a $\tau \in \{1, \ldots, \min\{\theta, t^1+1\}\}$ such that $V(x_{t^1+1}) \leq \eta^\tau V(x_{t^1+1-\tau})$. In the case that $t^0 \in \{t^1 + 1 - \tau, \ldots, t^1 + 1\}$, the choice of $t^{-1} = t^1 + 1 - \tau$ provides the inductive step. Assume now that $t^0 \in \{0, \ldots, t^1 - \tau\}$. By the induction assumption, noting that $t^1 + 1 - \tau \in \{0, \ldots, t^1\}$, there exists a $\hat{t}^{-1} \in \{t^0 - \theta + 1, \ldots, t^0\} \cap \mathbb{N}$ such that $V(x_{t^1+1-\tau}) \leq \eta^{t^1+1-\tau-\hat{t}^{-1}} V(x_{\hat{t}^{-1}})$. Hence, it holds that $V(x_{t^1+1}) \leq \eta^\tau V(x_{t^1+1-\tau}) \leq \eta^\tau \eta^{t^1+1-\tau-\hat{t}^{-1}} V(x_{\hat{t}^{-1}}) = \eta^{t^1+1-\hat{t}^{-1}} V(x_{\hat{t}^{-1}})$, completing the inductive step.

This result implies that for all $t^1 \in \mathbb{N}$ and all $t^0 \in \{0, \ldots, t^1\}$ there exists a $t^{-1} \in \{t^0 - \theta + 1, \ldots, t^0\} \cap \mathbb{N}$ such that

$$|x_{t^1}|_{\mathcal{X}} \leq \left(\frac{V(x_{t^1})}{c_1}\right)^{\frac{1}{a}} \leq \left(\eta^{t^1-t^{-1}} \frac{V(x_{t^{-1}})}{c_1}\right)^{\frac{1}{a}} \leq (\eta^{\frac{1}{a}})^{t^1-t^{-1}} \left(\frac{c_2}{c_1}\right)^{\frac{1}{a}} |x_{t^{-1}}|_{\mathcal{X}}, \qquad \text{(B.1)}$$

thereby completing the proof. $\qquad \square$

## B.2. Proof of Theorem 2.3

We prove the statement by establishing the existence of a function $V : \mathcal{X}_{\mathrm{f}} \to \mathbb{R}^{n_{\mathrm{x}}}$ with the properties stated in Theorem 2.1. The analysis is similar to the proof of Lemma 3.2 in [Ghaemi et al., 2008].

As $A + BK$ was assumed stable, there exists a matrix $P \in \mathbb{R}^{r \times n_{\mathrm{x}}}$ and scalars $c_1, c_2 \in (0, \infty)$, $\eta \in [0, 1)$ such that for all $x \in \mathcal{X}_{\mathrm{f}} \subseteq \mathbb{R}^{n_{\mathrm{x}}}$ it holds that $c_1|x| \leq |Px| \leq c_2|x|$ and

$|P(A + BK)x| \leq \eta |Px|$, see [Molchanov and Pyatnitskiy, 1989, Lazar, 2010]. Let now $V : x \mapsto \min_{y \in \gamma \mathscr{F}_\infty} |P(x - y)|$. It holds that

$$c_1 |x|_{\gamma \mathscr{F}_\infty} = \min_{y \in \gamma \mathscr{F}_\infty} c_1 |x - y| \leq \min_{y \in \gamma \mathscr{F}_\infty} |P(x - y)| \leq \min_{y \in \gamma \mathscr{F}_\infty} c_2 |x - y| = c_2 |x|_{\gamma \mathscr{F}_\infty}. \quad \text{(B.2)}$$

Let now $t \in \mathbb{N}$ be arbitrary. Considering that $x_{t+1} \in \{(A + BK)^\tau x_{t+1-\tau}\} \oplus \gamma \mathscr{F}_\tau$ by assumption, there exists an $f \in \gamma \mathscr{F}_\tau$ such that $x_{t+1} = (A + BK)^\tau x_{t+1-\tau} + f$, which implies

$$
\begin{aligned}
V(x_{t+1}) &= \min_{y \in \gamma \mathscr{F}_\infty} |P(x_{t+1} - y)| \\
&= \min_{y \in \gamma \mathscr{F}_\infty} |P((A + BK)^\tau x_{t+1-\tau} + f - y)| \\
&\overset{(2.4)}{=} \min_{y \in \gamma (A+BK)^\tau \mathscr{F}_\infty \oplus \gamma \mathscr{F}_\tau} |P((A + BK)^\tau x_{t+1-\tau} + f - y)| \\
&= \min_{\substack{y_1 \in \gamma (A+BK)^\tau \mathscr{F}_\infty \\ y_2 \in \gamma \mathscr{F}_\tau}} |P((A + BK)^\tau x_{t+1-\tau} - y_1 + f - y_2)| \\
&\overset{f \in \gamma \mathscr{F}_\tau}{\leq} \min_{y_1 \in \gamma (A+BK)^\tau \mathscr{F}_\infty} |P((A + BK)^\tau x_{t+1-\tau} - y_1)| \\
&= \min_{y_3 \in \gamma \mathscr{F}_\infty} |P(A + BK)^\tau (x_{t+1-\tau} - y_3)| \\
&\leq \min_{y_3 \in \gamma \mathscr{F}_\infty} \eta^\tau |P(x_{t+1-\tau} - y_3)| \\
&= \eta^\tau V(x_{t+1-\tau}). \quad \text{(B.3)}
\end{aligned}
$$

Hence, $V$ satisfies the requirements of Theorem 2.1 with $a = 1$, which completes the proof. $\qquad \square$

## B.3. Proof of Proposition 2.1

We will establish that for all $j \in \{0, \dots, M\}$ it holds that $\{\hat{x}_{t-M+j}\} \oplus \mathscr{E}_{t-M+j}^j = \mathcal{X}'_{j|t-M}$. The case $j = M$ then yields the result.

The proof is by induction, for which the base case follows from the fact that $\{\hat{x}_{t-M}\} \oplus \mathscr{E}_{t-M}^0 = \{\hat{x}_{t-M}\} \oplus \mathscr{E}_{t-M} = \mathcal{X}'_{0|t-M}$ holds by definition. Assume now that $\{\hat{x}_{t-M+j}\} \oplus \mathscr{E}_{t-M+j}^j = \mathcal{X}'_{j|t-M}$ holds for an arbitrary $j \in \{0, \dots, M-1\}$. Define $\tilde{y}_{t+1} := C(A\hat{x}_t + Bu_t) - y_{t+1}$ for all $t \in \mathbb{N}$. We have $\hat{x}_{t-M+j+1} = A\hat{x}_{t-M+j} + Bu_{t-M+j} + L\tilde{y}_{t-M+j+1}$ and

$$
\mathscr{E}_{t-M+j+1}^{j+1} = \left\{ (A + LCA)\tilde{x} + (I + LC)w + Lv \;\middle|\; \begin{array}{l} \tilde{x} \in \mathscr{E}_{t-M+j}^j, \; w \in \mathscr{W}, \; v \in \mathscr{V}, \\ \tilde{y}_{t-M+j+1} = -CA\tilde{x} - Cw - v \end{array} \right\},
$$
$$\text{(B.4)}$$

which follows directly from the definitions.

Hence, using $\tilde{y}_{t-M+j+1} = -CA\tilde{x}_{t-M+j} - Cw_{t-M+j} - v_{t-M+j+1}$, it holds that

$$
\{\hat{x}_{t-M+j+1}\} \oplus \mathscr{E}_{t-M+j+1}^{j+1}
$$
$$
= \left\{ \begin{array}{l} (A+LCA)\tilde{x} + (I+LC)w + Lv \\ \quad + A\hat{x}_{t-M+j} + Bu_{t-M+j} + L\tilde{y}_{t-M+j+1} \end{array} \middle| \begin{array}{l} \tilde{x} \in \mathscr{E}_{t-M+j}^{j}, \ w \in \mathscr{W}, \ v \in \mathscr{V}, \\ \tilde{y}_{t-M+j+1} = -CA\tilde{x} - Cw - v \end{array} \right\}
$$
$$
= \left\{ A(\tilde{x} + \hat{x}_{t-M+j}) + Bu_{t-M+j} + w \ \middle| \ \begin{array}{l} \tilde{x} \in \mathscr{E}_{t-M+j}^{j}, \ w \in \mathscr{W}, \ v \in \mathscr{V}, \\ \tilde{y}_{t-M+j+1} = -CA\tilde{x} - Cw - v \end{array} \right\}
$$
$$
= \left\{ A(\tilde{x} + \hat{x}_{t-M+j}) + Bu_{t-M+j} + w \ \middle| \ \begin{array}{l} \tilde{x} \in \mathscr{E}_{t-M+j}^{j}, \ w \in \mathscr{W}, \ v \in \mathscr{V}, \\ C(A\hat{x}_{t-M+j} + Bu_{t-M+j}) - \tilde{y}_{t-M+j+1} \\ \quad = C(A\hat{x}_{t-M+j} + Bu_{t-M+j}) \\ \qquad + CA\tilde{x} + Cw + v \end{array} \right\}.
$$

$$(B.5)$$

Further, using $y_{t-M+j+1} = C(A\hat{x}_{t-M+j} + Bu_{t-M+j}) - \tilde{y}_{t-M+j+1}$, we obtain

$$
\{\hat{x}_{t-M+j+1}\} \oplus \mathscr{E}_{t-M+j+1}^{j+1}
$$
$$
= \left\{ A(\tilde{x} + \hat{x}_{t-M+j}) + Bu_{t-M+j} + w \ \middle| \ \begin{array}{l} \tilde{x} \in \mathscr{E}_{t-M+j}^{j}, \ w \in \mathscr{W}, \ v \in \mathscr{V}, \\ y_{t-M+j+1} \\ \quad = C(A(\tilde{x}+\hat{x}_{t-M+j}) + Bu_{t-M+j} + w) + v \end{array} \right\}
$$
$$
= \left( A\left(\{\hat{x}_{t-M+j}\} \oplus \mathscr{E}_{t-M}^{j}\right) \oplus \{Bu_{t-M+j}\} \oplus \mathscr{W} \right) \cap C^{-1}\left(\{y_{t-M+j+1}\} \oplus (-\mathscr{V})\right)
$$
$$
= \left( A\mathscr{X}_{j|t-M}' \oplus \{Bu_{t-M+j}\} \oplus \mathscr{W} \right) \cap C^{-1}\left(\{y_{t-M+j+1}\} \oplus (-\mathscr{V})\right)
$$
$$
= \mathscr{X}_{j+1|t-M}',
$$

$$(B.6)$$

where the second-to-last line follows from the induction assumption and the last line from the definition of $\mathscr{X}_{j+1|t-M}'$. This completes the inductive step, and, thereby, the proof. $\qquad\square$

# B.4. Proof of Lemma 3.1

First note that, by the convexity of $\mathscr{W}$ and the fact that $0 \in \mathscr{W}$, it holds that $(A + BK)\gamma\mathscr{F}_\infty \oplus \mathscr{W} \subseteq (A + BK)\gamma\mathscr{F}_\infty \oplus \gamma\mathscr{W} \subseteq \gamma\mathscr{F}_\infty$ for all $\gamma \in [1,\infty)$, compare for example [Ong and Gilbert, 2006, Theorem 1], [Raković and Kouramas, 2006b, Theorem 1], [Raković and Kouramas, 2006a, Lemma 2], [Schulze Darup et al., 2017] with reference to [Blanchini and Miani, 2008] (see also page 203 therein), and [Heemels et al., 2008] with reference to [Blanchini, 1994]. Let $x \in \mathbb{R}^{n_x}$ and $w \in \mathscr{W}$ be arbitrary. It holds that

$$
\begin{aligned}
\alpha_1(|x|_{\gamma \mathscr{F}_\infty}) &= \alpha_1 \left( \min_{y \in \gamma \mathscr{F}_\infty} |x - y| \right) \\
&= \min_{y \in \gamma \mathscr{F}_\infty} \alpha_1(|x - y|) \\
&\leq \min_{y \in \gamma \mathscr{F}_\infty} W(x - y) = V(x) \\
&\leq \min_{y \in \gamma \mathscr{F}_\infty} \alpha_2(|x - y|) \\
&= \alpha_2 \left( \min_{y \in \gamma \mathscr{F}_\infty} |x - y| \right) \\
&= \alpha_2(|x|_{\gamma \mathscr{F}_\infty}),
\end{aligned}
\tag{B.7}
$$

where the second and the second-to-last lines follow from the monotonicity of $\alpha_1$ and $\alpha_2$. Further, it holds that

$$
\begin{aligned}
V((A + BK)x + w) &= \min_{y \in \gamma \mathscr{F}_\infty} W((A + BK)x + w - y) \\
&\stackrel{(A+BK)\gamma \mathscr{F}_\infty \oplus \mathscr{W} \subseteq \gamma \mathscr{F}_\infty}{\leq} \min_{y' \in \gamma \mathscr{F}_\infty} W((A + BK)x + w - ((A + BK)y' + w)) \\
&= \min_{y' \in \gamma \mathscr{F}_\infty} W((A + BK)(x - y')) \\
&\leq \min_{y' \in \gamma \mathscr{F}_\infty} \lambda W(x - y') \\
&= \lambda V(x),
\end{aligned}
\tag{B.8}
$$

thereby completing the proof. $\qquad\square$

## B.5. Proof of Lemma 3.2

It holds that

$$
\begin{aligned}
\min_{y \in c\mathscr{Y}} W(x - y) &= \min\{\zeta(\min\{r \in [0, \infty) \mid x - y \in r\mathscr{Y}\}) \mid y \in c\mathscr{Y}\} \\
&\stackrel{\zeta \text{ monotonic}}{=} \zeta(\min\{\min\{r \in [0, \infty) \mid x - y \in r\mathscr{Y}\} \mid y \in c\mathscr{Y}\}) \\
&= \zeta(\min\{r \in [0, \infty) \mid \exists y \in c\mathscr{Y}, x - y \in r\mathscr{Y}\}) \\
&= \zeta(\min\{r \in [0, \infty) \mid x \in r\mathscr{Y} \oplus c\mathscr{Y}\}) \\
&\stackrel{\text{Lemma A.1(iii)}}{=} \zeta(\min\{r \in [0, \infty) \mid x \in (r + c)\mathscr{Y}\}) \\
&= \zeta(\max\{0, \Phi_{\mathscr{Y}}(x) - c\}) \\
&= \zeta(\max\{0, \zeta^{-1}(W(x)) - c\}),
\end{aligned}
\tag{B.9}
$$

where the second-to-last line follows by distinguishing between the two cases $x \in c\mathscr{Y}$ and $x \notin c\mathscr{Y}$, thereby completing the proof. $\qquad\square$

# B.6. Proof of Theorem 3.5

We will show that for all $t \in \{t_i + 1, \ldots, t_{i+1}\}$, $i \in \{0, \ldots, i_{\max}\}$ it holds that $x_t \in \{(A + BK)^{t-t_i} x_{t_i}\} \oplus \gamma \mathscr{F}_{t-t_i}$. The first part of the statement then follows from Theorem 2.3. First, we use induction to characterize the reachable sets for the closed-loop system under the event-triggered control scheme, that is, we show that

$$
\begin{aligned}
\mathcal{S}_{t-t_i}(x_{t_i}) := &\left\{ A^{t-t_i} x_{t_i} + \sum_{k=0}^{t-t_i-1} A^{t-t_i-1-k}(BK(A+BK)^k x_{t_i} + w_{t_i+k}) \right| \\
&\quad w_{t_i+k} \in \mathscr{W}, k \in \{0, \ldots, t - t_i - 1\}, \\
&\quad A^j x_{t_i} + \sum_{k=0}^{j-1} A^{j-1-k}(BK(A+BK)^k x_{t_i} + w_{t_i+k}) \in \{(A+BK)^j x_{t_i}\} \oplus \mathscr{T}_j, \\
&\quad j \in \{0, \ldots, t - t_i - 1\} \Big\} \\
= &\{(A+BK)^{t-t_i} x_{t_i}\} \oplus \mathscr{H}_{t-t_i}
\end{aligned}
\tag{B.10}
$$

for all $t \in \{t_i, \ldots, t_{i+1}\}$, $i \in \{0, \ldots, i_{\max}\}$. If $t = t_i$, the claimed equality holds trivially, proving the base case. Assume now that $t \in \{t_i, \ldots, t_{i+1} - 1\}$ and $\mathcal{S}_{t-t_i}(x_{t_i}) = \{(A + BK)^{t-t_i} x_{t_i}\} \oplus \mathscr{H}_{t-t_i}$. It holds that

$$
\begin{aligned}
\mathcal{S}_{t+1-t_i}(x_{t_i}) &= \left\{ Ax_t + BK(A+BK)^{t-t_i} x_{t_i} + w_t \mid \right. \\
&\qquad \left. x_t \in \mathcal{S}_{t-t_i}(x_{t_i}),\ w_t \in \mathscr{W},\ x_t \in \{(A+BK)^{t-t_i} x_{t_i}\} \oplus \mathscr{T}_{t-t_i} \right\} \\
&= \left\{ Ax_t + BK(A+BK)^{t-t_i} x_{t_i} + w_t \mid \right. \\
&\qquad \left. x_t \in \{(A+BK)^{t-t_i} x_{t_i}\} \oplus (\mathscr{H}_{t-t_i} \cap \mathscr{T}_{t-t_i}),\ w_t \in \mathscr{W}, \right\} \\
&= A\{(A+BK)^{t-t_i} x_{t_i}\} \oplus A(\mathscr{H}_{t-t_i} \cap \mathscr{T}_{t-t_i}) \oplus \{BK(A+BK)^{t-t_i} x_{t_i}\} \oplus \mathscr{W} \\
&= \{(A+BK)^{t+1-t_i} x_{t_i}\} \oplus A(\mathscr{H}_{t-t_i} \cap \mathscr{T}_{t-t_i}) \oplus \mathscr{W} \\
&= \{(A+BK)^{t+1-t_i} x_{t_i}\} \oplus \mathscr{H}_{t+1-t_i},
\end{aligned}
\tag{B.11}
$$

completing the inductive step. If $\gamma = \min\{c \in [1, \infty) \mid \mathscr{H}_\tau \subseteq c\mathscr{F}_\tau,\ \tau \in \mathbb{N}\}$ it holds that $\mathscr{H}_\tau \subseteq \gamma \mathscr{F}_\tau$ for all $\tau \in \mathbb{N}$, which completes the proof for the first part of the statement.

For the second part, assume that $c < \gamma$ which implies that there exists a $\tau \in \mathbb{N}$ with $\mathscr{H}_\tau \not\subseteq c\mathscr{F}_\tau$. Therefore, it holds that

$$
\begin{aligned}
(A+BK)^\tau c\mathscr{F}_\infty \oplus \mathscr{H}_\tau &\not\subseteq (A+BK)^\tau c\mathscr{F}_\infty \oplus c\mathscr{F}_\tau \\
&= c\mathscr{F}_\infty,
\end{aligned}
\tag{B.12}
$$

where the first line follows from the properties of the Minkowski sum and the convexity of $\mathscr{F}_\tau$, see [Schneider, 2013]. Hence, with the first part of the statement proven above, there exist $x_0 \in c\mathscr{F}_\infty$ and $w_j \in \mathscr{W}$, $j \in \{0, \ldots, \tau - 1\}$ with $x_\tau \notin c\mathscr{F}_\infty$, such that $c\mathscr{F}_\infty$ is not $\theta$-uniformly asymptotically stable for any $\theta \in \mathbb{N} \cup \{\infty\}$. $\qquad\square$

## B.7. Proof of Lemma 4.1

The first part of the statement follows directly from the definition of $\bar{P}_i$. For the second part of the statement, consider that

$$\sum_{i=1}^{N} iP_i = \sum_{i=1}^{N} i(\bar{P}_{i-1} - \bar{P}_i)$$

$$= \bar{P}_0 + \sum_{i=1}^{N-1}(i+1)\bar{P}_i - \sum_{i=1}^{N-1} i\bar{P}_i - N\bar{P}_N$$

$$= 1 + \sum_{i=1}^{N-1} \bar{P}_i, \tag{B.13}$$

where we have used $\bar{P}_0 = 1$ and $\bar{P}_N = 0$. With (B.13), the second part of the statement follows from the first, thereby completing the proof. $\qquad\square$

## B.8. Proof of Lemma 4.2

By the definition of $\bar{P}_i$ in (4.5) (and the short discussion afterwards), Lemma 4.2 is equivalent to the statement that for all $i \in \{1, \ldots, N-1\}$, $\bar{P}_i$ is a monotonously non-decreasing continuous function of $r_i$, $\bar{P}_i = 0$ for $r_i = 0$, and that $\bar{P}_i$ converges to $\bar{P}_{i-1}$ for $r_i$ approaching infinity. The continuity equivalence follows from the fact that the sum of continuous functions is again continuous, the equivalence of the monotonicity and the respective limits from the relation $P_i = \bar{P}_{i-1} - \bar{P}_i$ and the independence of $\bar{P}_{i-1}$ of $r_i$.

Due to the assumption that the set $\mathscr{T}$ is convex and contains the origin, it holds that $r \leq r' \Leftrightarrow r\mathscr{T} \subseteq r'\mathscr{T}$, such that, by Lemma 4.1, $\bar{P}_i$ is a monotonously non-decreasing function of $r_i$. Let in the following $i \in \{1, \ldots, N-1\}$ and $r_j \in [0, \infty)$ for all $j \in \{1, \ldots, i-1\}$ be arbitrary but fixed and define

$$\bar{P}_i(r) := \int_{r_1\mathscr{T} \times r_2\mathscr{T} \times \cdots \times r_{i-1}\mathscr{T} \times r\mathscr{T}} \rho_{\mathbf{v}}^i(\mathbf{v})\, d\mathbf{v} \tag{B.14}$$

for all $r \in [0, \infty)$. Let $r \in [0, \infty)$ and $r' \in [r, \infty)$. It holds that

$$\bar{P}_i(r') = \int_{r_1\mathscr{T} \times \cdots \times r\mathscr{T}} \rho_{\mathbf{v}}^i(\mathbf{v})\, d\mathbf{v} + \int_{r_1\mathscr{T} \times \cdots \times (r'\mathscr{T} \backslash r\mathscr{T})} \rho_{\mathbf{v}}^i(\mathbf{v})\, d\mathbf{v}$$

$$\leq \int_{r_1\mathscr{T} \times \cdots \times r\mathscr{T}} \rho_{\mathbf{v}}^i(\mathbf{v})\, d\mathbf{v} + \sup_{\mathbf{v} \in \mathbb{R}^{in_x}} \rho_{\mathbf{v}}^i(\mathbf{v}) \operatorname{vol}(r_1\mathscr{T} \times \cdots \times (r'\mathscr{T} \backslash r\mathscr{T})), \tag{B.15}$$

where $\hat{p} := \sup_{\mathbf{v} \in \mathbb{R}^{in_x}} \mathbf{p}_{\mathbf{v}}^i(\mathbf{v}) \in [0, \infty)$ due to the assumption that $\rho_{\mathrm{w}}$ is bounded.

Using $\bar{P}_i(r') \geq \bar{P}_i(r)$, we obtain

$$
\begin{aligned}
|\bar{P}_i(r') - \bar{P}_i(r)| &\leq \hat{p}\,\mathrm{vol}(r_1\mathscr{T} \times \cdots \times (r'\mathscr{T} \setminus r\mathscr{T})) \\
&= \hat{p}\prod_{j=1}^{i-1}\mathrm{vol}(r_j\mathscr{T})\,\mathrm{vol}(r'\mathscr{T} \setminus r\mathscr{T}) \\
&= \hat{p}\prod_{j=1}^{i-1}\mathrm{vol}(r_j\mathscr{T})(\mathrm{vol}(r'\mathscr{T}) - \mathrm{vol}(r\mathscr{T})) \\
&= \hat{p}(\mathrm{vol}(\mathscr{T}))^i\prod_{j=1}^{i-1}r_j^{n_\mathrm{x}}(r'^{n_\mathrm{x}} - r^{n_\mathrm{x}}).
\end{aligned}
\tag{B.16}
$$

By Lemma A.4, it follows that $\bar{P}_i$, as a function of $r$, is continuous, thereby proving the first part of the statement. As $\rho_\mathrm{w}$ was assumed to be bounded, it follows from (B.14) that $\bar{P}_i = 0$ if $r = 0$, proving also the second part of the statement. It holds that

$$
\begin{aligned}
\bar{P}_{i-1} &= \mathbb{P}\left(\sum_{k=0}^{j-1}A^{j-1-k}w_k \in r_j\mathscr{T}, j \in \{1,\ldots,i-1\}\right) \\
&= \mathbb{P}\left(\sum_{k=0}^{j-1}A^{j-1-k}w_k \in r_j\mathscr{T}, j \in \{1,\ldots,i-1\}, \sum_{k=0}^{i-1}A^{i-1-k}w_k \in \mathbb{R}^{n_\mathrm{x}}\right) \\
&= \mathbb{P}\left(v_j \in r_j\mathscr{T}, j \in \{1,\ldots,i-1\}, v_i \in \mathbb{R}^{n_\mathrm{x}}\right) \\
&= \int_{r_1\mathscr{T} \times r_2\mathscr{T} \times \cdots \times r_{i-1}\mathscr{T} \times \mathbb{R}^{n_\mathrm{x}}}\rho_\mathbf{v}^i(\boldsymbol{v})\,\mathrm{d}\boldsymbol{v} \quad .
\end{aligned}
\tag{B.17}
$$

Further, as we assumed the origin to be contained in the interior of $\mathscr{T}$, we have $r_1\mathscr{T} \times r_2\mathscr{T} \times \cdots \times r_{i-1}\mathscr{T} \times \mathbb{R}^{n_\mathrm{x}} = \bigcup_{r\in[0,\infty)}r_1\mathscr{T} \times r_2\mathscr{T} \times \cdots \times r_{i-1}\mathscr{T} \times r\mathscr{T}$. Hence, the claimed convergence property follows from Lemma A.5, thereby completing the proof for the third part of the statement. $\qquad\square$

# B.9. Proof of Theorem 4.1

Define $\mathscr{H}_0' = \{0\}$ and $\mathscr{H}_{\tau+1}' = A(\mathscr{H}_\tau' \cap \mathscr{T}_\tau') \oplus \mathscr{W}$, $\tau \in \mathbb{N}$. Following the first part of the proof of Theorem 3.5, it holds that

$$
\begin{aligned}
\bar{P}_i &= \mathbb{P}\left(\sum_{k=0}^{j-1}A^{j-1-k}w_k \in \mathscr{T}_j, j \in \{1,\ldots,i\}\right) \\
&= \mathbb{P}\left(\sum_{k=0}^{j-1}A^{j-1-k}w_k \in \mathscr{H}_j \cap \mathscr{T}_j, j \in \{1,\ldots,i\}\right)
\end{aligned}
\tag{B.18}
$$

for all $i \in \{1, \dots\}$, analogously for $\bar{P}_i'$.

Hence, if $\mathcal{H}_\tau \cap \mathcal{T}_\tau \subseteq \mathcal{H}_\tau' \cap \mathcal{T}_\tau'$ for all $\tau \in \{0, \dots, N-1\}$, then $\bar{P}_i \leq \bar{P}_i'$, $i \in \{1, \dots, N-1\}$, and, following the proof of Lemma 4.1, also $\sum_{i=1}^{N} iP_i \leq \sum_{i=1}^{N} iP_i'$.

Due to $\mathcal{H}_0 = \mathcal{H}_0' = \{0\}$ and $0 \in \mathcal{T}_0'$, the inclusion $\mathcal{H}_0 \cap \mathcal{T}_0 \subseteq \mathcal{H}_0' \cap \mathcal{T}_0'$ holds, providing the base case for the following induction on $\tau$. Assume now that $\mathcal{H}_\tau \cap \mathcal{T}_\tau \subseteq \mathcal{H}_\tau' \cap \mathcal{T}_\tau'$ for an arbitrary $\tau \in \{0, \dots, N-2\}$. It follows that $\mathcal{H}_{\tau+1} = A(\mathcal{H}_\tau \cap \mathcal{T}_\tau) \oplus \mathcal{W} \subseteq A(\mathcal{H}_\tau' \cap \mathcal{T}_\tau') \oplus \mathcal{W} = \mathcal{H}_{\tau+1}'$. By assumption, it holds that $A(\mathcal{H}_{\tau+1} \cap \mathcal{T}_{\tau+1}) \oplus \mathcal{W} = \mathcal{H}_{\tau+2} \subseteq \gamma \mathcal{F}_{\tau+2}$, such that $\mathcal{H}_{\tau+1} \cap \mathcal{T}_{\tau+1} \subseteq A^{-1}(\gamma \mathcal{F}_{\tau+2} \ominus \mathcal{W}) = \mathcal{T}_{\tau+1}'$, which, together with $\mathcal{H}_{\tau+1} \subseteq \mathcal{H}_{\tau+1}'$ implies $\mathcal{H}_{\tau+1} \cap \mathcal{T}_{\tau+1} \subseteq \mathcal{H}_{\tau+1}' \cap \mathcal{T}_{\tau+1}'$. This completes the inductive step and, thereby, the proof. $\qquad\square$

# B.10. Proof of Lemma 4.3

By assumption, $\mathcal{W}$ is a convex set containing the origin and the same also holds for $\mathcal{F}_{\tau+1}$, $\tau \in \mathbb{N}$. Hence, $r \leq r'$ implies $r\mathcal{F}_{\tau+1} \subseteq r'\mathcal{F}_{\tau+1}$ and therefore also $A^{-1}(r\mathcal{F}_{\tau+1} \ominus \mathcal{W}) \subseteq A^{-1}(r'\mathcal{F}_{\tau+1} \ominus \mathcal{W})$. Consequently, Lemma 4.1 implies that $\sum_{i=1}^{N} iP_i$ is a monotonically non-decreasing function of $r$.

Continuity of $\sum_{i=1}^{N} iP_i$ in $r$ is implied by the continuity of $\bar{P}_i$, $i \in \{1, \dots, N-1\}$ in $r$, which we establish next. Let $i \in \{1, \dots, N-1\}$ and define

$$\bar{P}_i(r) := \int_{A^{-1}(r\mathcal{F}_2 \ominus \mathcal{W}) \times A^{-1}(r\mathcal{F}_3 \ominus \mathcal{W}) \times \cdots \times A^{-1}(r\mathcal{F}_{i+1} \ominus \mathcal{W})} \rho_{\mathbf{v}}^i(\boldsymbol{v}) \, \mathrm{d}\boldsymbol{v} \tag{B.19}$$

for all $r \in [0, \infty)$. Let $r \in [0, \infty)$ and $r' \in [r, \infty)$. It holds that

$$
\begin{aligned}
\bar{P}_i(r') &= \int_{A^{-1}(r'\mathcal{F}_2 \ominus \mathcal{W}) \times \cdots \times A^{-1}(r'\mathcal{F}_{i+1} \ominus \mathcal{W})} \rho_{\mathbf{v}}^i(\boldsymbol{v}) \, \mathrm{d}\boldsymbol{v} \\
&= \int_{A^{-1}(r\mathcal{F}_2 \ominus \mathcal{W}) \times \cdots \times A^{-1}(r\mathcal{F}_{i+1} \ominus \mathcal{W})} \rho_{\mathbf{v}}^i(\boldsymbol{v}) \, \mathrm{d}\boldsymbol{v} \\
&\quad + \int_{(A^{-1}(r'\mathcal{F}_2 \ominus \mathcal{W}) \times \cdots \times A^{-1}(r'\mathcal{F}_{i+1} \ominus \mathcal{W})) \setminus (A^{-1}(r\mathcal{F}_2 \ominus \mathcal{W}) \times \cdots \times A^{-1}(r\mathcal{F}_{i+1} \ominus \mathcal{W}))} \rho_{\mathbf{v}}^i(\boldsymbol{v}) \, \mathrm{d}\boldsymbol{v} \\
&\leq \bar{P}_i(r) + \sup_{\boldsymbol{v} \in \mathbb{R}^{in_x}} \rho_{\mathbf{v}}^i(\boldsymbol{v}) \, \mathrm{vol}\big((A^{-1}(r'\mathcal{F}_2 \ominus \mathcal{W}) \times \cdots \times A^{-1}(r'\mathcal{F}_{i+1} \ominus \mathcal{W})) \\
&\qquad\qquad\qquad\qquad\qquad\qquad \setminus (A^{-1}(r\mathcal{F}_2 \ominus \mathcal{W}) \times \cdots \times A^{-1}(r\mathcal{F}_{i+1} \ominus \mathcal{W}))\big) \\
&= \bar{P}_i(r) + g(r', r), \tag{B.20}
\end{aligned}
$$

where $g : [0, \infty) \times [0, \infty) \to \mathbb{R}$ with

$$
\begin{aligned}
g(r', r) &= \sup_{\boldsymbol{v} \in \mathbb{R}^{in_x}} \rho_{\mathbf{v}}^i(\boldsymbol{v}) \left( \prod_{j=1}^{i} \mathrm{vol}(A^{-1}(r'\mathcal{F}_{j+1} \ominus \mathcal{W})) - \prod_{j=1}^{i} \mathrm{vol}(A^{-1}(r\mathcal{F}_{j+1} \ominus \mathcal{W})) \right) \\
&= \sup_{\boldsymbol{v} \in \mathbb{R}^{in_x}} \rho_{\mathbf{v}}^i(\boldsymbol{v}) \frac{1}{|\det(A)|^i} \left( \prod_{j=1}^{i} \mathrm{vol}(r'\mathcal{F}_{j+1} \ominus \mathcal{W}) - \prod_{j=1}^{i} \mathrm{vol}(r\mathcal{F}_{j+1} \ominus \mathcal{W}) \right). \tag{B.21}
\end{aligned}
$$

By Lemma A.6, $g$ is a continuous function satisfying $g(r, r) = 0$ for all $r \in [0, \infty)$, such that with Lemma A.4 we conclude that $\bar{P}_i$ is continuous in $r$, proving the first part of the statement. As the interior of $\mathscr{W}$ was assumed to be nonempty, it follows that $r\mathscr{F}_{i+1} \ominus \mathscr{W} = \emptyset$ for $r = 0$, proving also the second part of the statement.

Finally, it holds that $\bigcup_{r \in [0, \infty)} A^{-1}(r\mathscr{F}_{i+1} \ominus \mathscr{W}) = \mathbb{R}^{n_x}$, $i \in \{1, \ldots, N-1\}$, due to the assumption that $\mathscr{W}$ contained the origin in its interior. By Lemma A.5 it follows that $\lim_{r \to \infty} \bar{P}_i(r) = 1$, $i \in \{1, \ldots, N-1\}$, such that, with (B.13), $\lim_{r \to \infty} \sum_{i=1}^{N} P_i = 1 + \lim_{r \to \infty} \sum_{i=1}^{N-1} \bar{P}_i(r) = N$, thereby completing the proof. $\qquad\square$

# B.11. Proof of Lemma 4.4

The proof is based on similar arguments as the proofs of Lemma 4 and Lemma 5 in [Wu et al., 2016a]. Consider first a normally distributed random variable $z \in \mathbb{R}^{n_z}$ with $z \sim \mathcal{N}(0, \Sigma)$ where $\Sigma \in \mathbb{R}^{n_z \times n_z}$ is a nonzero symmetric positive semi-definite matrix. Let further another random variable $r \in \mathbb{R}$, uniformly distributed on $[0, 1]$, and a scalar $\lambda > 0$ be given and let $z$ and $r$ be independent. Let a third random variable be defined by

$$
\delta := \begin{cases} 0 & \frac{1}{2} z^{\mathsf{T}} \Sigma^{\dagger} z \le -\frac{1}{\lambda} \ln(r) \\ 1 & \text{else.} \end{cases} \tag{B.22}
$$

It holds that $z = Vy$ where $y \sim N(0, V^{\mathsf{T}} \Sigma V)$ and the columns of $V \in \mathbb{R}^{n_z \times \mathrm{rk}(\Sigma)}$ are orthonormal and span $\ker(\Sigma)^{\perp}$ (using the standard scalar product as the inner product on $\mathbb{R}^{n_z}$), compare [Rao, 1973]. Hence, the probability of $\delta = 0$ is given by

$$
\begin{aligned}
\mathbb{P}(\delta = 0) &= \int\limits_{y \in \mathbb{R}^{\mathrm{rk}(\Sigma)}} \int\limits_{r=0}^{e^{-\lambda \frac{1}{2} y^{\mathsf{T}} V^{\mathsf{T}} \Sigma^{\dagger} V y}} \frac{1}{\det^{\dagger}(2\pi\Sigma)^{\frac{1}{2}}} e^{-\frac{1}{2} y^{\mathsf{T}} V^{\mathsf{T}} \Sigma^{\dagger} V y} \, \mathrm{d}r \, \mathrm{d}y \\
&= \int\limits_{y \in \mathbb{R}^{\mathrm{rk}(\Sigma)}} \frac{1}{\det^{\dagger}(2\pi\Sigma)^{\frac{1}{2}}} e^{-\frac{1}{2} y^{\mathsf{T}} V^{\mathsf{T}} \Sigma^{\dagger} V y} e^{-\lambda \frac{1}{2} y^{\mathsf{T}} V^{\mathsf{T}} \Sigma^{\dagger} V y} \, \mathrm{d}y \\
&= \int\limits_{y \in \mathbb{R}^{\mathrm{rk}(\Sigma)}} \frac{1}{\det^{\dagger}(2\pi\Sigma)^{\frac{1}{2}}} e^{-\frac{1}{2} y^{\mathsf{T}} V^{\mathsf{T}} (1+\lambda) \Sigma^{\dagger} V y} \, \mathrm{d}y \\
&= \frac{\det^{\dagger}(2\pi \frac{1}{1+\lambda} \Sigma)^{\frac{1}{2}}}{\det^{\dagger}(2\pi\Sigma)^{\frac{1}{2}}} = (1+\lambda)^{-\frac{\mathrm{rk}(\Sigma)}{2}}.
\end{aligned} \tag{B.23}
$$

Further, the distribution function of $y$ under the condition that $\delta = 0$ is given by

$$\mathbb{P}(y \le y' | \delta = 0) = \frac{\mathbb{P}(y \le y', \delta = 0)}{\mathbb{P}(\delta = 0)}$$

$$= \frac{1}{\mathbb{P}(\delta = 0)} \int\limits_{y \le y'} \int\limits_{r=0}^{e^{-\lambda \frac{1}{2} y^{\mathsf{T}} V^{\mathsf{T}} \Sigma^{\dagger} V y}} \frac{1}{\det^{\dagger}(2\pi\Sigma)^{\frac{1}{2}}} e^{-\frac{1}{2} y^{\mathsf{T}} V^{\mathsf{T}} \Sigma^{\dagger} V y} \, dr \, dy \qquad (B.24)$$

and the corresponding conditional density is given by

$$p_y(y' | \delta = 0) = \frac{1}{\mathbb{P}(\delta = 0)} \int\limits_{r=0}^{e^{-\lambda \frac{1}{2} y'^{\mathsf{T}} V^{\mathsf{T}} \Sigma^{\dagger} V y'}} \frac{1}{\det^{\dagger}(2\pi\Sigma)^{\frac{1}{2}}} e^{-\frac{1}{2} y'^{\mathsf{T}} V^{\mathsf{T}} \Sigma^{\dagger} V y'} \, dr$$

$$= \frac{(1+\lambda)^{\frac{\text{rk}(\Sigma)}{2}}}{\det^{\dagger}(2\pi\Sigma)^{\frac{1}{2}}} e^{-\frac{1}{2} y'^{\mathsf{T}} (1+\lambda) V^{\mathsf{T}} \Sigma^{\dagger} V y'} \,. \qquad (B.25)$$

That is, it holds that $y \sim \mathcal{N}\left(0, \frac{1}{1+\lambda} V^{\mathsf{T}} \Sigma V\right)$ and $z \sim \mathcal{N}\left(0, \frac{1}{1+\lambda} \Sigma\right)$ under the condition that $\delta = 0$.

Let $t_i$ be known and consider a time point $t \in \{t_i + 1, \dots, t_{i+1}\}$ where $t_{i+1}$ is unknown. It holds that $z_{t_i+1} = w_{t_i}$, such that $z_{t_i+1} \sim \mathcal{N}(0, Q_{t_i})$. Assume now that $z_t \sim \mathcal{N}(0, \hat{\Sigma}_t)$ under the condition that $t \in \{t_i + 1, \dots, t_{i+1}\}$ for a matrix $\hat{\Sigma}_t$. By the discussion above it follows that $z_t \sim \mathcal{N}\left(0, \frac{1}{1+\lambda_t} \hat{\Sigma}_t\right)$ under the condition that $t \in \{t_i + 1, \dots, t_{i+1} - 1\}$, as then $\delta_t = 0$. As $z_{t+1} = A z_t + w_t$ it follows that $z_{t+1} \sim \mathcal{N}\left(0, \frac{1}{1+\lambda_t} A \hat{\Sigma}_t A^{\mathsf{T}} + Q_t\right)$ under the condition that $t \in \{t_i + 1, \dots, t_{i+1} - 1\}$. By induction, it follows that $z_t \sim \mathcal{N}(0, \Sigma_t)$ under the condition that $t \in \{t_i + 1, \dots, t_{i+1}\}$, for all $i \in \{1, \dots, i_{\max}\}$, where $\Sigma_t$ is computed according to (4.20). Further, using (B.23), an event occurs at time $t$ with probability $1 - (1 + \lambda_t)^{-\frac{\text{rk}(\Sigma_t)}{2}}$.

In the following discussion, we omit the additional conditioning of most probabilities on $t \in \{t_i + 1, \dots, t_{i+1}\}$ and $t_i$ for the sake of simpler exposition. Further, we use the notational convention $\mathbb{P}(a_j, \ j \in \mathcal{M}) := \mathbb{P}(\forall j \in \mathcal{M}, \ a_j)$. It holds that $\mathbb{P}(\delta_{t_i+1} = 0) = 1 - \mathbb{P}(\delta_{t_i+1} = 1)$ and, for all $t \in \{t_i + 2, \dots, t_{i+1}\}$, that

$$\mathbb{P}(\delta_j = 0, \ j \in \{t_i + 1, \dots, t\})$$
$$= \mathbb{P}(\delta_j = 0, \ j \in \{t_i + 1, \dots, t - 1\}) - \mathbb{P}(\delta_t = 1, \ \delta_j = 0, \ j \in \{t_i + 1, \dots, t - 1\}), \qquad (B.26)$$

where we emphasize that $t_{i+1}$ is not known and the set membership $t \in \{t_i + 2, \dots, t_{i+1}\}$ only states that $t_i$ was the last transmission instant before $t$ and $t \ge t_i + 2$.

Hence, for all $t \in \{t_i + 2, \dots, t_{i+1}\}$, it holds that

$$\mathbb{P}(\delta_j = 0, \ j \in \{t_i + 1, \dots, t\}) = 1 - \sum_{k=t_i+1}^{t} \mathbb{P}(\delta_k = 1, \ \delta_j = 0, \ j \in \{t_i + 1, \dots, k - 1\}).$$

$$(B.27)$$

Further, we have

$$
\begin{aligned}
\mathbb{P}(t_{i+1} - t_i = \tau) &= \mathbb{P}(\delta_{t_i+\tau} = 1, \ \delta_j = 0, \ j \in \{t_i + 1, \ldots, t_i + \tau - 1\}) \\
&= \mathbb{P}(\delta_{t_i+\tau} = 1 \mid \delta_j = 0, \ j \in \{t_i + 1, \ldots, t_i + \tau - 1\}) \\
&\qquad\qquad\qquad \cdot \mathbb{P}(\delta_j = 0, \ j \in \{t_i + 1, \ldots, t_i + \tau - 1\}) \\
&\stackrel{(B.23)}{=} \left(1 - (1 + \lambda_{t_i+\tau})^{-\frac{\mathrm{rk}(\Sigma_{t_i+\tau})}{2}}\right) \mathbb{P}(\delta_j = 0, \ j \in \{t_i + 1, \ldots, t_i + \tau - 1\}) \\
&\stackrel{(B.27)}{=} \left(1 - (1 + \lambda_{t_i+\tau})^{-\frac{\mathrm{rk}(\Sigma_{t_i+\tau})}{2}}\right) \left(1 - \sum_{j=1}^{\tau-1} \mathbb{P}(t_{i+1} - t_i = j)\right). \quad (B.28)
\end{aligned}
$$

With (4.21), the last line in turn implies

$$
\mathbb{P}(t_{i+1} - t_i = \tau) = \frac{p_\Delta(\tau)}{1 - \sum_{j=1}^{\tau-1} p_\Delta(j)} \left(1 - \sum_{j=1}^{\tau-1} \mathbb{P}(t_{i+1} - t_i = j)\right). \quad (B.29)
$$

Via strong induction, (B.29) implies $\mathbb{P}(t_{i+1} - t_i = \tau) = p_\Delta(\tau)$ for all $\tau \in \{1, \ldots\}$, thereby completing the proof. $\qquad\square$

# B.12. Proof of Theorem 4.2

Let $\Theta_t = \{0, 1\}^t$ denote the set of sequences of length $t$ of all possible realizations of the values of the event-generating function, that is $(\delta_0, \ldots, \delta_{t-1}) = \theta$ for some $\theta \in \Theta_t$. It holds that

$$
\mathbb{E}[x_t^\mathsf{T} x_t] = \sum_{\theta \in \Theta_t} \mathbb{P}((\delta_0, \ldots, \delta_{t-1}) = \theta) \mathbb{E}[x_t^\mathsf{T} x_t \mid \theta]. \quad (B.30)
$$

We disregard the probabilities $\mathbb{P}((\delta_0, \ldots, \delta_{t-1}) = \theta)$ and compute an upper bound of the expression $\mathbb{E}[x_t^\mathsf{T} x_t \mid \theta]$ which then also serves as a upper bound[1] for $\mathbb{E}[x_t^\mathsf{T} x_t]$. Let $\theta = (\delta_0, \ldots, \delta_{t-1})$ be arbitrary but fixed with $\mathbb{P}((\delta_0, \ldots, \delta_{t-1}) = \theta) > 0$, let $t_0 = 0$, $t_{j+1} = \inf\{t' \in \{t_j + 1, \ldots, t - 1\} \mid \delta_{t'} = 1\}$ for $j \in \mathbb{N}$, and let $i = \max\{j \in \mathbb{N} \mid t_j < t\}$. Note that $\mathbb{E}[x^\mathsf{T} x] = \mathrm{tr}(\mathbb{E}[xx^\mathsf{T}])$. It holds that $\mathbb{E}[x_{t_0} x_{t_0}^\mathsf{T} \mid \theta] = \mathbb{E}[x_0 x_0^\mathsf{T} \mid \theta] = x_0 x_0^\mathsf{T}$. Let $j \in \{1, \ldots, i\}$. We have $x_{t_j} = x_{t_j} - (A + BK)^{t_j - t_{j-1}} x_{t_{j-1}} + (A + BK)^{t_j - t_{j-1}} x_{t_{j-1}}$ and

---

[1] This introduces conservatism.

$x_{t_j} - (A + BK)^{t_j - t_{j-1}} x_{t_{j-1}} = \sum_{k=t_{j-1}}^{t_j - 1} A^{t_j - 1 - k} w_k$. Hence it holds that

$$
\begin{aligned}
\mathbb{E}[x_{t_j} x_{t_j}^\mathsf{T} \mid \theta] = & \mathbb{E}\left[ \left( \sum_{k=t_{j-1}}^{t_j - 1} A^{t_j - 1 - k} w_k \right) \left( \sum_{k=t_{j-1}}^{t_j - 1} A^{t_j - 1 - k} w_k \right)^\mathsf{T} \Bigg| \theta \right] \\
& + \mathbb{E}\left[ (A + BK)^{t_j - t_{j-1}} x_{t_{j-1}} \left( \sum_{k=t_{j-1}}^{t_j - 1} A^{t_j - 1 - k} w_k \right)^\mathsf{T} \Bigg| \theta \right] \\
& + \mathbb{E}\left[ \left( \sum_{k=t_{j-1}}^{t_j - 1} A^{t_j - 1 - k} w_k \right) \left( (A + BK)^{t_j - t_{j-1}} x_{t_{j-1}} \right)^\mathsf{T} \Bigg| \theta \right] \\
& + \mathbb{E}\left[ (A + BK)^{t_j - t_{j-1}} x_{t_{j-1}} x_{t_{j-1}}^\mathsf{T} \left( (A + BK)^{t_j - t_{j-1}} \right)^\mathsf{T} \Big| \theta \right]. \quad \text{(B.31)}
\end{aligned}
$$

As the variables $(A + BK)^{t_j - t_{j-1}} x_{t_{j-1}}$ and $\sum_{k=t_{j-1}}^{t_j - 1} A^{t_j - 1 - k} w_k$ are independent and furthermore the (conditional) distribution of the latter is symmetric about the origin, the second and third addends in the expression are zero. Along the sames lines as the proof of Lemma 4.4, considering in particular the derivation of (B.25), it follows that $\sum_{k=t_{j-1}}^{t_j - 1} A^{t_j - 1 - k} w_k = Vy$ for some matrix $V \in \mathbb{R}^{n_x \times p}$ where $y$ is a random variable with conditional density

$$
\begin{aligned}
p_y(y' \mid \theta) = & \frac{1}{\mathbb{P}(\delta_{t_j} = 1 \mid \delta_{t_{j-1}} = 1, \delta_k = 0, k \in \{t_{j-1} + 1, \ldots, t_j - 1\})} \\
& \cdot \int_{r = e^{-\lambda_{t_j} \frac{1}{2} y'^\mathsf{T} V^\mathsf{T} \Sigma_{t_j}^\dagger V y'}}^{1} \frac{1}{\det^\dagger (2\pi \Sigma_{t_j})^{\frac{1}{2}}} e^{-\frac{1}{2} y'^\mathsf{T} V^\mathsf{T} \Sigma_{t_j}^\dagger V y'} \, dr \\
= & \frac{1}{\left( 1 - (1 + \lambda_{t_j})^{-\frac{\mathrm{rk}(\Sigma_{t_j})}{2}} \right) \det^\dagger (2\pi \Sigma_{t_j})^{\frac{1}{2}}} \left( e^{-\frac{1}{2} y'^\mathsf{T} V^\mathsf{T} \Sigma_{t_j}^\dagger V y'} - e^{-\frac{1}{2} y'^\mathsf{T} (1 + \lambda_{t_j}) V^\mathsf{T} \Sigma_{t_j}^\dagger V y'} \right) \\
= & \frac{1}{\left( 1 - (1 + \lambda_{t_j})^{-\frac{\mathrm{rk}(\Sigma_{t_j})}{2}} \right)} \frac{e^{-\frac{1}{2} y'^\mathsf{T} V^\mathsf{T} \Sigma_{t_j}^\dagger V y'}}{\det^\dagger (2\pi \Sigma_{t_j})^{\frac{1}{2}}} \\
& - \frac{(1 + \lambda_{t_j})^{-\frac{\mathrm{rk}(\Sigma_{t_j})}{2}}}{\left( 1 - (1 + \lambda_{t_j})^{-\frac{\mathrm{rk}(\Sigma_{t_j})}{2}} \right) (1 + \lambda_{t_j})^{-\frac{\mathrm{rk}(\Sigma_{t_j})}{2}} \det^\dagger (2\pi \Sigma_{t_j})^{\frac{1}{2}}}. \quad \text{(B.32)}
\end{aligned}
$$

As this density function is a linear combination of two normal densities, the conditional covariance $\mathbb{E}[yy^\mathsf{T} \mid \theta]$ is given by

$$\mathbb{E}[yy^\mathsf{T} \mid \theta] = \frac{1}{\left(1 - (1 + \lambda_{t_j})^{-\frac{\mathrm{rk}(\Sigma_{t_j})}{2}}\right)} V^\mathsf{T} \Sigma_{t_j} V - \frac{(1 + \lambda_{t_j})^{-\frac{\mathrm{rk}(\Sigma_{t_j})}{2}}}{\left(1 - (1 + \lambda_{t_j})^{-\frac{\mathrm{rk}(\Sigma_{t_j})}{2}}\right)} \frac{1}{1 + \lambda_{t_j}} V^\mathsf{T} \Sigma_{t_j} V$$

$$= g(\lambda_{t_j}, \mathrm{rk}(\Sigma_{t_j})) V^\mathsf{T} \Sigma_{t_j} V. \tag{B.33}$$

This implies

$$\mathbb{E}\left[\left(\sum_{k=t_{j-1}}^{t_j-1} A^{t_j-1-k} w_k\right) \left(\sum_{k=t_{j-1}}^{t_j-1} A^{t_j-1-k} w_k\right)^\mathsf{T} \;\middle|\; \theta\right] = g(\lambda_{t_j}, \mathrm{rk}(\Sigma_{t_j})) \Sigma_{t_j}. \tag{B.34}$$

By induction, we obtain

$$\mathbb{E}[x_{t_i} x_{t_i}^\mathsf{T} \mid \theta] = \sum_{j=1}^{i} (A + BK)^{t_i-t_j} g(\lambda_{t_j}, \mathrm{rk}(\Sigma_{t_j})) \Sigma_{t_j} \left((A + BK)^{t_i-t_j}\right)^\mathsf{T}$$

$$+ (A + BK)^{t_i} x_0 x_0^\mathsf{T} \left((A + BK)^{t_i}\right)^\mathsf{T}. \tag{B.35}$$

By similar reasoning, and noting that $\theta$ does not determine whether an event occurs at time $t$, we have

$$\mathbb{E}[x_t x_t^\mathsf{T} \mid \theta]$$

$$= \Sigma_t + \sum_{j=1}^{i} (A + BK)^{t-t_j} g(\lambda_{t_j}, \mathrm{rk}(\Sigma_{t_j})) \Sigma_{t_j} \left((A + BK)^{t-t_j}\right)^\mathsf{T}$$

$$+ (A + BK)^{t} x_0 x_0^\mathsf{T} \left((A + BK)^{t}\right)^\mathsf{T}$$

$$\preceq g(\lambda_t, \mathrm{rk}(\Sigma_t)) \Sigma_t + \sum_{k=1}^{t-1} (A + BK)^{t-k} g(\lambda_k, \mathrm{rk}(\Sigma_k)) \Sigma_k \left((A + BK)^{t-k}\right)^\mathsf{T}$$

$$+ (A + BK)^{t} x_0 x_0^\mathsf{T} \left((A + BK)^{t}\right)^\mathsf{T}$$

$$\preceq S - (A + BK) S \left((A + BK)\right)^\mathsf{T}$$

$$+ \sum_{k=1}^{t-1} \left((A + BK)^{t-k} S \left((A + BK)^{t-k}\right)^\mathsf{T} - (A + BK)^{t-k+1} S \left((A + BK)^{t-k+1}\right)^\mathsf{T}\right)$$

$$+ (A + BK)^{t} x_0 x_0^\mathsf{T} \left((A + BK)^{t}\right)^\mathsf{T}$$

$$\preceq S + (A + BK)^{t} x_0 x_0^\mathsf{T} \left((A + BK)^{t}\right)^\mathsf{T}. \tag{B.36}$$

Hence, it holds that

$$\mathbb{E}[x_t^\mathsf{T} x_t \mid \theta] = \mathrm{tr}(\mathbb{E}[x_t x_t^\mathsf{T} \mid \theta])$$
$$\leq \mathrm{tr}(S) + x_0^\mathsf{T} \left((A + BK)^t\right)^\mathsf{T} (A + BK)^t x_0$$
$$\leq \mathrm{tr}(S) + \frac{\lambda_{\max}(P)}{\lambda_{\min}(P)} \mu^t x_0^\mathsf{T} x_0, \tag{B.37}$$

thereby completing the proof. $\qquad\square$

## B.13. Proof of Corollary 4.1

First, we discuss how the conditions in (4.24) guarantee $g(\lambda_t, \mathrm{rk}(\Sigma_t))\Sigma_t \preceq \gamma \sum_{j=0}^{t-1-t_i}(A+BK)^j Q\left((A+BK)^j\right)^\mathsf{T}$ for all $t \in \{t_i + 1, \ldots, t_{i+1}\}$, $i \in \{0, \ldots, i_{\max}\}$. Of particular interest is only the case $\lambda_t = \infty$, which implies $t = t_{i+1}$. It holds that $g(\infty, k) = 1$ for all $k \in \{1, \ldots\}$, such that $\lambda_{t_i+1} = \infty$ implies $g(\lambda_{t_i+1}, \mathrm{rk}(\Sigma_{t_i+1}))\Sigma_{t_i+1} = Q \preceq \gamma Q$ for all $\gamma \in [1, \infty)$, by the definition of $\Sigma_t$ in (4.20). Assume now $\lambda_t = \infty$ for some $t \in \{t_i + 2, \ldots, t_{i+1}\}$, and $\lambda_{t-1} < \infty$. By the first of the two matrix inequalities in (4.24) and the recursion in (4.20), it holds that

$$g(\lambda_t, \mathrm{rk}(\Sigma_t))\Sigma_t = \Sigma_t \preceq \gamma \sum_{j=0}^{t-1-t_i} (A + BK)^j Q \left((A+BK)^j\right)^\mathsf{T}, \tag{B.38}$$

completing the proof of the preliminary statement.

Following the the discussion in the proof of Theorem 4.2, and noting that $g(\lambda, k) \geq 1$ for all $\lambda \in [\,, \infty) \cup \{\infty\}$ and all $k \in \{1, \ldots\}$, we hence obtain

$$\mathbb{E}[x_t^\mathsf{T} x_t] = \mathrm{tr}(\mathbb{E}[x_t x_t^\mathsf{T}])$$
$$\leq \mathrm{tr}(\mathbb{E}[x_t x_t^\mathsf{T} \mid \theta])$$
$$\leq \mathrm{tr}\left(\Sigma_t + \sum_{j=1}^{i}(A + BK)^{t-t_j} g(\lambda_{t_j}, \mathrm{rk}(\Sigma_{t_j}))\Sigma_{t_j}\left((A+BK)^{t-t_j}\right)^\mathsf{T}\right)$$
$$\quad + \mathrm{tr}\left((A + BK)^t x_0 x_0^\mathsf{T}\left((A+BK)^t\right)^\mathsf{T}\right)$$
$$\leq \gamma\, \mathrm{tr}\left(\sum_{k=0}^{t-1-t_i}(A + BK)^k Q\left((A+BK)^k\right)^\mathsf{T}\right.$$
$$\quad + \sum_{j=1}^{i}(A + BK)^{t-t_j}\left(\sum_{k=0}^{t_j-1-t_{j-1}}(A + BK)^k Q\left((A+BK)^k\right)^\mathsf{T}\right)\left((A+BK)^{t-t_j}\right)^\mathsf{T}\right)$$
$$\quad + x_0^\mathsf{T}\left((A + BK)^t\right)^\mathsf{T}(A + BK)^t x_0$$

$$
\begin{aligned}
=&\gamma \operatorname{tr}\left(\sum_{k=0}^{t-1-t_i}(A+BK)^k Q\left((A+BK)^k\right)^{\mathsf{T}}\right.\\
&\left.+\sum_{j=1}^{i}\sum_{k=0}^{t_j-1-t_{j-1}}(A+BK)^{t-t_j+k} Q\left((A+BK)^{t-t_j+k}\right)^{\mathsf{T}}\right)\\
&+x_0^{\mathsf{T}}\left((A+BK)^t\right)^{\mathsf{T}}(A+BK)^t x_0\\
=&\gamma \operatorname{tr}\left(\sum_{k=0}^{t-1-t_i}(A+BK)^k Q\left((A+BK)^k\right)^{\mathsf{T}}+\sum_{j=1}^{i}\sum_{k=t-t_j}^{t-t_{j-1}-1}(A+BK)^k Q\left((A+BK)^k\right)^{\mathsf{T}}\right)\\
&+x_0^{\mathsf{T}}\left((A+BK)^t\right)^{\mathsf{T}}(A+BK)^t x_0\\
=&\gamma \operatorname{tr}\left(\sum_{k=0}^{t-1}(A+BK)^k Q\left((A+BK)^k\right)^{\mathsf{T}}\right)+x_0^{\mathsf{T}}\left((A+BK)^t\right)^{\mathsf{T}}(A+BK)^t x_0\\
\leq&\gamma \operatorname{tr}(\bar{S})+\mu^t\frac{\lambda_{\max}(P)}{\lambda_{\min}(P)}x_0^{\mathsf{T}}x_0, \tag{B.39}
\end{aligned}
$$

which is the desired result. $\qquad\square$

## B.14. Proof of Lemma 5.1

By assumption, $\ell$ is positive definite and, hence, $J(x)$ is non-negative. Therefore it holds that

$$
V_\gamma(x) \geq V_\gamma^{\mathrm{s}}(x) = \min_{y\in\gamma\mathscr{F}_\infty}\bar{V}^{\mathrm{s}}(x-y) \overset{\text{Assumption 5.1}}{\geq} \min_{y\in\gamma\mathscr{F}_\infty}\bar{c}_1|x-y|^a = \bar{c}_1|x|_{\gamma\mathscr{F}_\infty}^a \tag{B.40}
$$

which establishes the lower bound. For the decrease of the Lyapunov function, let $y_t\in\gamma\mathscr{F}_\infty$ such that $V_\gamma^{\mathrm{s}}(x_t)=\bar{V}^{\mathrm{s}}(x_t-y_t)$. Exploiting Theorem 2.4 and using $J(x_{t+1})\leq J(x_t)-\ell(u_t-Kx_t)$, we obtain

$$
\begin{aligned}
V_\gamma(x_{t+1}) &= V_\gamma^{\mathrm{s}}(Ax_t+Bu_t+w_t)+J(x_{t+1})\\
&= V_\gamma^{\mathrm{s}}((A+BK)x_t+B(u_t-Kx_t)+w_t)+J(x_{t+1})\\
&= \min_{y\in\gamma\mathscr{F}_\infty}\bar{V}^{\mathrm{s}}((A+BK)x_t+B(u_t-Kx_t)+w_t-y)+J(x_{t+1})\\
&\overset{(2.4),\gamma\geq 1}{\leq} \min_{y\in\gamma(A+BK)\mathscr{F}_\infty\oplus\mathscr{W}}\bar{V}^{\mathrm{s}}((A+BK)x_t+B(u_t-Kx_t)+w_t-y)+J(x_{t+1})\\
&\overset{w_t\in\mathscr{W}}{\leq} \min_{y\in\gamma(A+BK)\mathscr{F}_\infty}\bar{V}^{\mathrm{s}}((A+BK)x_t+B(u_t-Kx_t)-y)+J(x_{t+1})\\
&= \min_{y\in\gamma\mathscr{F}_\infty}\bar{V}^{\mathrm{s}}((A+BK)(x_t-y)+B(u_t-Kx_t))+J(x_{t+1})\\
&\leq \bar{V}^{\mathrm{s}}((A+BK)(x_t-y_t)+B(u_t-Kx_t))+J(x_{t+1})
\end{aligned}
$$

$$\leq \bar{V}^{\mathrm{s}}((A + BK)(x_t - y_t) + B(u_t - Kx_t)) + J(x_t) - \ell(u_t - Kx_t)$$
$$= V_\gamma(x_t) - \left(\bar{V}^{\mathrm{s}}(x_t - y_t) - \bar{V}^{\mathrm{s}}((A + BK)(x_t - y_t) + B(u_t - Kx_t)) + \ell(u_t - Kx_t)\right)$$

Assumption 5.1
$$\leq V_\gamma(x_t) - \bar{c}_3 |x_t - y_t|^a$$
$$\leq V_\gamma(x_t) - \min_{y \in \gamma \mathscr{F}_\infty} \bar{c}_3 |x_t - y|^a$$
$$= V_\gamma(x_t) - \bar{c}_3 |x_t|_{\gamma \mathscr{F}_\infty}^a. \tag{B.41}$$

The upper bound is established by similar arguments as in the proof of Theorem III.2 in [Lazar et al., 2006]: By assumption, for all $x \in \mathbb{R}^{n_x}$ with $|x|_{\gamma \mathscr{F}_\infty} \leq \epsilon$ it holds that $x \in \bar{\mathscr{X}}_0$, hence $J(x) = 0$ and

$$V_\gamma(x) = V_\gamma^{\mathrm{s}}(x) = \min_{y \in \gamma \mathscr{F}_\infty} \bar{V}^{\mathrm{s}}(x - y) \leq \min_{y \in \gamma \mathscr{F}_\infty} \bar{c}_2 |x - y|^a = \bar{c}_2 |x|_{\gamma \mathscr{F}_\infty}^a. \tag{B.42}$$

As the set $\mathscr{X}_{\mathrm{f}}$ is compact and $\ell$ is continuous, there exists a $\Gamma \in [0, \infty)$ such that $V_\gamma(x) \leq \Gamma$ for all $x \in \mathscr{X}_{\mathrm{f}}$. Hence, for all $x \in \mathscr{X}_{\mathrm{f}}$ it holds that

$$V_\gamma(x) \leq \max \left\{ 1, \frac{\Gamma}{\bar{c}_2 \epsilon^a} \right\} \bar{c}_2 |x|_{\gamma \mathscr{F}_\infty}^a. \tag{B.43}$$

This completes the proof. □

## B.15. Proof of Lemma 5.2

Define the expressions

$$W_i := \bar{V}^{\mathrm{s}}(x[i] - (A + BK)^i y) + \sum_{j=i}^{N-1} \ell(u[j] - Kx[j]) \tag{B.44}$$

for all $i \in \mathbb{N}$. It follows that

$$W_{i+1} = W_i$$
$$+ \bar{V}^{\mathrm{s}}(x[i+1] - (A + BK)^{i+1} y) - \bar{V}^{\mathrm{s}}(x[i] - (A + BK)^i y) - \ell(u[i] - Kx[i]) \tag{B.45}$$

and

$$W_i = V_\gamma(x) + \bar{V}^{\mathrm{s}} \left( x[i] - (A + BK)^i y \right) - \bar{V}^{\mathrm{s}} \left( x - y \right) - \sum_{j=0}^{i-1} \ell(u[j] - Kx[j]) \tag{B.46}$$

for all $i \in \mathbb{N}$.

Further, by optimality of $(u[0], \ldots, u[N-1])$, it follows that $u[j] - Kx[j] = 0$ for $j \in \{i, \ldots, N-1\}$ if $x[i] \in \bar{\mathscr{X}}_i$ as then $u[j] - Kx[j] = 0$, $j \in \{i, \ldots, N-1\}$, is feasible.

Hence, considering that $y \in \gamma \mathscr{F}_\infty$, if $x[i] - (A + BK)^i y \in \bar{\mathcal{X}}_i \ominus \gamma(A + BK)^i \mathscr{F}_\infty$ for some $i \in \{0, \ldots, N - 1\}$, or if $i \in \{N, \ldots\}$, then $W_i = \bar{V}^s(x[i] - (A + BK)^i y) \leq \bar{c}_2 |x[i] - (A + BK)^i y|^a$. As $\gamma(A + BK)^i \mathscr{F}_\infty \oplus \epsilon \mathscr{B}_{n_x} \subseteq \bar{\mathcal{X}}_i$ for $i \in \{0, \ldots, N - 1\}$, $|x[i] - (A + BK)^i y| \leq \epsilon$ implies $x[i] \in \bar{\mathcal{X}}_i$, such that

$$W_i \leq \underbrace{\max\left\{1, \frac{\bar{\Gamma}}{\bar{c}_2 \epsilon^a}\right\} \bar{c}_2}_{:=\bar{c}_4} |x[i] - (A + BK)^i y|^a \tag{B.47}$$

for all $i \in \mathbb{N}$ and some $\bar{\Gamma} \in [0, \infty)$ (independent of $i$), similar to the proof of Lemma 5.1. Hence, with (5.4b), (B.45) implies

$$\begin{aligned} W_{i+1} &\leq W_i - \bar{c}_3 |x[i] - (A + BK)^i y|^a \\ &\leq W_i - \frac{\bar{c}_3}{\bar{c}_4} W_i \\ &= \lambda W_i \end{aligned} \tag{B.48}$$

for all $i \in \mathbb{N}$, where $\lambda = 1 - \bar{c}_3/\bar{c}_4 \in [0, 1)$. Therefore, it holds that $W_i \leq \lambda^i W_0$ and, hence, $W_0 - W_i \geq (1 - \lambda^i) W_0$. Realizing that $W_0 = V_\gamma(x)$ and that $W_0 - W_i$ is equal to the left-hand side of (5.5) completes the proof. □

## B.16. Proof of Proposition 5.1

For $\tau = 0$ the inequality holds for every $\mu \in [0, 1)$ (using the convention $0^0 = 1$). If $\min\{\nu, \lambda\} = 0$, then $\mu = \max\{\nu, \lambda\} \in [0, 1)$ is a sufficient choice for all $\tau \in \mathbb{N}$. Assume now that $\nu, \lambda \in (0, 1)$. The derivative of the left-hand side of (5.6) (interpreted as a function mapping $(0, \infty)$ to $\mathbb{R}$) with respect to $\tau$ is equal to $\ln(\nu)\nu^\tau(1 - \lambda^\tau) + \ln(\lambda)\lambda^\tau(1 - \nu^\tau)$, which is negative for all $\tau > 0$. Thus, the the left-hand side of (5.6) is strictly monotonically decreasing in $\tau$ and, furthermore, less than 1 for all $\tau \in \{1, \ldots, \}$. Further, it holds that $2 \max\{\nu, \lambda\}^\tau$ is an upper bound for the left-hand side of (5.6). Let now $\bar{\tau} \in \mathbb{N}$ be chosen such that $\bar{\mu} := 2^{\frac{1}{\bar{\tau}}} \max\{\nu, \lambda\} < 1$. It holds that $\mu^\tau \geq 2 \max\{\nu, \lambda\}^\tau$ if $\mu \geq \bar{\mu}$ and $\tau \geq \bar{\tau}$. Define now

$$\mu := \min\{\hat{\mu} \in [\bar{\mu}, \infty) \mid 1 - (1 - \nu^\tau)(1 - \lambda^\tau) \leq \hat{\mu}^\tau, \tau \in \{1, \ldots, \bar{\tau} - 1\}\} \tag{B.49}$$

By the discussion above it holds that $\mu \in [0, 1)$ and $1 - (1 - \nu^\tau)(1 - \lambda^\tau) \leq \mu^\tau$ for all $\tau \in \mathbb{N}$, which is the desired result. □

## B.17. Proof of Theorem 5.3

With the assumptions on $\kappa$, it is ensured that $u_t \in \mathcal{U}$ if the execution of the Algorithm is well defined. In the following, we will show that the prerequisites of Theorem 5.2 are satisfied, which is sufficient to prove the statement.

First, we prove via strong induction on $t$ that $W_t^{\text{et}}$ is finite for all $t \in \mathbb{N}$. Note that if $x_{t^{-1}} \in \mathcal{X}_{\text{f}}$ for all $t^{-1} \in \{0, \ldots, t\}$, then $W_t^{\text{et}}$ is guaranteed to be finite. It holds that $x_0 \in \mathcal{X}_{\text{f}}$ by assumption. Let now $t \in \mathbb{N}$ be arbitrary and assume that $x_{t^{-1}} \in \mathcal{X}_{\text{f}}$ for all $t^{-1} \in \{0, \ldots, t\}$. By the definition of the algorithm, it is either explicitly ensured that $V_\gamma(x_{t+1}) \leq W_t^{\text{et}}$, in which case $x_{t+1} \in \mathcal{X}_{\text{f}}$ as $V_\gamma(x_{t+1})$ will be finite, or it holds that $u_t \in \kappa_{\text{MPC}}(x_t)$, in which case it also holds that $x_{t+1} \in \mathcal{X}_{\text{f}}$. Hence, $W_t^{\text{et}}$ is finite for all $t \in \mathbb{N}$.

Considering the definition of $W_t^{\text{et}}$, whenever $V_\gamma(x_{t+1}) \leq W_t^{\text{et}}$, then (5.7) in Theorem 5.2 is satisfied for some $\tau \in \{1, \ldots, \min\{\theta, t+1\}\}$, some $((x_{t+1-\tau}[i])_{i \in \mathbb{N}}, (u_{t+1-\tau}[i])_{i \in \mathbb{N}}) \in \mathcal{T}(x_{t+1-\tau})$ and some $y_{t+1-\tau} \in \mathcal{Y}^\gamma(x_{t+1-\tau})$. Again, by the definition of the algorithm, it is either explicitly ensured that $V_\gamma(x_{t+1}) \leq W_t^{\text{et}}$, or it holds that $u_t \in \kappa_{\text{MPC}}(x_t)$. In the latter case, using the proof of Lemma 5.1, in particular (B.41), it holds that

$V_\gamma(x_{t+1})$
$$\leq V_\gamma(x_t) - \left( \bar{V}^{\text{s}}(x_t - y_t) - \bar{V}^{\text{s}}((A + BK)(x_t - y_t) + B(u_t - Kx_t)) + \ell(u_t - Kx_t) \right)$$
$$= V_\gamma(x_t) - \left( \bar{V}^{\text{s}}(x_t - y_t) - \bar{V}^{\text{s}}(Ax_t + Bu_t - (A + BK)y_t) + \ell(u_t - Kx_t) \right)$$
$$= V_\gamma(x_t) - \left( \bar{V}^{\text{s}}(x_t - y_t) - \bar{V}^{\text{s}}(x_t[1] - (A + BK)y_t) + \ell(u_t[0] - Kx_t[0]) \right) \quad \text{(B.50)}$$

for some $((x_t[i])_{i \in \mathbb{N}}, (u_t[i])_{i \in \mathbb{N}}) \in \mathcal{T}(x_t)$ and some $y_t \in \mathcal{Y}^\gamma(x_t)$. That is, (5.7) in Theorem 5.2 holds with $\nu = 0$ and $\tau = 1$, and, hence, the inequality is satisfied for all $t \in \mathbb{N}$, thereby completing the proof. $\qquad \square$

# B.18. Proof of Theorem 5.5

Let $u_t = u[t - t_i]$ for all $t \in \{t_i, \ldots, t_{i+1} - 1\}$, $i \in \mathbb{N}$, where $(u[i])_{i \in \{0, \ldots, t_{i+1} - t_i - 1\}} \in \mathcal{D}_{t_{i+1} - t_i}^{\text{L}}(x_{t_i}, W_{i, t_i + 1}^{\text{st}}, \ldots, W_{i, t_{i+1}}^{\text{st}})$ and assume that $W_{i, t_i + 1}^{\text{st}}, \ldots, W_{i, t_{i+1}}^{\text{st}}$ are finite. It immediately follows that $u_t \in \mathcal{U}$ for all $t \in \mathbb{N}$. Further, it holds that

$$x_t \in \{x[t - t_i]\} \oplus \bigoplus_{j=0}^{t - t_i - 1} A^j \mathcal{W} \quad \text{(B.51)}$$

for $t \in \{t_i, \ldots, t_{i+1}\}$, where $x[0] = x_{t_i}$ and $x[k+1] = Ax[k] + Bu[k]$ for all $k \in \{0, \ldots, t - t_i - 1\}$. Therefore, by the definition of $\mathcal{D}_M^{\text{L}}$ and $W_{i,k}^{\text{st}}$, considering in particular (5.11d), the prerequisites of Theorem 5.2 are satisfied and $x_t \in \mathcal{X}_{\text{f}} \subseteq \mathcal{X}$ for all $t \in \mathbb{N}$, such that the quantities $W_{i,k}^{\text{st}}$ are well-defined and finite for all $i \in \mathbb{N}$.

Hence, in order to complete the proof, it is sufficient to establish that $M^\star > 0$ for all $t_i$, $i \in \mathbb{N}$, in step 11 of Algorithm 5. In the following, we show that $\mathcal{D}_1^{\text{L}}(x_{t_i}, W_{i,t_i+1}^{\text{st}}) \neq \emptyset$ for all $i \in \mathbb{N}$ with $t_i \in \mathbb{N}$, based on the assumption that $x_{t_i} \in \mathcal{X}_{\text{f}}$. Let $x \in \mathcal{X}_{\text{f}}$, $((x[i])_{i \in \mathbb{N}}, (u[i])_{i \in \mathbb{N}}) \in \mathcal{T}(x)$, $y \in \gamma \mathcal{F}_\infty$ such that $\bar{V}^{\text{s}}(x - y) = V_\gamma^{\text{s}}(x)$ and let $W \geq V_\gamma(x) - (1 - \nu) \left( \bar{V}^{\text{s}}(x - y) - \bar{V}(x[1] - (A + BK)y) + \ell(u[0] - Kx[0]) \right)$. Following the proof of Lemma 5.1, it holds that $V_\gamma(Ax + Bu[0] + w) \leq W$ for all $w \in \mathcal{W}$, such that $u[0] \in \mathcal{D}_1^{\text{L}}(x, W)$. This completes the proof. $\qquad \square$

# B.19. Proof of Lemma 5.3

Let $((u[j|k])_{j\in\{0,\ldots,N-k-1\}}, (x[j|k])_{j\in\{0,\ldots,N-k\}})_{k\in\{1,\ldots,M\}}$ satisfy the constraints in (5.18). Let further $\bar{x} = x[k] + \bar{w}$ for some $k \in \{1,\ldots,M\}$ and some $\bar{w} \in \bigoplus_{j=0}^{k-1} A^j \mathscr{W}$. Define $u[i|k] := K(A+BK)^{i-(N-k)}x[N-k|k]$ for all $i \in \{N-k,\ldots,N-1\}$ and $\tilde{u}[i] := u[i|k] + K(A+BK)^i\bar{w}$ for $i \in \{0,\ldots,N-1\}$. Define further $\tilde{x}[0] := \bar{x}$ and $\tilde{x}[i+1] = A\tilde{x}[i] + B\tilde{u}[i]$, $i \in \{0,\ldots,N-1\}$, which implies that $\tilde{u}[i] = K\tilde{x}[i]$ for $i \in \{N-k,\ldots,N-1\}$. It holds that

$$\{\tilde{x}[i]\} \oplus \mathscr{F}_i \subseteq \{x[i|k]\} \oplus \mathscr{F}_i \oplus (A+BK)^i \bigoplus_{l=0}^{k-1} A^l \mathscr{W}$$

$$\stackrel{(5.18g)}{\subseteq} \mathscr{X} \tag{B.52}$$

for all $i \in \{0,\ldots,N-1-k\}$. Further, we have

$$\{\tilde{x}[i]\} \oplus \mathscr{F}_i = (A+BK)^{i-(N-k)}\{\tilde{x}[N-k]\} \oplus \mathscr{F}_i$$

$$\subseteq (A+BK)^{i-(N-k)}\left(\{x[N-k|k]\} \oplus (A+BK)^{N-k}\bigoplus_{l=0}^{k-1} A^l \mathscr{W}\right) \oplus \mathscr{F}_i$$

$$\stackrel{(5.18h)}{\subseteq} (A+BK)^{i-(N-k)}\mathscr{X}_T^k \oplus \mathscr{F}_i$$

$$\stackrel{(5.19)}{\subseteq} \mathscr{X} \tag{B.53}$$

for $i \in \{N-k,\ldots,N-1\}$ and, similarly,

$$\tilde{x}[N] = (A+BK)^k \tilde{x}[N-k]$$

$$\stackrel{(5.18h)}{\in} (A+BK)^k \mathscr{X}_T^k$$

$$\stackrel{(5.19)}{\subseteq} \mathscr{X}_T. \tag{B.54}$$

By the same reasoning, it holds that $\{\tilde{u}[i]\} \oplus K\mathscr{F}_i \subseteq \mathscr{U}$ for $i \in \{0,\ldots,N-1\}$. Hence,

$$J(\bar{x}) \le \bar{J}(\bar{x}, (\tilde{u}[0],\ldots,\tilde{u}[N-1]))$$

$$= \sum_{i=0}^{N-1} \ell(\tilde{u}[i] - K\tilde{x}[i])$$

$$= \sum_{j=0}^{N-k-1} \ell(u[j|k] - Kx[j|k]). \tag{B.55}$$

Together with the discussion surrounding (5.14), this inequality implies, via (5.18i), that $V_\gamma^s(\bar{x}) + J(\bar{x}) \le W_k$, which completes the proof. $\qquad\square$

# B.20. Proof of Lemma 5.4

In the following, let $((x[i])_{i\in\mathbb{N}}, (u[i])_{i\in\mathbb{N}}) := ((x_{t+1-\tau}[i])_{i\in\mathbb{N}}, (u_{t+1-\tau}[i])_{i\in\mathbb{N}})$. Under the stated assumptions, there exists an $f_\tau \in \gamma\mathscr{F}_\tau$ such that $x_{t+1} = x[\tau] + f_\tau$. Hence, we have

$$V_\gamma^s(x_{t+1}) = \min_{y\in\gamma\mathscr{F}_\infty} \bar{V}^s(x[\tau] + f_\tau - y)$$

$$\overset{\gamma\mathscr{F}_\infty = \gamma(A+BK)^\tau \mathscr{F}_\infty \oplus \gamma\mathscr{F}_\tau}{\leq} \bar{V}^s(x[\tau] - (A+BK)^\tau y_{t+1-\tau}). \tag{B.56}$$

Further, it holds that

$$\bar{V}^s(x[\tau] - (A+BK)^\tau y_{t+1-\tau}) = \bar{V}^s(x_{t+1-\tau} - y_{t+1-\tau})$$

$$+ \sum_{k=0}^{\tau-1} \bar{V}^s(x[k+1] - (A+BK)^{k+1} y_{t+1-\tau}) - \bar{V}^s(x[k] - (A+BK)^k y_{t+1-\tau}) \tag{B.57}$$

and

$$\bar{V}^s(x[k+1] - (A+BK)^{k+1} y_{t+1-\tau}) - \bar{V}^s(x[k] - (A+BK)^k y_{t+1-\tau})$$

$$= \bar{V}^s\left((A+BK)(x[k] - (A+BK)^k y_{t+1-\tau}) + B(u[k] - Kx[k])\right)$$

$$- \bar{V}^s(x[k] - (A+BK)^k y_{t+1-\tau}). \tag{B.58}$$

From (5.20), we obtain

$$\bar{V}^s((A+BK)z + Bv) - \bar{V}^s(z) - (1-\nu^{\tau+1})\ell(v)$$

$$= -\nu^\tau \underbrace{\left(\bar{V}^s(z) - \bar{V}^s((A+BK)z + Bv) + (1-\nu)\ell(v)\right)}_{\geq 0}$$

$$- (1-\nu^\tau)\left(\bar{V}^s(z) - \bar{V}^s((A+BK)z + Bv) + \ell(v)\right)$$

$$\leq -(1-\nu^\tau)\left(\bar{V}^s(z) - \bar{V}^s((A+BK)z + Bv) + \ell(v)\right) \tag{B.59}$$

for all $z \in \mathbb{R}^{n_x}$ and all $v \in \mathbb{R}^{n_u}$. Thus, collecting (B.57), (B.58), and (B.59), we conclude

$$\bar{V}^s(x[\tau] - (A+BK)^\tau y_{t+1-\tau}) - (1-\nu^{\tau+1})\sum_{k=0}^{\tau-1} \ell(u[k] - Kx[k])$$

$$\overset{(B.57)}{=} \bar{V}^s(x_{t+1-\tau} - y_{t+1-\tau}) + \sum_{k=0}^{\tau-1} \left(\bar{V}^s(x[k+1] - (A+BK)^{k+1} y_{t+1-\tau})\right.$$

$$\left. - \bar{V}^s(x[k] - (A+BK)^k y_{t+1-\tau}) - (1-\nu^{\tau+1})\ell(u[k] - Kx[k])\right)$$

$$\overset{(B.58),\ (B.59)}{\leq} \bar{V}^s(x_{t+1-\tau} - y_{t+1-\tau}) - (1-\nu^\tau)\sum_{k=0}^{\tau-1} \left(\bar{V}^s(x[k] - (A+BK)^k y_{t+1-\tau})\right.$$

$$\left. - \bar{V}^s(x[k+1] - (A+BK)^{k+1} y_{t+1-\tau}) + \ell(u[k] - Kx[k])\right)$$

$$\overset{(B.57)}{=} \bar{V}^{\mathrm{s}}(x_{t+1-\tau} - y_{t+1-\tau}) - (1-\nu^{\tau})\bigg(\bar{V}^{\mathrm{s}}(x_{t+1-\tau} - y_{t+1-\tau})$$

$$- \bar{V}^{\mathrm{s}}(x[\tau] - (A+BK)^{\tau}y_{t+1-\tau}) + \sum_{k=0}^{\tau-1} \ell(u[k] - Kx[k])\bigg). \qquad (B.60)$$

Using $V_{\gamma}(x_{t+1}) = V_{\gamma}^{\mathrm{s}}(x_{t+1}) + J(x_{t+1})$, the assumption that $J(x_{t+1}) \leq J(x_{t+1-\tau}) - (1 - \nu^{\tau+1})\sum_{k=0}^{\tau-1}\ell(u[k] - Kx[k])$, (B.56), and (B.60), we arrive at (5.7) , which completes the proof. $\qquad\square$

## B.21. Proof of Theorem 5.6

With the assumptions on $\kappa$, it holds that $u_t \in \mathcal{U}$ if the execution of the algorithm is well defined. Analogously to the proof of Theorem 5.3, it is ensured that $x_t \in \mathcal{X}_{\mathrm{f}} \subseteq \mathcal{X}$ for all $t \in \mathbb{N}$, due to the finiteness of $J(x_t)$. Further, the algorithm either explicitly or implicitly (if $u_t = \kappa(x_t, 0)$) guarantees that the prerequisites of Lemma 5.4 are satisfied, such that also the stated stability properties of the closed-loop system are ensured. $\quad\square$

## B.22. Proof of Theorem 5.8

Following the reasoning in the proof of Theorem 5.5, it is ensured that $u_t \in \mathcal{U}$ and $x_t \in \mathcal{X}_{\mathrm{f}} \subseteq \mathcal{X}$ for all $t \in \mathbb{N}$, in particular because of constraint (5.22d), as long as $M^{\star} > 0$ in step 9 of Algorithm 8, whenever this step is executed. By the same reasoning, and considering Lemma 5.4, the claimed stability property of the closed-loop system is established.

Analogously to the proof of Theorem 5.5, if $x_t \in \mathcal{X}_{\mathrm{f}}$ and $((x[i])_{i\in\mathbb{N}}, (u[i])_{i\in\mathbb{N}}) \in \mathcal{T}(x_t)$, then $u[0] \in \mathscr{D}_1^{\mathrm{m}}(t, x_t, (\mathbf{x}_{t-1}, \mathbf{u}_{t-1})_{t-1\in\mathscr{R}}, \mathscr{R})$ if $t \in \mathscr{R}$. Considering step 8 of Algorithm 8, $t_i \in \mathscr{R}_i$ is ensured, and by the discussion above $x_{t_i} \in \mathcal{X}_{\mathrm{f}}$ holds, such that $M^{\star} \geq 1$. This completes the proof. $\qquad\square$

## B.23. Proof of Lemma 5.5

It can be readily seen that the modified algorithm ensures (if the optimization problems are feasible at $t_i$ for some arbitrary $i \in \mathbb{N}$) that for all $t \in \{t_i + 1, \ldots, t_{i+1}\}$, it holds that $u_{t-1} \in \mathcal{U}$ and

$$x_t \in \{x[t - t_i]\} \oplus \bigoplus_{j=0}^{t-t_i-1} A^j \mathcal{W}, \qquad (B.61)$$

where

$$x[t - t_i] \in \left\{ \sum_{t^{-1} \in \mathscr{R}} \lambda_{t^{-1}} x_{t^{-1}}[t - t^{-1}] \right\} \oplus \bigoplus_{t^{-1} \in \mathscr{R}} \lambda_{t^{-1}} \bar{\mathscr{F}}_{t-t_i, t-t^{-1}}, \tag{B.62}$$

for some $((x_{t^{-1}}[i])_{i \in \mathbb{N}}, (u_{t^{-1}}[i])_{i \in \mathbb{N}}) \in \mathscr{T}(x_{t^{-1}})$, $\lambda_{t^{-1}} \in [0, \infty)$, $t^{-1} \in \mathscr{R}$, $\sum_{t^{-1} \in \mathscr{R}} \lambda_{t^{-1}} = 1$, and $\mathscr{R} \subseteq \{t - \min\{t, \theta\}, \dots, t - 1\}$. Using the derivations in the proof of Lemma 5.3, we also have $x_t \in \mathscr{X}_f$, implying that the algorithm is well-defined and $x_t \in \mathscr{X}$ for all $t \in \mathbb{N}$. Furthermore, again following the proof of Lemma 5.3, in particular considering (B.55), we obtain

$$J(x_t) \leq \sum_{t^{-1} \in \mathscr{R}} \lambda_{t^{-1}} \left( J(x_{t^{-1}}) - (1 - \nu^{t-t^{-1}+1}) \sum_{l=0}^{t-t^{-1}-1} \ell(u_{t^{-1}}[l] - K x_{t^{-1}}[l]) \right). \tag{B.63}$$

By exploiting the convexity of all involved sets, we have

$$x_t \in \left\{ \sum_{t^{-1} \in \mathscr{R}} \lambda_\tau x_{t^{-1}}[t - t^{-1}] \right\} \oplus \bigoplus_{t^{-1} \in \mathscr{R}} \lambda_{t^{-1}} \bar{\mathscr{F}}_{t-t_i, t-t^{-1}} \oplus \bigoplus_{j=0}^{t-t_i-1} A^j \mathscr{W}$$

$$= \left\{ \sum_{t^{-1} \in \mathscr{R}} \lambda_\tau x_{t^{-1}}[t - t^{-1}] \right\} \oplus \bigoplus_{t^{-1} \in \mathscr{R}} \lambda_{t^{-1}} \left( \bar{\mathscr{F}}_{t-t_i, t-t^{-1}} \oplus \bigoplus_{j=0}^{t-t_i-1} A^j \mathscr{W} \right)$$

$$\subseteq \left\{ \sum_{t^{-1} \in \mathscr{R}} \lambda_\tau x_{t^{-1}}[t - t^{-1}] \right\} \oplus \bigoplus_{t^{-1} \in \mathscr{R}} \lambda_{t^{-1}} \gamma \mathscr{F}_{t-t^{-1}}. \tag{B.64}$$

Here it is important to note that if $\gamma \mathscr{F}_{t-t^{-1}} \ominus \bigoplus_{j=0}^{t-t_i-1} A^j \mathscr{W} = \emptyset$, then, by the definition of $\bar{\mathscr{R}}_k$ in (5.24), it holds that $t^{-1} \notin \mathscr{R}$. Let now $y_{t^{-1}} \in \mathscr{Y}^\gamma(x_{t^{-1}})$ and $f_{t^{-1}} \in \gamma \mathscr{F}_{t-t^{-1}}$ for $t^{-1} \in \mathscr{R}$ such that

$$x_t = \sum_{t^{-1} \in \mathscr{R}} \lambda_{t^{-1}} (x_{t^{-1}}[t - t^{-1}] + f_{t^{-1}}). \tag{B.65}$$

From (B.65) and the convexity of $\bar{V}^s$, which implies the convexity of $V_\gamma^s$, we obtain

$$V_\gamma^s(x_t) \leq \sum_{t^{-1} \in \mathscr{R}} \lambda_{t^{-1}} V_\gamma^s(x_{t^{-1}}[t - t^{-1}] + f_{t^{-1}})$$

$$= \sum_{t^{-1} \in \mathscr{R}} \lambda_{t^{-1}} \min_{y \in \gamma \mathscr{F}_\infty} \bar{V}^s(x_{t^{-1}}[t - t^{-1}] + f_{t^{-1}} - y)$$

$$= \sum_{t^{-1} \in \mathscr{R}} \lambda_{t^{-1}} \min_{y \in (A+BK)^{t-t^{-1}} \gamma \mathscr{F}_\infty \oplus \gamma \mathscr{F}_{t-t^{-1}}} \bar{V}^s(x_{t^{-1}}[t - t^{-1}] + f_{t^{-1}} - y)$$

$$\overset{f_{t^{-1}} \in \gamma \mathscr{F}_{t-t^{-1}}}{\leq} \sum_{t^{-1} \in \mathscr{R}} \lambda_{t^{-1}} \min_{y \in \gamma \mathscr{F}_\infty} \bar{V}^s(x_{t^{-1}}[t - t^{-1}] - (A + BK)^{t-t^{-1}} y)$$

$$\leq \sum_{t^{-1} \in \mathscr{R}} \lambda_{t^{-1}} \bar{V}^s(x_{t^{-1}}[t - t^{-1}] - (A + BK)^{t-t^{-1}} y_{t^{-1}}) \tag{B.66}$$

Hence,

$$
V_\gamma(x_t) = V_\gamma^{\mathrm{s}}(x_t) + J(x_t)
$$

$$
\leq \sum_{t^{-1}\in\mathscr{R}} \lambda_{t^{-1}} \Bigg( \bar{V}^{\mathrm{s}}(x_{t^{-1}}[t-t^{-1}] - (A+BK)^{t-t^{-1}} y_{t^{-1}}) + J(x_{t^{-1}})
$$

$$
- (1-\nu^{t-t^{-1}+1}) \sum_{l=0}^{t-t^{-1}-1} \ell(u_{t^{-1}}[l] - Kx_{t^{-1}}[l]) \Bigg)
$$

$$
\leq \sum_{t^{-1}\in\mathscr{R}} \lambda_{t^{-1}} \Bigg( V_\gamma(x_{t^{-1}}) - (1-\nu^{t-t^{-1}}) \bigg( \bar{V}^{\mathrm{s}}(x_{t^{-1}} - y_{t^{-1}})
$$

$$
- \bar{V}^{\mathrm{s}}(x_{t^{-1}}[t-t^{-1}] - (A+BK)^{t-t^{-1}} y_{t^{-1}}) + \sum_{l=0}^{t-t^{-1}-1} \ell(u_{t^{-1}}[l] - Kx_{t^{-1}}[l]) \bigg) \Bigg).
$$

$$(B.67)$$

where the third line follows by the same arguments as the proof of Lemma 5.4, in particular by using (B.60). Considering that $\mathscr{R} \subseteq \{t - \min\{t,\theta\}, \ldots, t-1\}$ this last inequality implies

$$
V_\gamma(x_t) \leq \max_{t^{-1}\in\{t-\min\{t,\theta\},\ldots,t-1\}} \Bigg( V_\gamma(x_{t^{-1}}) - (1-\nu^{t-t^{-1}}) \bigg( \bar{V}^{\mathrm{s}}(x_{t^{-1}} - y_{t^{-1}})
$$

$$
- \bar{V}^{\mathrm{s}}(x_{t^{-1}}[t-t^{-1}] - (A+BK)^{t-t^{-1}} y_{t^{-1}}) + \sum_{l=0}^{t-t^{-1}-1} \ell(u_{t^{-1}}[l] - Kx_{t^{-1}}[l]) \bigg) \Bigg). \quad (B.68)
$$

With Theorem 5.2 we conclude that the claimed stability properties hold, which completes the proof. $\qquad\square$

# B.24. Proof of Theorem 5.9

By the definition of $\mathscr{D}^\gamma$, it holds that $x[\tau] \in \mathscr{X} \ominus \gamma\mathscr{F}_\tau$, implying that $x_{t+1} \in \mathscr{X}$ for all $t \in \mathbb{N}$. Further, we have $x_0 \in \mathscr{X}_{\mathrm{f}}^\gamma \subseteq \mathscr{X}$ by assumption, proving the first part of the statement.

For the second part, consider that $x_{t+1} \in \{x[\tau]\} \oplus \gamma\mathscr{F}_\tau$ implies $x_{t+1} = x[\tau] + f_\tau$ for some $f_\tau \in \gamma\mathscr{F}_\tau$. Let now $((x[i])_{i\in\mathbb{N}}, (u[i])_{i\in\mathbb{N}}) \in \mathscr{T}^\gamma(x_{t+1-\tau})$ and define $\tilde{x}[0] := x_{t+1}$ and $\tilde{x}[i+1] = A\tilde{x}[i] + B\tilde{u}[i]$ for $i \in \mathbb{N}$, where $\tilde{u}[i] := u[i+\tau] + K(\tilde{x}[i] - x[i+\tau])$. It follows that $\tilde{x}[i] = x[i+\tau] + (A+BK)^i f_\tau$ and, hence,

$$
\{\tilde{x}[i]\} \oplus \gamma\mathscr{F}_i \subseteq \{x[i+\tau]\} \oplus (A+BK)^i \gamma\mathscr{F}_\tau \oplus \gamma\mathscr{F}_\tau
$$

$$
\overset{(2.4)}{=} \{x[i+\tau]\} \oplus \gamma\mathscr{F}_{i+\tau}
$$

$$
\subseteq \mathscr{X} \qquad\qquad (B.69)
$$

for $i \in \{0, \ldots, N-1\}$. In particular, the last line follows for $i \in \{0, \ldots, N-1-\tau\}$ directly from the definition of $\mathscr{D}^\gamma$ and for $i \in \{N-\tau, \ldots, N-1\}$ from the fact that $x[N] \in \mathscr{X}_T^\gamma$, $x[i] = (A + BK)^{i-N} x[N]$ for $i \in \{N, \ldots\}$, and Lemma A.8. Similarly, it holds that $\{\tilde{u}[i]\} \oplus K\gamma\mathscr{F}_i \subseteq \mathscr{U}$, $i \in \{0, \ldots, N-1\}$ and $\tilde{x}[N] \in \mathscr{X}_T^\gamma$. Hence, $\tilde{\mathbf{u}} := (\tilde{u}[0], \ldots, \tilde{u}[N-1]) \in \mathscr{D}^\gamma(x_{t+1})$ and

$$J^\gamma(x_{t+1}) \leq \bar{J}^\gamma(x_{t+1}, \tilde{\mathbf{u}}) = J^\gamma(x_{t+1-\tau}) - \sum_{k=0}^{\tau-1} \ell(u[k] - Kx[k]), \tag{B.70}$$

analogously[2] to the result in [Chisci et al., 2001].

Applying now first Lemma 5.4 with $\nu = 0$ and $J$ replaced[3] with $J^\gamma$ and then Theorem 5.2, where $V_\gamma(x)$ is replaced with $V_\gamma^s(x) + J^\gamma(x)$, we obtain the claimed stability result, thereby completing the proof. $\qquad\square$

# B.25. Proof of Lemma 5.6

The proof mostly follows the same arguments as the proof of Theorem 5.5.

It can readily be seen that the modified algorithm ensures (if the optimization problems are feasible at $t_i$ for some arbitrary $i \in \mathbb{N}$) that for all $t \in \{t_i + 1, \ldots, t_{i+1}\}$, it holds that $u_{t-1} \in \mathscr{U}$ and

$$x_t \in \{x[t - t_i]\} \oplus \bigoplus_{j=0}^{t-t_i-1} A^j \mathscr{W}, \tag{B.71}$$

where

$$x[t - t_i] \in \left\{ \sum_{t^{-1} \in \mathscr{R}} \lambda_{t-1} x_{t-1}[t - t^{-1}] \right\} \oplus \bigoplus_{t^{-1} \in \mathscr{R}} \lambda_{t-1} \bar{\mathscr{F}}_{t-t_i, t-t^{-1}}, \tag{B.72}$$

for some $((x_{t-1}[i])_{i \in \mathbb{N}}, (u_{t-1}[i])_{i \in \mathbb{N}}) \in \mathscr{T}^\gamma(x_{t-1})$, $\lambda_{t-1} \in [0, \infty)$, $t^{-1} \in \mathscr{R}$, $\sum_{t^{-1} \in \mathscr{R}} \lambda_{t-1} = 1$, and $\mathscr{R} \subseteq \{t - \min\{t, \theta\}, \ldots, t-1\}$. In the following, we will show that this implies $x_t \in \mathscr{X}_f^\gamma$ (implying that the algorithm is well-defined and $x_t \in \mathscr{X}$ for all $t \in \mathbb{N}$) and, furthermore, that

$$V_\gamma^\gamma(x_t) \leq \max_{t^{-1} \in \{t - \min\{t, \theta\}, \ldots, t-1\}} \left( V_\gamma^\gamma(x_{t-1}) - \left( \bar{V}^s(x_{t-1} - y_{t-1}) \right.\right.$$
$$\left.\left. - \bar{V}^s(x_{t-1}[t - t^{-1}]) - (A + BK)^{t-t^{-1}} y_{t-1}) + \sum_{k=0}^{t-t^{-1}-1-k} \ell(u_{t-1}[k] - Kx_{t-1}[k]) \right) \right), \tag{B.73}$$

---

[2] As we have made no assumptions on the actual input $u_j$, applied to the system for $j \in \{t+1-\tau, \ldots, t\}$, we cannot directly follow the arguments in [Chisci et al., 2001], which are based on observing that (B.70) implies $\lim_{k \to \infty} u_t - Kx_t = 0$ if $u_t = u_t^\star[0]$ for all $t \in \mathbb{N}$, where $(\mathbf{x}_t, (u_t^\star[i])_{i \in \mathbb{N}}) \in \mathscr{T}(x_t)$.

[3] This replacement is equivalent to replacing $\mathscr{W}$ with $\gamma\mathscr{W}$ and $\gamma$ with 1.

where $y_{t^{-1}} \in \mathscr{Y}^\gamma(x_{t^{-1}})$, $t^{-1} \in \mathscr{R}$ and $V_\gamma^\gamma(x) := V_\gamma^s(x) + J^\gamma(x)$ for all $x \in \mathscr{X}_f^\gamma$. As $t \in \{t_i, \dots, t_{i+1-1}\}$ and $i \in \mathbb{N}$ were arbitrary, this yields the claimed stability properties of the closed-loop system via Theorem 5.2.

By exploiting the convexity of all sets involved, we have

$$
x_t \in \left\{ \sum_{t^{-1} \in \mathscr{R}} \lambda_\tau x_{t^{-1}}[t - t^{-1}] \right\} \oplus \bigoplus_{t^{-1} \in \mathscr{R}} \lambda_{t^{-1}} \bar{\mathscr{F}}_{t-t_i, t-t^{-1}} \oplus \bigoplus_{j=0}^{t-t_i-1} A^j \mathscr{W}
$$

$$
= \left\{ \sum_{t^{-1} \in \mathscr{R}} \lambda_\tau x_{t^{-1}}[t - t^{-1}] \right\} \oplus \bigoplus_{t^{-1} \in \mathscr{R}} \lambda_{t^{-1}} \left( \bar{\mathscr{F}}_{t-t_i, t-t^{-1}} \oplus \bigoplus_{j=0}^{t-t_i-1} A^j \mathscr{W} \right)
$$

$$
\subseteq \left\{ \sum_{t^{-1} \in \mathscr{R}} \lambda_\tau x_{t^{-1}}[t - t^{-1}] \right\} \oplus \bigoplus_{t^{-1} \in \mathscr{R}} \lambda_{t^{-1}} \gamma \mathscr{F}_{t-t^{-1}}, \tag{B.74}
$$

where we emphasize that $t^{-1} \notin \mathscr{R}$ if $\gamma \mathscr{F}_{t-t^{-1}} \ominus \bigoplus_{j=0}^{t-t_i-1} A^j \mathscr{W} = \emptyset$, by the definition of $\bar{\mathscr{R}}_k$ in (5.37). Let now $f_{t^{-1}} \in \gamma \mathscr{F}_{t-t^{-1}}$ for $t^{-1} \in \mathscr{R}$ such that

$$
x_t = \sum_{t^{-1} \in \mathscr{R}} \lambda_{t^{-1}} (x_{t^{-1}}[t - t^{-1}] + f_{t^{-1}}) \tag{B.75}
$$

and define

$$
\tilde{x}[i] := \sum_{t^{-1} \in \mathscr{R}} \lambda_{t^{-1}} \left( x_{t^{-1}}[t - t^{-1} + i] + (A + BK)^i f_{t^{-1}} \right) \tag{B.76}
$$

and

$$
\tilde{u}[i] := \sum_{t^{-1} \in \mathscr{R}} \lambda_{t^{-1}} \left( u_{t^{-1}}[t - t^{-1} + i] + K(A + BK)^i f_{t^{-1}} \right) \tag{B.77}
$$

for $i \in \mathbb{N}$. It holds that $\tilde{x}[0] = x_t$ and $\tilde{x}[i + 1] = A\tilde{x}[i] + B\tilde{u}[i]$ for $i \in \mathbb{N}$, such that $\tilde{\mathbf{u}} := (\tilde{u}[0], \dots, \tilde{u}[N-1])$ satisfies (5.28a) and (5.28b) in the definition of $\mathscr{D}^\gamma(x_t)$. Further, exploiting again convexity, it holds that

$$
\{\tilde{x}[i]\} \oplus \gamma \mathscr{F}_i = \{\tilde{x}[i]\} \oplus \bigoplus_{t^{-1} \in \mathscr{R}} \lambda_{t^{-1}} \gamma \mathscr{F}_i
$$

$$
= \bigoplus_{t^{-1} \in \mathscr{R}} \lambda_{t^{-1}} \left( \{x_{t^{-1}}[t - t^{-1} + i] + (A + BK)^i f_{t^{-1}}\} \oplus \gamma \mathscr{F}_i \right)
$$

$$
\subseteq \bigoplus_{t^{-1} \in \mathscr{R}} \lambda_{t^{-1}} \left( \{x_{t^{-1}}[t - t^{-1} + i]\} \oplus \gamma(A + BK)^i \mathscr{F}_{t-t^{-1}} \oplus \gamma \mathscr{F}_i \right)
$$

$$
= \bigoplus_{t^{-1} \in \mathscr{R}} \lambda_{t^{-1}} \left( \{x_{t^{-1}}[t - t^{-1} + i]\} \oplus \gamma \mathscr{F}_{t-t^{-1}+i} \right)
$$

$$
\subseteq \bigoplus_{t^{-1} \in \mathscr{R}} \lambda_{t^{-1}} \mathscr{X}
$$

$$
= \mathscr{X}, \tag{B.78}
$$

for all $i \in \mathbb{N}$, where the second-to-last line follows from the definition of $\mathscr{D}^\gamma$, the assumptions on the terminal set $\mathcal{X}_T^\gamma$, and Lemma A.8. By similar reasoning, it follows that $\{\tilde{u}[i]\} \oplus \gamma K \mathscr{F}_i \subseteq \mathcal{U}$, $i \in \mathbb{N}$, and $\tilde{x}[N] \in \mathcal{X}_T^\gamma$. Hence, it holds that $\tilde{\mathbf{u}} \in \mathscr{D}^\gamma(x_t)$ and $x_t \in \mathcal{X}_f^\gamma$, completing the first part of the proof.

It remains to show the claimed decrease of the Lyapunov function. Using the convexity of $\ell$, we have

$$
\begin{aligned}
J^\gamma(x_t) &\leq \bar{J}(x_t, \tilde{\mathbf{u}}) \\
&= \sum_{i=0}^{N-1} \ell\left(\tilde{u}[i] - K\tilde{x}[i]\right) \\
&= \sum_{i=0}^{N-1} \ell\left(\sum_{t^{-1} \in \mathscr{R}} \lambda_{t^{-1}}\left(u_{t^{-1}}[t - t^{-1} + i] - K x_{t^{-1}}[t - t^{-1} + i]\right)\right) \\
&\leq \sum_{t^{-1} \in \mathscr{R}} \lambda_{t^{-1}} \sum_{i=0}^{N-1} \ell\left(u_{t^{-1}}[t - t^{-1} + i] - K x_{t^{-1}}[t - t^{-1} + i]\right) \\
&= \sum_{t^{-1} \in \mathscr{R}} \lambda_{t^{-1}}\left(J^\gamma(x_{t^{-1}}) - \sum_{i=0}^{t - t^{-1} - 1} \ell\left(u_{t^{-1}}[i] - K x_{t^{-1}}[i]\right)\right).
\end{aligned}
\tag{B.79}
$$

From (B.75) and the convexity of $\bar{V}^{\mathrm{s}}$, which implies the convexity of $V_\gamma^{\mathrm{s}}$, we obtain

$$
\begin{aligned}
V_\gamma^{\mathrm{s}}(x_t) &\leq \sum_{t^{-1} \in \mathscr{R}} \lambda_{t^{-1}} V_\gamma^{\mathrm{s}}(x_{t^{-1}}[t - t^{-1}] + f_{t^{-1}}) \\
&= \sum_{t^{-1} \in \mathscr{R}} \lambda_{t^{-1}} \min_{y \in \gamma \mathscr{F}_\infty} \bar{V}^{\mathrm{s}}(x_{t^{-1}}[t - t^{-1}] + f_{t^{-1}} - y) \\
&= \sum_{t^{-1} \in \mathscr{R}} \lambda_{t^{-1}} \min_{y \in (A+BK)^{t-t^{-1}} \gamma \mathscr{F}_\infty \oplus \gamma \mathscr{F}_{t-t^{-1}}} \bar{V}^{\mathrm{s}}(x_{t^{-1}}[t - t^{-1}] + f_{t^{-1}} - y) \\
&\stackrel{f_{t^{-1}} \in \gamma \mathscr{F}_{t-t^{-1}}}{\leq} \sum_{t^{-1} \in \mathscr{R}} \lambda_{t^{-1}} \min_{y \in \gamma \mathscr{F}_\infty} \bar{V}^{\mathrm{s}}(x_{t^{-1}}[t - t^{-1}] - (A + BK)^{t-t^{-1}} y) \\
&\leq \sum_{t^{-1} \in \mathscr{R}} \lambda_{t^{-1}} \bar{V}^{\mathrm{s}}(x_{t^{-1}}[t - t^{-1}] - (A + BK)^{t-t^{-1}} y_{t^{-1}}).
\end{aligned}
\tag{B.80}
$$

Hence,

$$
\begin{aligned}
V_\gamma^\gamma(x_t) &= V_\gamma^{\mathrm{s}}(x_t) + J^\gamma(x_t) \\
&\leq \sum_{t^{-1} \in \mathscr{R}} \lambda_{t^{-1}}\left(\bar{V}^{\mathrm{s}}(x_{t^{-1}}[t - t^{-1}] - (A + BK)^{t-t^{-1}} y_{t^{-1}}) + J^\gamma(x_{t^{-1}}) \right. \\
&\qquad\qquad\qquad\qquad\qquad \left. - \sum_{l=0}^{t - t^{-1} - 1} \ell\left(u_{t^{-1}}[l] - K x_{t^{-1}}[l]\right)\right)
\end{aligned}
$$

$$= \sum_{t^{-1} \in \mathcal{R}} \lambda_{t^{-1}} \left( V_\gamma^\gamma(x_{t^{-1}}) - \left( \bar{V}^s(x_{t^{-1}} - y_{t^{-1}}) \right. \right.$$

$$\left. \left. - \bar{V}^s(x_{t^{-1}}[t - t^{-1}] - (A + BK)^{t-t^{-1}} y_{t^{-1}}) + \sum_{l=0}^{t-t^{-1}-1} \ell(u_{t^{-1}}[l] - Kx_{t^{-1}}[l]) \right) \right)$$

$$\leq \max_{t^{-1} \in \mathcal{R}} \left( V_\gamma^\gamma(x_{t^{-1}}) - \left( \bar{V}^s(x_{t^{-1}} - y_{t^{-1}}) \right. \right.$$

$$\left. \left. - \bar{V}^s(x_{t^{-1}}[t - t^{-1}] - (A + BK)^{t-t^{-1}} y_{t^{-1}}) + \sum_{l=0}^{t-t^{-1}-1} \ell(u_{t^{-1}}[l] - Kx_{t^{-1}}[l]) \right) \right) \tag{B.81}$$

Considering that $\mathcal{R} \subseteq \{t - \min\{t, \theta\}, \ldots, t - 1\}$ this last inequality implies (B.73), thereby completing the proof. $\qquad \square$

## B.26. Proof of Lemma 5.7

Following along the lines of the proof of Theorem 3.5, it holds that

$$\left\{ A^j x_{t_i} + \sum_{k=0}^{j-1} A^{j-1-k}(Bu[k] + w_k) \ \middle| \ w_k \in \mathcal{W}, k \in \{0, \ldots, j-1\}, \right.$$

$$A^k x_{t_i} + \sum_{l=0}^{k-1} A^{k-1-l}(Bu[l] + w_l) \in \{x[k]\} \oplus \mathcal{T}_k,$$

$$\left. k \in \{0, \ldots, j-1\} \right\}$$

$$= \{x[j]\} \oplus \mathcal{H}_j \tag{B.82}$$

for any $x_{t_i} \in \mathcal{X}_f^\gamma$ and all $j \in \mathbb{N}$, where $((x[k])_{k \in \mathbb{N}}, (u[k])_{k \in \mathbb{N}}) \in \mathcal{T}(x_{t_i})$ and $\mathcal{H}_j$ is defined as in Theorem 3.5.

Hence, if $\gamma = \min\{c \in [1, \infty) \mid \mathcal{H}_\tau \subseteq c\mathcal{F}_\tau, \ \tau \in \mathbb{N}\}$, then it holds that $x_t \in \{x[t - t_i]\} \oplus \gamma\mathcal{F}_{t-t_i}$ for all $t \in \{t_i + 1, \ldots, t_{i+1}\}$, $i \in \mathbb{N}$ for the closed-loop system defined by Algorithm 12. This completes the proof. $\qquad \square$

## B.27. Proof of Theorem 6.1

Assume that $((\hat{x}[i])_{i \in \mathbb{N}}, (u[i])_{i \in \mathbb{N}}) \in \mathcal{T}_t(\hat{x}_t)$ and $u_t = u[0]$. We have $\hat{x}_{t+1} = A\hat{x}_t + Bu_t + L\tilde{y}_{t+1} = \hat{x}[1] + L\tilde{y}_{t+1}$. Define $\hat{x}'[i] := \hat{x}[i+1] + (A+BK)^i L\tilde{y}_{t+1}$ and $u'[i] := u[i+1] + K(A + BK)^i L\tilde{y}_{t+1}$, for $i \in \mathbb{N}$. It follows immediately that $((\hat{x}'[0], \ldots, \hat{x}'[N]), (u'[0], \ldots, u'[N - 1]))$ satisfies constraints (6.8a) and (6.8b) for $\hat{x} = \hat{x}_{t+1}$.

Further, for all $i \in \{0, \ldots, N-2\}$ it holds that

$$
\left\{ \begin{bmatrix} \hat{x}'[i] \\ u'[i] \end{bmatrix} \right\} \oplus \begin{bmatrix} I & I \\ 0 & K \end{bmatrix} \hat{\mathcal{X}}[i|t+1]
$$

$$
= \left\{ \begin{bmatrix} \hat{x}[i+1] \\ u[i+1] \end{bmatrix} \right\} \oplus \begin{bmatrix} I & I \\ 0 & K \end{bmatrix} \left( \hat{\mathcal{X}}[i|t+1] \oplus \left\{ \begin{bmatrix} 0 \\ I \end{bmatrix} (A+BK)^i L\tilde{y}_{t+1} \right\} \right)
$$

$$
\overset{\text{Assumption 6.1(ii)}}{\subseteq} \left\{ \begin{bmatrix} \hat{x}[i+1] \\ u[i+1] \end{bmatrix} \right\} \oplus \begin{bmatrix} I & I \\ 0 & K \end{bmatrix} \hat{\mathcal{X}}[i+1|t]
$$

$$
\subseteq \mathcal{X} \times \mathcal{U}, \tag{B.83}
$$

last inequality readily following from the assumption that $(u[0], \ldots, u[N-1])) \in \mathcal{D}_t(\hat{x}_t)$. Further, using $u[N] = K\hat{x}[N]$, we have

$$
\left\{ \begin{bmatrix} \hat{x}'[N-1] \\ u'[N-1] \end{bmatrix} \right\} \oplus \begin{bmatrix} I & I \\ 0 & K \end{bmatrix} \hat{\mathcal{X}}[N-1|t+1]
$$

$$
= \left\{ \begin{bmatrix} I \\ K \end{bmatrix} \hat{x}[N] \right\} \oplus \begin{bmatrix} I & I \\ 0 & K \end{bmatrix} \left( \hat{\mathcal{X}}[N-1|t+1] \oplus \left\{ \begin{bmatrix} 0 \\ I \end{bmatrix} (A+BK)^{N-1} L\tilde{y}_{t+1} \right\} \right)
$$

$$
\overset{\hat{x}[N] \in \hat{\mathcal{X}}_{\mathrm{T},t}}{\subseteq} \begin{bmatrix} I \\ K \end{bmatrix} \hat{\mathcal{X}}_{\mathrm{T},t} \oplus \begin{bmatrix} I & I \\ 0 & K \end{bmatrix} \hat{\mathcal{X}}[N|t]
$$

$$
\overset{\text{Assumption 6.1(iv)}}{\subseteq} \mathcal{X} \times \mathcal{U}. \tag{B.84}
$$

From (B.83) and (B.84) it follows that (6.8c) is satisfied at time $t+1$. Consider finally that $\hat{x}[N+1] \in (A+BK)\hat{\mathcal{X}}_{\mathrm{T},t}$ and $\hat{x}'[N] = \hat{x}[N+1] + (A+BK)^N L\tilde{y}_{t+1}$. Assumption 6.1(iv) then implies that $\hat{x}'[N] \in \hat{\mathcal{X}}_{\mathrm{T},t+1}$, such that (6.8d) is satisfied. Hence, $((\hat{x}'[0], \ldots, \hat{x}'[N]), (u'[0], \ldots, u'[N-1]))$ satisfies the constraints in (6.8), implying that $\hat{x}_{t+1} \in \hat{\mathcal{X}}_{\mathrm{f},t+1}$. This completes the proof.

$\square$

# B.28. Proof of Lemma 6.2

The proof largely follows the arguments used to prove Lemma 5.1 and Theorem 5.1. Define the functions $V_t : \mathbb{R}^{n_x} \times \hat{\mathcal{X}}_{\mathrm{f},t} \to \mathbb{R}$, $t \in \mathbb{N}$, by

$$
V_t(\tilde{x}, \hat{x}) := \min_{f \in \hat{\mathscr{F}}_\infty} \bar{V}^{\mathrm{s}}((\tilde{x}, \hat{x}) - f) + J_t(\hat{x}). \tag{B.85}
$$

For all $t \in \mathbb{N}$ it holds that $V_t(\tilde{x}, \hat{x}) \geq \bar{c}_1 |(\tilde{x}, \hat{x})|_{\hat{\mathscr{F}}_\infty}^a$ for all $(\tilde{x}, \hat{x}) \in \mathbb{R}^{n_x} \times \hat{\mathcal{X}}_{\mathrm{f},t}$ and $\min_{f \in \hat{\mathscr{F}}_\infty} \bar{V}^{\mathrm{s}}((\tilde{x}, \hat{x}) - f) \leq \bar{c}_2 |(\tilde{x}, \hat{x})|_{\hat{\mathscr{F}}_\infty}^a$ for all $(\tilde{x}, \hat{x}) \in \mathbb{R}^{2n_x}$. Further, it holds that $J_t(\hat{x}) = 0$ for all $\hat{x} \in \bar{\mathcal{X}}_0$ and, by the boundedness of $\mathcal{X}$, $\mathcal{U}$, and $\hat{\mathcal{X}}[i|t]$, that there exists a $\Gamma \in [0, \infty)$ such that $J_t(\hat{x}) \leq \Gamma$ for all $\hat{x} \in \hat{\mathcal{X}}_{\mathrm{f},t}$, for all all $t \in \mathbb{N}$. Since

$\hat{\mathscr{F}}_{\infty} \oplus \epsilon \mathscr{B}_{2n_x} \subseteq \mathbb{R}^{n_x} \times \hat{\mathcal{X}}_0$, this implies that $J_t(\hat{x}) \leq (\Gamma/\epsilon)|(\tilde{x}, \hat{x})|_{\hat{\mathscr{F}}_{\infty}}$ for all $t \in \mathbb{N}$ and all $(\tilde{x}, \hat{x}) \in \mathbb{R}^{n_x} \times \hat{\mathcal{X}}_{\mathrm{f},t}$. Hence, for all $t \in \mathbb{N}$ and all $(\tilde{x}, \hat{x}) \in \mathbb{R}^{n_x} \times \hat{\mathcal{X}}_{\mathrm{f},t}$ it holds that $V_t(\tilde{x}, \hat{x}) \leq \bar{c}_4|(\tilde{x}, \hat{x})|^a_{\hat{\mathscr{F}}_{\infty}}$, with $\bar{c}_4 := \bar{c}_2 + \Gamma/\epsilon$. Finally, as remarked in the proof of Corollary 6.1, it holds that $J_{t+1}(\hat{x}_{t+1}) \leq J_t(\hat{x}_t) - \ell(u_t - K\hat{x}_t)$ for all $t \in \mathbb{N}$ in the closed-loop system. Together with (6.17b), this implies $V_{t+1}(\tilde{x}_{t+1}, \hat{x}_{t+1}) \leq V_t(\tilde{x}_t, \hat{x}_t) - \bar{c}_3|(\tilde{x}_t, \hat{x}_t)|^a_{\hat{\mathscr{F}}_{\infty}}$ for all $t \in \mathbb{N}$, using the same derivations used to obtain (B.41) in the proof of Lemma 5.1. Hence, in the closed-loop system it holds that

$$|(\tilde{x}_t, \hat{x}_t)|_{\hat{\mathscr{F}}_{\infty}} \leq c\mu^{t-t_0}|(\tilde{x}_{t_0}, \hat{x}_{t_0})|_{\hat{\mathscr{F}}_{\infty}} \tag{B.86}$$

for all $t \in \mathbb{N}$ and $t_0 \in \{0, \ldots, t\}$ if $(\tilde{x}_0, \hat{x}_0) \in [I\ 0]\hat{\mathcal{X}}[0|0] \times \hat{\mathcal{X}}_{\mathrm{f},0}$, where $c = (\bar{c}_4/\bar{c}_1)^{1/a}$ and $\mu = (1 - \bar{c}_3/\bar{c}_1)^{1/a}$. This completes the proof. $\square$

## B.29. Proof of Lemma 6.3

Assumption 6.1(i) holds by the definition of the sets.

Assumption 6.1(ii) is shown in three steps: First, we show that

$$\hat{\mathcal{X}}^{\mathrm{I}}[i-1|t+1] \oplus \left\{ \begin{bmatrix} 0 \\ I \end{bmatrix} (A + BK)^{i-1}L\tilde{y}_{t+1} \right\} \subseteq \hat{\mathcal{X}}^{\mathrm{I}}[i|t] \tag{B.87}$$

for all $t \in \mathbb{N}$ and $i \in \{1, \ldots, \hat{M}\}$. Second, we show that

$$\hat{\mathcal{X}}^{\mathrm{II}}[i-1|t+1] \oplus \left\{ \begin{bmatrix} 0 \\ I \end{bmatrix} (A + BK)^{i-1}L\tilde{y}_{t+1} \right\} \subseteq \hat{\mathcal{X}}^{\mathrm{II}}[i|t] \tag{B.88}$$

for all $t \in \mathbb{N}$ and $i \in \{\hat{M}+1, \ldots, N\}$, and third, we recall the fact that

$$\hat{\mathcal{X}}^{\mathrm{I}}[\hat{M}|t] \subseteq \hat{\mathcal{X}}^{\mathrm{II}}[\hat{M}|t] \tag{B.89}$$

for all $t \in \mathbb{N}$ from the discussion leading up to Lemma 6.3. The combination of these three claims then readily yields the result stated in the lemma.

For the first claim, considering the definition of $\hat{A}$, $\hat{\mathscr{F}}_i$, and $\mathscr{E}^j_t$, it holds that

$$\hat{\mathcal{X}}^{\mathrm{I}}[i-1|t+1] = \left\{ \hat{A}^{i-1} \begin{bmatrix} \tilde{x} \\ 0 \end{bmatrix} + \sum_{k=0}^{i-2} \hat{A}^{i-2-k} \begin{bmatrix} I + LC & L \\ -LC & -L \end{bmatrix} \begin{bmatrix} w_{t+1+k} \\ v_{t+2+k} \end{bmatrix} \right|$$

$$(w_{t+1+k}, v_{t+2+k}) \in \mathscr{W} \times \mathscr{V},\ k \in \{0, \ldots, i-2\},\ \tilde{x} \in \mathscr{E}^{\min\{t+1, M-i+1\}}_{t+1} \Big\}$$

$$
\begin{aligned}
&= \left\{ \hat{A}^{i-1} \begin{bmatrix} (A+LCA)\tilde{x} + (I+LC)w_t + Lv_{t+1} \\ 0 \end{bmatrix} \right.\\
&\qquad + \sum_{k=0}^{i-2} \hat{A}^{i-2-k} \begin{bmatrix} I+LC & L \\ -LC & -L \end{bmatrix} \begin{bmatrix} w_{t+1+k} \\ v_{t+2+k} \end{bmatrix} \Bigg| \\
&\qquad (w_{t+1+k}, v_{t+2+k}) \in \mathscr{W} \times \mathscr{V}, k \in \{-1,0,\dots,i-2\}, \ \tilde{x} \in \mathscr{E}_t^{\min\{t,M-i\}}, \\
&\qquad\qquad\qquad \left. \tilde{y}_{t+1} = -CA\tilde{x} - Cw_t - v_{t+1} \right\}
\end{aligned}
$$

$$
\begin{aligned}
&= \left\{ \hat{A}^i \begin{bmatrix} \tilde{x} \\ 0 \end{bmatrix} + \sum_{k=0}^{i-1} \hat{A}^{i-1-k} \begin{bmatrix} I+LC & L \\ -LC & -L \end{bmatrix} \begin{bmatrix} w_{t+k} \\ v_{t+1+k} \end{bmatrix} \right.\\
&\qquad - \hat{A}^{i-1} \begin{bmatrix} 0 & 0 \\ -LC & -L \end{bmatrix} \begin{bmatrix} w_t \\ v_{t+1} \end{bmatrix} - \hat{A}^{i-1} \begin{bmatrix} 0 \\ -LCA\tilde{x} \end{bmatrix} \Bigg| \\
&\qquad (w_{t+k}, v_{t+1+k}) \in \mathscr{W} \times \mathscr{V}, \ k \in \{0,\dots,i-1\}, \ \tilde{x} \in \mathscr{E}_t^{\min\{t,M-i\}}, \\
&\qquad\qquad\qquad \left. \tilde{y}_{t+1} = -CA\tilde{x} - Cw_t - v_{t+1} \right\}
\end{aligned}
$$

$$
\begin{aligned}
&= \left\{ \hat{A}^i \begin{bmatrix} \tilde{x} \\ 0 \end{bmatrix} + \sum_{k=0}^{i-1} \hat{A}^{i-1-k} \begin{bmatrix} I+LC & L \\ -LC & -L \end{bmatrix} \begin{bmatrix} w_{t+k} \\ v_{t+1+k} \end{bmatrix} \right.\\
&\qquad - \begin{bmatrix} 0 \\ I \end{bmatrix} (A+BK)^{i-1} L(-CA\tilde{x} - Cw_t - v_{t+1}) \Bigg| \\
&\qquad (w_{t+k}, v_{t+1+k}) \in \mathscr{W} \times \mathscr{V}, \ k \in \{0,\dots,i-1\}, \ \tilde{x} \in \mathscr{E}_t^{\min\{t,M-i\}}, \\
&\qquad\qquad\qquad \left. \tilde{y}_{t+1} = -CA\tilde{x} - Cw_t - v_{t+1} \right\}
\end{aligned}
$$

$$
\begin{aligned}
&\subseteq \left\{ \hat{A}^i \begin{bmatrix} \tilde{x} \\ 0 \end{bmatrix} + \sum_{k=0}^{i-1} \hat{A}^{i-1-k} \begin{bmatrix} I+LC & L \\ -LC & -L \end{bmatrix} \begin{bmatrix} w_{t+k} \\ v_{t+1+k} \end{bmatrix} - \begin{bmatrix} 0 \\ I \end{bmatrix} (A+BK)^{i-1} L\tilde{y}_{t+1} \right. \Bigg| \\
&\qquad (w_{t+k}, v_{t+1+k}) \in \mathscr{W} \times \mathscr{V}, \ k \in \{0,\dots,i-1\}, \ \tilde{x} \in \mathscr{E}_t^{\min\{t,M-i\}} \Bigg\}
\end{aligned}
$$

$$
= \hat{\mathscr{X}}^{\mathrm{I}}[i|t] \oplus (-1) \left\{ \begin{bmatrix} 0 \\ I \end{bmatrix} (A+BK)^{i-1} L\tilde{y}_{t+1} \right\}, \tag{B.90}
$$

which completes the proof for the first statement.

For the second claim, consider that by definition

$$\mathscr{G}[i-1|t+1] \oplus (A+BK)^{i-1}\Big((-LCA)\hat{\mathscr{E}}_t \oplus (-LC)\mathscr{W} \oplus (-L)\mathscr{V}\Big) = \mathscr{G}[i|t] \quad \text{(B.91)}$$

and that $\tilde{y}_{t+1} \in (-CA)\hat{\mathscr{E}}_t \oplus (-C)\mathscr{W} \oplus (-)\mathscr{V}$. This readily yields the second statement, thereby completing the proof. $\qquad\square$

## B.30. Proof of Lemma 6.4

Under the stated assumptions, there exists an $\hat{f}_\gamma \in \gamma\hat{\mathscr{F}}_\infty$ such that

$$z_{t+1} = \begin{bmatrix} 0 \\ \hat{x}[\tau] \end{bmatrix} + \hat{A}^\tau \begin{bmatrix} \tilde{x}_{t+1-\tau} \\ 0 \end{bmatrix} + \hat{f}_\gamma. \tag{B.92}$$

Further, it holds that

$$\begin{bmatrix} 0 \\ \hat{x}[\tau] \end{bmatrix} = \hat{A}^\tau \begin{bmatrix} 0 \\ \hat{x}_{t+1-\tau} \end{bmatrix} + \sum_{j=0}^{\tau-1} \hat{A}^{\tau-1-j} \begin{bmatrix} 0 \\ B \end{bmatrix} (u[j] - K\hat{x}[j]). \tag{B.93}$$

Hence,

$$V_\gamma^{\mathrm{s}}(z_{t+1}) = \min_{\substack{z' \in \gamma\hat{A}^\tau\hat{\mathscr{F}}_\infty \oplus \gamma\hat{\mathscr{F}}_\tau \\ z'_{t+1-\tau} \in \gamma\hat{\mathscr{F}}_\infty}} \bar{V}^{\mathrm{s}}\left( \hat{A}^\tau \begin{bmatrix} \tilde{x}_{t+1-\tau} \\ \hat{x}_{t+1-\tau} \end{bmatrix} + \hat{f}_\tau - z' + \sum_{j=0}^{\tau-1} \hat{A}^{\tau-1-j} \begin{bmatrix} 0 \\ B \end{bmatrix} (u[j] - K\hat{x}[j]) \right)$$

$$\leq \bar{V}^{\mathrm{s}}\left( \hat{A}^\tau(z_{t+1-\tau} - z'_{t+1-\tau}) + \sum_{j=0}^{\tau-1} \hat{A}^{\tau-1-j} \begin{bmatrix} 0 \\ B \end{bmatrix} (u[j] - K\hat{x}[j]) \right) \tag{B.94}$$

$$= \bar{V}^{\mathrm{s}}(z_{t+1-\tau} - z'_{t+1-\tau})$$
$$+ \sum_{k=0}^{\tau-1} \bar{V}^{\mathrm{s}}\left( \hat{A}^{k+1}(z_{t+1-\tau} - z'_{t+1-\tau}) + \sum_{j=0}^{k} \hat{A}^{k-j} \begin{bmatrix} 0 \\ B \end{bmatrix} (u[j] - K\hat{x}[j]) \right)$$
$$- \bar{V}^{\mathrm{s}}\left( \hat{A}^k(z_{t+1-\tau} - z'_{t+1-\tau}) + \sum_{j=0}^{k-1} \hat{A}^{k-1-j} \begin{bmatrix} 0 \\ B \end{bmatrix} (u[j] - K\hat{x}[j]) \right)$$

$$= \bar{V}^{\mathrm{s}}(z_{t+1-\tau} - z'_{t+1-\tau})$$
$$+ \sum_{k=0}^{\tau-1} \bar{V}^{\mathrm{s}}\left( \hat{A}\left( \hat{A}^k(z_{t+1-\tau} - z'_{t+1-\tau}) + \sum_{j=0}^{k-1} \hat{A}^{k-1-j} \begin{bmatrix} 0 \\ B \end{bmatrix} (u[j] - K\hat{x}[j]) \right) \right.$$
$$\left. + \begin{bmatrix} 0 \\ B \end{bmatrix} (u[k] - K\hat{x}[k]) \right)$$
$$- \bar{V}^{\mathrm{s}}\left( \hat{A}^k(z_{t+1-\tau} - z'_{t+1-\tau}) + \sum_{j=0}^{k-1} \hat{A}^{k-1-j} \begin{bmatrix} 0 \\ B \end{bmatrix} (u[j] - K\hat{x}[j]) \right). \tag{B.95}$$

From (6.38b), we have for all $z \in \mathbb{R}^{2n_x}$ and all $v \in \mathbb{R}^{n_u}$ that

$$
\bar{V}^{\mathrm{s}}\left(\hat{A}z + \begin{bmatrix} 0 \\ B \end{bmatrix} v\right) - \bar{V}^{\mathrm{s}}(z) - (1 - \nu^{\tau+1})\ell(v)
$$

$$
= -\nu^{\tau}\underbrace{\left(\bar{V}^{\mathrm{s}}(z) - \bar{V}^{\mathrm{s}}\left(\hat{A}z + \begin{bmatrix} 0 \\ B \end{bmatrix} v\right) + (1 - \nu)\ell(v)\right)}_{\geq 0}
$$

$$
- (1 - \nu^{\tau})\left(\bar{V}^{\mathrm{s}}(z) - \bar{V}^{\mathrm{s}}\left(\hat{A}z + \begin{bmatrix} 0 \\ B \end{bmatrix} v\right) + \ell(v)\right)
$$

$$
\leq -(1 - \nu^{\tau})\left(\bar{V}^{\mathrm{s}}(z) - \bar{V}^{\mathrm{s}}\left(\hat{A}z + \begin{bmatrix} 0 \\ B \end{bmatrix} v\right) + \ell(v)\right). \tag{B.96}
$$

Thus,

$$
\bar{V}^{\mathrm{s}}\left(\hat{A}^{\tau}(z_{t+1-\tau} - z'_{t+1-\tau}) + \sum_{j=0}^{\tau-1} \hat{A}^{\tau-1-j} \begin{bmatrix} 0 \\ B \end{bmatrix} (u[j] - K\hat{x}[j])\right) - (1 - \nu^{\tau+1}) \sum_{k=0}^{\tau-1} \ell(u[k] - K\hat{x}[k])
$$

$$
\stackrel{\text{(B.95)}}{=} \bar{V}^{\mathrm{s}}(z_{t+1-\tau} - z'_{t+1-\tau}) + \sum_{k=0}^{\tau-1} \left(\bar{V}^{\mathrm{s}}\left(\hat{A}\left(\hat{A}^{k}(z_{t+1-\tau} - z'_{t+1-\tau})\right.\right.\right.
$$

$$
\left.\left. + \sum_{j=0}^{k-1} \hat{A}^{k-1-j} \begin{bmatrix} 0 \\ B \end{bmatrix} (u[j] - K\hat{x}[j])\right) + \begin{bmatrix} 0 \\ B \end{bmatrix} (u[k] - K\hat{x}[k])\right)
$$

$$
\left. - \bar{V}^{\mathrm{s}}\left(\hat{A}^{k}(z_{t+1-\tau} - z'_{t+1-\tau}) + \sum_{j=0}^{k-1} \hat{A}^{k-1-j} \begin{bmatrix} 0 \\ B \end{bmatrix} (u[j] - K\hat{x}[j])\right) - (1 - \nu^{\tau+1})\ell(u[k] - K\hat{x}[k])\right)
$$

$$
\stackrel{\text{(B.96)}}{\leq} \bar{V}^{\mathrm{s}}(z_{t+1-\tau} - z'_{t+1-\tau})
$$

$$
- (1 - \nu^{\tau}) \sum_{k=0}^{\tau-1} \left(\bar{V}^{\mathrm{s}}\left(\hat{A}^{k}(z_{t+1-\tau} - z'_{t+1-\tau}) + \sum_{j=0}^{k-1} \hat{A}^{k-1-j} \begin{bmatrix} 0 \\ B \end{bmatrix} (u[j] - K\hat{x}[j])\right)\right)
$$

$$
- \bar{V}^{\mathrm{s}}\left(\hat{A}\left(\hat{A}^{k}(z_{t+1-\tau} - z'_{t+1-\tau}) + \sum_{j=0}^{k-1} \hat{A}^{k-1-j} \begin{bmatrix} 0 \\ B \end{bmatrix} (u[j] - K\hat{x}[j])\right)\right.
$$

$$
\left. + \begin{bmatrix} 0 \\ B \end{bmatrix} (u[k] - K\hat{x}[k])\right) + \ell(u[k] - K\hat{x}[k])\right)
$$

$$
\leq \bar{V}^{\mathrm{s}}(z_{t+1-\tau} - z'_{t+1-\tau}) - (1 - \nu^{\tau})\left(\bar{V}^{\mathrm{s}}(z_{t+1-\tau} - z'_{t+1-\tau})\right.
$$

$$
\left. - \bar{V}^{\mathrm{s}}\left(\hat{A}^{\tau}(z_{t+1-\tau} - z'_{t+1-\tau}) + \sum_{j=0}^{\tau-1} \hat{A}^{\tau-1-j} \begin{bmatrix} 0 \\ B \end{bmatrix} (u[j] - K\hat{x}[j])\right) + \sum_{k=0}^{\tau-1} \ell(u[k] - K\hat{x}[k])\right). \tag{B.97}
$$

With (B.94) and $J_{t+1}(\hat{x}_{t+1}) \leq J_{t+1-\tau}(\hat{x}_{t+1-\tau}) - (1 - \nu^{\tau+1}) \sum_{k=0}^{\tau-1} \ell(u[k] - K\hat{x}[k])$, (B.97) implies (6.41), which completes the proof. □

# B.31. Proof of Theorem 6.2

By the definition of the algorithm, if $\hat{x}_0 \in \hat{\mathcal{X}}_{\mathrm{f},0}$, then $J_{t+1}(\hat{x}_{t+1})$ is guaranteed to be finite either because this condition is ensured in step 8 (for $t > t_i$ and $u_t = \kappa(\hat{x}_{t_i}, t - t_i)$) or it holds by the recursive feasibility property of the MPC scheme (if $u_t = \kappa(\hat{x}_{t_i}, 0) \in \kappa_{\mathrm{MPC},t}(\hat{x}_t)$). Hence, it follows by induction that $\hat{x}_t \in \hat{\mathcal{X}}_{\mathrm{f},t}$ for all $t \in \mathbb{N}$. By the definition of the constraints in the MPC problem, it follows that $(x_t, u_t) \in \mathcal{X} \times \mathcal{U}$ for $t \in \mathbb{N}$, which completes the proof for the first part of the statement.

For the second part, we first adapt the proof of Lemma 5.2 to the output-feedback case: Let $t \in \mathbb{N}$, $\hat{x}_t \in \hat{\mathcal{X}}_{\mathrm{f},t}$, $\tilde{x}_t \in \mathbb{R}^{n_x}$, $((\hat{x}[i])_{i \in \mathbb{N}}, (u[i])_{i \in \mathbb{N}}) \in \mathcal{T}_t(\hat{x}_t)$, and $z'_t \in \mathcal{Z}^{\gamma}(z_t)$, arbitrary (but consistent with the assumptions on the set-valued estimator), where $z_t = (\tilde{x}_t, \hat{x}_t)$. Define

$$W_i := \bar{V}^{\mathrm{s}}\left(\hat{A}^i(z_t - z'_t) + \sum_{k=0}^{i-1} \hat{A}^{i-1-k} \begin{bmatrix} 0 \\ B \end{bmatrix} (u[k] - K\hat{x}[k])\right) + \sum_{j=i}^{N-1} \ell(u[j] - K\hat{x}[j]) \quad (B.98)$$

for $i \in \mathbb{N}$. It follows that

$$W_{i+1} = W_i + \bar{V}^{\mathrm{s}}\left(\hat{A}^{i+1}(z_t - z'_t) + \sum_{k=0}^{i} \hat{A}^{i-k} \begin{bmatrix} 0 \\ B \end{bmatrix} (u[k] - K\hat{x}[k])\right)$$

$$- \bar{V}^{\mathrm{s}}\left(\hat{A}^i(z_t - z'_t) + \sum_{k=0}^{i-1} \hat{A}^{i-1-k} \begin{bmatrix} 0 \\ B \end{bmatrix} (u[k] - K\hat{x}[k])\right) - \ell(u[i] - K\hat{x}[i]) \quad (B.99)$$

and

$$W_i = J_t(\hat{x}_t) - \sum_{j=0}^{i-1} \ell(u[j] - K\hat{x}[j])$$

$$+ \bar{V}^{\mathrm{s}}\left(\hat{A}^i(z_t - z'_t) + \sum_{k=0}^{i-1} \hat{A}^{i-1-k} \begin{bmatrix} 0 \\ B \end{bmatrix} (u[k] - K\hat{x}[k])\right). \quad (B.100)$$

By the optimality of $(u[0], \ldots, u[N-1])$ it follows that $u[j] - K\hat{x}[j] = 0$ for $j \in \{i, \ldots, N-1\}$ if $\hat{x}[i] \in \mathcal{X}_i$, as then $u[j] - K\hat{x}[j] = 0$, $j \in \{i, \ldots, N-1\}$, is feasible. Thus, considering that $z'_t \in \gamma \hat{\mathcal{F}}_\infty$, if

$$\hat{A}^i(z_t - z'_t) + \sum_{k=0}^{i-1} \hat{A}^{i-1-k} \begin{bmatrix} 0 \\ B \end{bmatrix} (u[k] - K\hat{x}[k]) \in (\mathbb{R}^{n_x} \times \bar{\mathcal{X}}_i) \ominus \gamma \hat{A}^i \hat{\mathcal{F}}_\infty \ominus \hat{A}^i(-\mathcal{E}_t \times \{0\})$$

$$\text{(B.101)}$$

then

$$
\begin{bmatrix} 0 \\ \hat{x}[i] \end{bmatrix} = \hat{A}^i(z_t - z_t') + \sum_{k=0}^{i-1} \hat{A}^{i-1-k} \begin{bmatrix} 0 \\ B \end{bmatrix} (u[k] - K\hat{x}[k]) + \hat{A}^i z_t' - \hat{A}^i \begin{bmatrix} \tilde{x}_t \\ 0 \end{bmatrix}
$$
$$
\in \mathbb{R}^{n_x} \times \bar{\mathcal{X}}_i \qquad\qquad (B.102)
$$

and

$$
W_i = \bar{V}^s \left( \hat{A}^i(z_t - z_t') + \sum_{k=0}^{i-1} \hat{A}^{i-1-k} \begin{bmatrix} 0 \\ B \end{bmatrix} (u[k] - K\hat{x}[k]) \right)
$$
$$
\leq \bar{c}_2 \left| \hat{A}^i(z_t - z_t') + \sum_{k=0}^{i-1} \hat{A}^{i-1-k} \begin{bmatrix} 0 \\ B \end{bmatrix} (u[k] - K\hat{x}[k]) \right|^a . \quad (B.103)
$$

Hence, as by assumption $\gamma \hat{A}^i \hat{\mathscr{F}}_\infty \oplus \hat{A}^i(-\mathscr{E}_t \times \{0\}) \oplus \epsilon \mathscr{B}_{2n_x} \subseteq \mathbb{R}^{n_x} \times \bar{\mathcal{X}}_i$ for all $i \in \{0, \dots, N-1\}$,

$$
\left| \hat{A}^i(z_t - z_t') + \sum_{k=0}^{i-1} \hat{A}^{i-1-k} \begin{bmatrix} 0 \\ B \end{bmatrix} (u[k] - K\hat{x}[k]) \right| \leq \epsilon \qquad (B.104)
$$

implies $\hat{x}[i] \in \bar{\mathcal{X}}_i$, such that

$$
W_i \leq \underbrace{\left( \bar{c}_2 + \frac{\bar{\Gamma}}{\epsilon^a} \right)}_{:=\bar{c}_4} \left| \hat{A}^i(z_t - z_t') + \sum_{k=0}^{i-1} \hat{A}^{i-1-k} \begin{bmatrix} 0 \\ B \end{bmatrix} (u[k] - K\hat{x}[k]) \right|^a \qquad (B.105)
$$

for all $i \in \mathbb{N}$ and some $\bar{\Gamma} \in [0, \infty)$ (independent of $i$ and $t$), similar to the proof of Lemma 6.2. Therefore, with (6.38b), (B.99) implies

$$
W_{i+1} \leq W_i - \bar{c}_3 \left| \hat{A}^i(z_t - z_t') + \sum_{k=0}^{i-1} \hat{A}^{i-1-k} \begin{bmatrix} 0 \\ B \end{bmatrix} (u[k] - K\hat{x}[k]) \right|^a
$$
$$
\leq W_i - \frac{\bar{c}_3}{\bar{c}_4} W_i
$$
$$
= \lambda W_i \qquad\qquad (B.106)
$$

for all $i \in \mathbb{N}$, where $\lambda := 1 - \bar{c}_3/\bar{c}_4 \in [0, 1)$. Thus, it holds that $W_0 - W_i \geq (1 - \lambda^i)W_0$ and, by the definition of $W_i$, that

$$
V_\gamma^s(z_t) - \bar{V}^s \left( \hat{A}^i(z_t - z_t') + \sum_{k=0}^{i-1} \hat{A}^{i-1-k} \begin{bmatrix} 0 \\ B \end{bmatrix} (u[k] - K\hat{x}[k]) \right) + \sum_{j=0}^{i-1} \ell(u[j] - K\hat{x}[j])
$$
$$
\geq (1 - \lambda^i)(V_\gamma^s(z_t) + J_t(\hat{x}_t)) \quad (B.107)
$$

for all $i \in \mathbb{N}$.

Considering now the definition of the trigger condition in step 8 of Algorithm 14, we arrive at the fact that in the closed-loop system for all $t \in \mathbb{N}$ there exists a $\tau \in \{1, \ldots, \min\{t+1, \theta\}\}$ such that

$$J_{t+1}(\hat{x}_{t+1}) \leq J_{t+1-\tau}(\hat{x}_{t+1-\tau}) - (1 - \nu^{\tau+1}) \sum_{k=0}^{\tau-1} \ell(u[k] - K\hat{x}[k]) \tag{B.108}$$

and

$$\begin{bmatrix} \tilde{x}_{t+1} \\ \hat{x}_{t+1} \end{bmatrix} \in \left\{ \begin{bmatrix} 0 \\ x[\tau] \end{bmatrix} \right\} \oplus \hat{A}^{\tau} \begin{bmatrix} \tilde{x}_{t+1-\tau} \\ 0 \end{bmatrix} \oplus \gamma \hat{\mathscr{F}}_{\tau} \tag{B.109}$$

where $((\hat{x}[j])_{j \in \mathbb{N}}, (u[j])_{j \in \mathbb{N}}) \in \mathscr{T}_{t+1-\tau}(\hat{x}_{t+1-\tau})$. Note that in the case that the trigger condition fails (that is, $t = t_i$), the inequality and inclusion above are satisfied by the properties of the MPC scheme. Lemma 6.4, the derivations leading up to (B.107), and Proposition 5.1 then imply that there exists a $\mu \in [0, 1)$ such that in the closed-loop system it holds that for every $t \in \mathbb{N}$ there exists a $\tau \in \{1, \ldots, \min\{t+1, \theta\}\}$ such that

$$J_{t+1}(\hat{x}_{t+1}) + V_{\gamma}^{\mathrm{s}}((\tilde{x}_{t+1}, \hat{x}_{t+1})) \leq \mu^{\tau}(J_{t+1-\tau}(\hat{x}_{t+1-\tau}) + V_{\gamma}^{\mathrm{s}}((\tilde{x}_{t+1-\tau}, \hat{x}_{t+1-\tau}))). \tag{B.110}$$

Furthermore, by the same reasoning as in the proof of Lemma 6.2, there exist $\hat{c}_1, \hat{c}_2 \in (0, \infty)$ such that

$$\hat{c}_1 |(\tilde{x}_t, \hat{x}_t)|_{\gamma \hat{\mathscr{F}}_{\infty}}^a \leq J_t(\hat{x}_t) + V_{\gamma}^{\mathrm{s}}((\tilde{x}_t, \hat{x}_t)) \leq \hat{c}_2 |(\tilde{x}_t, \hat{x}_t)|_{\gamma \hat{\mathscr{F}}_{\infty}}^a \tag{B.111}$$

for all $(\tilde{x}_0, \hat{x}_0) \in [I \ 0]\hat{\mathcal{X}}[0|0] \times \hat{\mathcal{X}}_{\mathrm{f},0}$ and all $t \in \mathbb{N}$. Hence, the conditions of Theorem 2.1 are met, $\gamma \hat{\mathscr{F}}_{\infty}$ is $\theta$-uniformly asymptotically stable for the dynamical system generating $((\tilde{x}_t, \hat{x}_t))_{t \in \mathbb{N}}$ in closed-loop, $[I \ 0]\hat{\mathcal{X}}[0|0] \times \hat{\mathcal{X}}_{\mathrm{f},0}$ belongs to the region of attraction, and the proof is complete. $\qquad \square$

# C. Numerical Data for the Examples in Chapter 5

In this appendix, we report the average transmission rates for various combinations of the parameters $\nu$, $\gamma$, and $\theta$. For the investigation of the transient behavior, we averaged the transmission rates over 30 simulations with 30 time points each, where the initial system state was a random point in the state space. For the investigation of the asymptotic behavior, we averaged the transmission rates over 10 simulations with 100 time points each. In Figure C.1, we depict the terminal set, an inner approximation of $\mathscr{F}_\infty$ and $\gamma\mathscr{F}_\infty$ for $\gamma = 5$, and the chosen initial conditions.

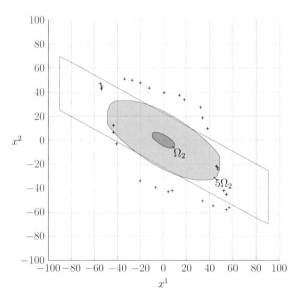

Figure C.1.: Terminal set $\mathscr{X}_\mathrm{T}$ (yellow), inner approximations $\Omega_2 \subseteq \mathscr{F}_\infty$ and $5\Omega_2 \subseteq 5\mathscr{F}_\infty$ of the stabilized sets (orange/red), and initial conditions for the simulations of the transient behavior (+).

Table C.1.: Lyapunov-based event-triggered control, transient behavior.

to-zero control

| | ν | | | | | | | | | | | |
|---|---|---|---|---|---|---|---|---|---|---|---|---|
| | 0.00 | | | | 0.50 | | | | 0.99 | | | |
| | θ | | | | θ | | | | θ | | | |
| $\gamma$ | 1 | 2 | 5 | 10 | $\gamma$ | 1 | 2 | 5 | 10 | $\gamma$ | 1 | 2 | 5 | 10 |
| 1.00 | 0.46 | 0.41 | 0.37 | 0.34 | 1.00 | 0.41 | 0.38 | 0.36 | 0.35 | 1.00 | 0.37 | 0.29 | 0.22 | 0.15 |
| 1.25 | 0.40 | 0.36 | 0.32 | 0.30 | 1.25 | 0.37 | 0.34 | 0.31 | 0.30 | 1.25 | 0.33 | 0.26 | 0.20 | 0.15 |
| 1.50 | 0.35 | 0.32 | 0.29 | 0.27 | 1.50 | 0.32 | 0.31 | 0.28 | 0.26 | 1.50 | 0.29 | 0.24 | 0.19 | 0.14 |
| 2.00 | 0.29 | 0.27 | 0.24 | 0.23 | 2.00 | 0.27 | 0.25 | 0.23 | 0.23 | 2.00 | 0.25 | 0.21 | 0.18 | 0.14 |
| 5.00 | 0.15 | 0.14 | 0.14 | 0.14 | 5.00 | 0.15 | 0.14 | 0.14 | 0.14 | 5.00 | 0.15 | 0.14 | 0.14 | 0.13 |

to-hold control

| | ν | | | | | | | | | | | |
|---|---|---|---|---|---|---|---|---|---|---|---|---|
| | 0.00 | | | | 0.50 | | | | 0.99 | | | |
| | θ | | | | θ | | | | θ | | | |
| $\gamma$ | 1 | 2 | 5 | 10 | $\gamma$ | 1 | 2 | 5 | 10 | $\gamma$ | 1 | 2 | 5 | 10 |
| 1.00 | 0.28 | 0.23 | 0.24 | 0.23 | 1.00 | 0.24 | 0.23 | 0.24 | 0.23 | 1.00 | 0.23 | 0.23 | 0.23 | 0.23 |
| 1.25 | 0.24 | 0.21 | 0.23 | 0.22 | 1.25 | 0.22 | 0.22 | 0.23 | 0.23 | 1.25 | 0.21 | 0.23 | 0.23 | 0.23 |
| 1.50 | 0.23 | 0.23 | 0.23 | 0.23 | 1.50 | 0.23 | 0.23 | 0.23 | 0.23 | 1.50 | 0.22 | 0.23 | 0.23 | 0.23 |
| 2.00 | 0.23 | 0.22 | 0.23 | 0.23 | 2.00 | 0.23 | 0.22 | 0.23 | 0.23 | 2.00 | 0.23 | 0.22 | 0.23 | 0.23 |
| 5.00 | 0.22 | 0.22 | 0.22 | 0.22 | 5.00 | 0.22 | 0.22 | 0.22 | 0.22 | 5.00 | 0.22 | 0.22 | 0.22 | 0.22 |

model-based control

| | ν | | | | | | | | | | | |
|---|---|---|---|---|---|---|---|---|---|---|---|---|
| | 0.00 | | | | 0.50 | | | | 0.99 | | | |
| | θ | | | | θ | | | | θ | | | |
| $\gamma$ | 1 | 2 | 5 | 10 | $\gamma$ | 1 | 2 | 5 | 10 | $\gamma$ | 1 | 2 | 5 | 10 |
| 1.00 | 0.24 | 0.11 | 0.10 | 0.09 | 1.00 | 0.13 | 0.10 | 0.10 | 0.09 | 1.00 | 0.11 | 0.10 | 0.09 | 0.07 |
| 1.25 | 0.12 | 0.08 | 0.07 | 0.07 | 1.25 | 0.08 | 0.08 | 0.07 | 0.07 | 1.25 | 0.08 | 0.07 | 0.07 | 0.07 |
| 1.50 | 0.08 | 0.07 | 0.07 | 0.07 | 1.50 | 0.07 | 0.07 | 0.07 | 0.07 | 1.50 | 0.07 | 0.07 | 0.07 | 0.06 |
| 2.00 | 0.06 | 0.06 | 0.06 | 0.06 | 2.00 | 0.06 | 0.06 | 0.06 | 0.06 | 2.00 | 0.06 | 0.06 | 0.06 | 0.05 |
| 5.00 | 0.04 | 0.04 | 0.04 | 0.04 | 5.00 | 0.04 | 0.04 | 0.04 | 0.04 | 5.00 | 0.04 | 0.04 | 0.04 | 0.04 |

Table C.2.: Lyapunov-based self-triggered control with predefined inputs, transient behavior.

to-zero control

| | | $\nu$ | | | | | | | | | | | | |
| | 0.00 | | | | | 0.50 | | | | | 0.99 | | | |
| | | $\theta$ | | | | | $\theta$ | | | | | $\theta$ | | |
| $\gamma$ | 1 | 2 | 5 | 10 | $\gamma$ | 1 | 2 | 5 | 10 | $\gamma$ | 1 | 2 | 5 | 10 |
|---|---|---|---|---|---|---|---|---|---|---|---|---|---|---|
| 1.00 | 1.00 | 0.73 | 0.68 | 0.64 | 1.00 | 1.00 | 0.71 | 0.67 | 0.64 | 1.00 | 1.00 | 0.64 | 0.51 | 0.36 |
| 1.25 | 1.00 | 0.67 | 0.58 | 0.55 | 1.25 | 1.00 | 0.65 | 0.57 | 0.55 | 1.25 | 1.00 | 0.60 | 0.46 | 0.35 |
| 1.50 | 1.00 | 0.64 | 0.52 | 0.50 | 1.50 | 1.00 | 0.62 | 0.51 | 0.49 | 1.50 | 1.00 | 0.58 | 0.42 | 0.34 |
| 2.00 | 1.00 | 0.59 | 0.42 | 0.41 | 2.00 | 1.00 | 0.59 | 0.42 | 0.41 | 2.00 | 1.00 | 0.55 | 0.36 | 0.31 |
| 5.00 | 1.00 | 0.53 | 0.27 | 0.23 | 5.00 | 1.00 | 0.52 | 0.27 | 0.23 | 5.00 | 1.00 | 0.52 | 0.26 | 0.22 |

to-hold control

| | | $\nu$ | | | | | | | | | | | | |
| | 0.00 | | | | | 0.50 | | | | | 0.99 | | | |
| | | $\theta$ | | | | | $\theta$ | | | | | $\theta$ | | |
| $\gamma$ | 1 | 2 | 5 | 10 | $\gamma$ | 1 | 2 | 5 | 10 | $\gamma$ | 1 | 2 | 5 | 10 |
|---|---|---|---|---|---|---|---|---|---|---|---|---|---|---|
| 1.00 | 1.00 | 0.60 | 0.49 | 0.45 | 1.00 | 1.00 | 0.54 | 0.46 | 0.43 | 1.00 | 1.00 | 0.53 | 0.44 | 0.33 |
| 1.25 | 1.00 | 0.52 | 0.35 | 0.33 | 1.25 | 1.00 | 0.51 | 0.35 | 0.33 | 1.25 | 1.00 | 0.51 | 0.34 | 0.29 |
| 1.50 | 1.00 | 0.50 | 0.34 | 0.32 | 1.50 | 1.00 | 0.50 | 0.33 | 0.32 | 1.50 | 1.00 | 0.50 | 0.32 | 0.28 |
| 2.00 | 1.00 | 0.50 | 0.27 | 0.27 | 2.00 | 1.00 | 0.50 | 0.27 | 0.27 | 2.00 | 1.00 | 0.50 | 0.27 | 0.25 |
| 5.00 | 1.00 | 0.50 | 0.23 | 0.23 | 5.00 | 1.00 | 0.50 | 0.23 | 0.23 | 5.00 | 1.00 | 0.50 | 0.23 | 0.23 |

model-based control

| | | $\nu$ | | | | | | | | | | | | |
| | 0.00 | | | | | 0.50 | | | | | 0.99 | | | |
| | | $\theta$ | | | | | $\theta$ | | | | | $\theta$ | | |
| $\gamma$ | 1 | 2 | 5 | 10 | $\gamma$ | 1 | 2 | 5 | 10 | $\gamma$ | 1 | 2 | 5 | 10 |
|---|---|---|---|---|---|---|---|---|---|---|---|---|---|---|
| 1.00 | 1.00 | 0.69 | 0.53 | 0.49 | 1.00 | 1.00 | 0.58 | 0.49 | 0.45 | 1.00 | 1.00 | 0.54 | 0.43 | 0.37 |
| 1.25 | 1.00 | 0.55 | 0.40 | 0.36 | 1.25 | 1.00 | 0.52 | 0.38 | 0.35 | 1.25 | 1.00 | 0.51 | 0.33 | 0.28 |
| 1.50 | 1.00 | 0.50 | 0.36 | 0.33 | 1.50 | 1.00 | 0.50 | 0.33 | 0.32 | 1.50 | 1.00 | 0.50 | 0.31 | 0.26 |
| 2.00 | 1.00 | 0.50 | 0.26 | 0.25 | 2.00 | 1.00 | 0.50 | 0.26 | 0.24 | 2.00 | 1.00 | 0.50 | 0.25 | 0.21 |
| 5.00 | 1.00 | 0.50 | 0.20 | 0.13 | 5.00 | 1.00 | 0.50 | 0.20 | 0.13 | 5.00 | 1.00 | 0.50 | 0.20 | 0.13 |

Table C.3.: Lyapunov-based self-triggered control with optimized inputs, transient behavior.

to-zero control

| | | $\nu$ | | | | | | | | | | | |
|---|---|---|---|---|---|---|---|---|---|---|---|---|---|
| | 0.00 | | | | | 0.50 | | | | | 0.99 | | |
| | $\theta$ | | | | | $\theta$ | | | | | $\theta$ | | |
| $\gamma$ | 1 | 2 | 5 | 10 | $\gamma$ | 1 | 2 | 5 | 10 | $\gamma$ | 1 | 2 | 5 | 10 |
| 1.00 | 1.00 | 0.56 | 0.52 | 0.49 | 1.00 | 1.00 | 0.54 | 0.50 | 0.48 | 1.00 | 1.00 | 0.52 | 0.41 | 0.31 |
| 1.25 | 1.00 | 0.55 | 0.40 | 0.38 | 1.25 | 1.00 | 0.54 | 0.38 | 0.37 | 1.25 | 1.00 | 0.52 | 0.33 | 0.28 |
| 1.50 | 1.00 | 0.55 | 0.38 | 0.37 | 1.50 | 1.00 | 0.54 | 0.37 | 0.36 | 1.50 | 1.00 | 0.52 | 0.33 | 0.28 |
| 2.00 | 1.00 | 0 54 | 0.29 | 0.28 | 2.00 | 1.00 | 0.53 | 0.29 | 0.28 | 2.00 | 1.00 | 0.52 | 0.27 | 0.24 |
| 5.00 | 1.00 | 0.52 | 0.25 | 0.19 | 5.00 | 1.00 | 0.52 | 0.25 | 0.18 | 5.00 | 1.00 | 0.51 | 0.25 | 0.17 |

to-hold control

| | | $\nu$ | | | | | | | | | | | |
|---|---|---|---|---|---|---|---|---|---|---|---|---|---|---|
| | 0.00 | | | | | 0.50 | | | | | 0.99 | | |
| | $\theta$ | | | | | $\theta$ | | | | | $\theta$ | | |
| $\gamma$ | 1 | 2 | 5 | 10 | $\gamma$ | 1 | 2 | 5 | 10 | $\gamma$ | 1 | 2 | 5 | 10 |
| 1.00 | 1.00 | 0.53 | 0.47 | 0.44 | 1.00 | 1.00 | 0.50 | 0.45 | 0.42 | 1.00 | 1.00 | 0.50 | 0.42 | 0.33 |
| 1.25 | 1.00 | 0.50 | 0.33 | 0.32 | 1.25 | 1.00 | 0.50 | 0.33 | 0.31 | 1.25 | 1.00 | 0.50 | 0.31 | 0.25 |
| 1.50 | 1.00 | 0.50 | 0.32 | 0.30 | 1.50 | 1.00 | 0.50 | 0.32 | 0.30 | 1.50 | 1.00 | 0.50 | 0.31 | 0.24 |
| 2.00 | 1.00 | 0.50 | 0.23 | 0.22 | 2.00 | 1.00 | 0.50 | 0.23 | 0.22 | 2.00 | 1.00 | 0.50 | 0.23 | 0.19 |
| 5.00 | 1.00 | 0.50 | 0.20 | 0.13 | 5.00 | 1.00 | 0.50 | 0.20 | 0.13 | 5.00 | 1.00 | 0.50 | 0.20 | 0.12 |

model-based control

| | | $\nu$ | | | | | | | | | | | |
|---|---|---|---|---|---|---|---|---|---|---|---|---|---|---|
| | 0.00 | | | | | 0.50 | | | | | 0.99 | | |
| | $\theta$ | | | | | $\theta$ | | | | | $\theta$ | | |
| $\gamma$ | 1 | 2 | 5 | 10 | $\gamma$ | 1 | 2 | 5 | 10 | $\gamma$ | 1 | 2 | 5 | 10 |
| 1.00 | 1.00 | 0.52 | 0.46 | 0.43 | 1.00 | 1.00 | 0.50 | 0.44 | 0.41 | 1.00 | 1.00 | 0.50 | 0.42 | 0.35 |
| 1.25 | 1.00 | 0.50 | 0.33 | 0.30 | 1.25 | 1.00 | 0.50 | 0.32 | 0.30 | 1.25 | 1.00 | 0.50 | 0.30 | 0.26 |
| 1.50 | 1.00 | 0.50 | 0.31 | 0.29 | 1.50 | 1.00 | 0.50 | 0.31 | 0.29 | 1.50 | 1.00 | 0.50 | 0.30 | 0.26 |
| 2.00 | 1.00 | 0.50 | 0.21 | 0.20 | 2.00 | 1.00 | 0.50 | 0.21 | 0.20 | 2.00 | 1.00 | 0.50 | 0.21 | 0.17 |
| 5.00 | 1.00 | 0.50 | 0.20 | 0.10 | 5.00 | 1.00 | 0.50 | 0.20 | 0.10 | 5.00 | 1.00 | 0.50 | 0.20 | 0.10 |

Table C.4.: Mixed set–Lyapunov-based event-triggered control, transient behavior.

to-zero control

| | | ν | | | | | | | | | | | | |
| | 0.00 | | | | | 0.50 | | | | | 0.99 | | | |
| | | θ | | | | | θ | | | | | θ | | |
| $\gamma$ | 1 | 2 | 5 | 10 | $\gamma$ | 1 | 2 | 5 | 10 | $\gamma$ | 1 | 2 | 5 | 10 |
|---|---|---|---|---|---|---|---|---|---|---|---|---|---|---|
| 1.00 | 1.00 | 0.66 | 0.47 | 0.40 | 1.00 | 1.00 | 0.66 | 0.47 | 0.40 | 1.00 | 1.00 | 0.66 | 0.47 | 0.40 |
| 1.25 | 0.73 | 0.52 | 0.39 | 0.35 | 1.25 | 0.73 | 0.52 | 0.39 | 0.35 | 1.25 | 0.73 | 0.52 | 0.39 | 0.35 |
| 1.50 | 0.59 | 0.44 | 0.34 | 0.31 | 1.50 | 0.59 | 0.44 | 0.34 | 0.31 | 1.50 | 0.59 | 0.44 | 0.34 | 0.31 |
| 2.00 | 0.45 | 0.36 | 0.28 | 0.26 | 2.00 | 0.45 | 0.36 | 0.28 | 0.26 | 2.00 | 0.45 | 0.36 | 0.28 | 0.26 |
| 5.00 | 0.22 | 0.19 | 0.15 | 0.14 | 5.00 | 0.22 | 0.19 | 0.15 | 0.14 | 5.00 | 0.22 | 0.19 | 0.16 | 0.14 |

to-hold control

| | | ν | | | | | | | | | | | | |
| | 0.00 | | | | | 0.50 | | | | | 0.99 | | | |
| | | θ | | | | | θ | | | | | θ | | |
| $\gamma$ | 1 | 2 | 5 | 10 | $\gamma$ | 1 | 2 | 5 | 10 | $\gamma$ | 1 | 2 | 5 | 10 |
|---|---|---|---|---|---|---|---|---|---|---|---|---|---|---|
| 1.00 | 0.99 | 0.55 | 0.33 | 0.30 | 1.00 | 0.99 | 0.55 | 0.33 | 0.30 | 1.00 | 0.99 | 0.55 | 0.33 | 0.30 |
| 1.25 | 0.62 | 0.37 | 0.29 | 0.27 | 1.25 | 0.62 | 0.37 | 0.29 | 0.27 | 1.25 | 0.62 | 0.37 | 0.29 | 0.27 |
| 1.50 | 0.42 | 0.31 | 0.24 | 0.25 | 1.50 | 0.42 | 0.31 | 0.24 | 0.25 | 1.50 | 0.42 | 0.31 | 0.24 | 0.25 |
| 2.00 | 0.32 | 0.25 | 0.22 | 0.22 | 2.00 | 0.32 | 0.25 | 0.22 | 0.22 | 2.00 | 0.32 | 0.25 | 0.22 | 0.22 |
| 5.00 | 0.20 | 0.21 | 0.21 | 0.21 | 5.00 | 0.20 | 0.21 | 0.21 | 0.21 | 5.00 | 0.20 | 0.21 | 0.21 | 0.21 |

model-based control

| | | ν | | | | | | | | | | | | |
| | 0.00 | | | | | 0.50 | | | | | 0.99 | | | |
| | | θ | | | | | θ | | | | | θ | | |
| $\gamma$ | 1 | 2 | 5 | 10 | $\gamma$ | 1 | 2 | 5 | 10 | $\gamma$ | 1 | 2 | 5 | 10 |
|---|---|---|---|---|---|---|---|---|---|---|---|---|---|---|
| 1.00 | 1.00 | 0.46 | 0.18 | 0.14 | 1.00 | 1.00 | 0.46 | 0.18 | 0.14 | 1.00 | 1.00 | 0.46 | 0.18 | 0.14 |
| 1.25 | 0.49 | 0.19 | 0.10 | 0.09 | 1.25 | 0.49 | 0.19 | 0.10 | 0.09 | 1.25 | 0.49 | 0.19 | 0.10 | 0.09 |
| 1.50 | 0.24 | 0.14 | 0.08 | 0.07 | 1.50 | 0.24 | 0.14 | 0.08 | 0.07 | 1.50 | 0.24 | 0.14 | 0.08 | 0.07 |
| 2.00 | 0.11 | 0.08 | 0.07 | 0.06 | 2.00 | 0.10 | 0.08 | 0.07 | 0.06 | 2.00 | 0.10 | 0.08 | 0.07 | 0.06 |
| 5.00 | 0.06 | 0.05 | 0.04 | 0.04 | 5.00 | 0.05 | 0.05 | 0.04 | 0.04 | 5.00 | 0.05 | 0.05 | 0.04 | 0.04 |

Table C.5.: Mixed set–Lyapunov-based self-triggered control with predefined inputs, transient behavior.

to-zero control

| | ν | | | | | | | | | | | |
|---|---|---|---|---|---|---|---|---|---|---|---|---|
| | 0.00 | | | | 0.50 | | | | 0.99 | | | |
| | θ | | | | θ | | | | θ | | | |
| γ | 1 | 2 | 5 | 10 | γ | 1 | 2 | 5 | 10 | γ | 1 | 2 | 5 | 10 |
| 1.00 | 1.00 | 1.00 | 0.88 | 0.73 | 1.00 | 1.00 | 1.00 | 0.88 | 0.73 | 1.00 | 1.00 | 1.00 | 0.88 | 0.73 |
| 1.25 | 1.00 | 1.00 | 0.71 | 0.63 | 1.25 | 1.00 | 1.00 | 0.71 | 0.63 | 1.25 | 1.00 | 1.00 | 0.71 | 0.63 |
| 1.50 | 1.00 | 0.82 | 0.62 | 0.54 | 1.50 | 1.00 | 0.82 | 0.62 | 0.54 | 1.50 | 1.00 | 0.82 | 0.62 | 0.54 |
| 2.00 | 1.00 | 0.65 | 0.51 | 0.45 | 2.00 | 1.00 | 0.65 | 0.51 | 0.45 | 2.00 | 1.00 | 0.65 | 0.51 | 0.45 |
| 5.00 | 1.00 | 0.54 | 0.29 | 0.25 | 5.00 | 1.00 | 0.54 | 0.29 | 0.25 | 5.00 | 1.00 | 0.54 | 0.29 | 0.25 |

to-hold control

| | ν | | | | | | | | | | | |
|---|---|---|---|---|---|---|---|---|---|---|---|---|
| | 0.00 | | | | 0.50 | | | | 0.99 | | | |
| | θ | | | | θ | | | | θ | | | |
| γ | 1 | 2 | 5 | 10 | γ | 1 | 2 | 5 | 10 | γ | 1 | 2 | 5 | 10 |
| 1.00 | 1.00 | 1.00 | 0.76 | 0.62 | 1.00 | 1.00 | 1.00 | 0.76 | 0.62 | 1.00 | 1.00 | 1.00 | 0.76 | 0.62 |
| 1.25 | 1.00 | 1.00 | 0.59 | 0.55 | 1.25 | 1.00 | 1.00 | 0.59 | 0.55 | 1.25 | 1.00 | 1.00 | 0.59 | 0.55 |
| 1.50 | 1.00 | 0.70 | 0.52 | 0.44 | 1.50 | 1.00 | 0.70 | 0.52 | 0.44 | 1.50 | 1.00 | 0.70 | 0.52 | 0.44 |
| 2.00 | 1.00 | 0.55 | 0.42 | 0.35 | 2.00 | 1.00 | 0.55 | 0.42 | 0.35 | 2.00 | 1.00 | 0.55 | 0.42 | 0.35 |
| 5.00 | 1.00 | 0.50 | 0.24 | 0.22 | 5.00 | 1.00 | 0.50 | 0.24 | 0.22 | 5.00 | 1.00 | 0.50 | 0.24 | 0.22 |

model-based control

| | ν | | | | | | | | | | | |
|---|---|---|---|---|---|---|---|---|---|---|---|---|
| | 0.00 | | | | 0.50 | | | | 0.99 | | | |
| | θ | | | | θ | | | | θ | | | |
| γ | 1 | 2 | 5 | 10 | γ | 1 | 2 | 5 | 10 | γ | 1 | 2 | 5 | 10 |
| 1.00 | 1.00 | 1.00 | 0.76 | 0.58 | 1.00 | 1.00 | 1.00 | 0.76 | 0.58 | 1.00 | 1.00 | 1.00 | 0.76 | 0.58 |
| 1.25 | 1.00 | 1.00 | 0.54 | 0.49 | 1.25 | 1.00 | 1.00 | 0.54 | 0.49 | 1.25 | 1.00 | 1.00 | 0.54 | 0.49 |
| 1.50 | 1.00 | 0.51 | 0.44 | 0.37 | 1.50 | 1.00 | 0.50 | 0.44 | 0.37 | 1.50 | 1.00 | 0.50 | 0.44 | 0.37 |
| 2.00 | 1.00 | 0.51 | 0.34 | 0.28 | 2.00 | 1.00 | 0.50 | 0.33 | 0.27 | 2.00 | 1.00 | 0.50 | 0.33 | 0.27 |
| 5.00 | 1.00 | 0.51 | 0.21 | 0.14 | 5.00 | 1.00 | 0.50 | 0.20 | 0.13 | 5.00 | 1.00 | 0.50 | 0.20 | 0.13 |

Table C.6.: Mixed set–Lyapunov-based self-triggered control with optimized inputs, transient behavior.

to-zero control

| | | | | | | | | | | | | | | |
|---|---|---|---|---|---|---|---|---|---|---|---|---|---|---|
| | | | | | | $\nu$ | | | | | | | | |
| | 0.00 | | | | | 0.50 | | | | | 0.99 | | | |
| | $\theta$ | | | | | $\theta$ | | | | | $\theta$ | | | |
| $\gamma$ | 1 | 2 | 5 | 10 | $\gamma$ | 1 | 2 | 5 | 10 | $\gamma$ | 1 | 2 | 5 | 10 |
| 1.00 | 1.00 | 0.57 | 0.57 | 0.57 | 1.00 | 1.00 | 0.57 | 0.57 | 0.57 | 1.00 | 1.00 | 0.57 | 0.57 | 0.57 |
| 1.25 | 1.00 | 0.56 | 0.45 | 0.45 | 1.25 | 1.00 | 0.56 | 0.45 | 0.45 | 1.25 | 1.00 | 0.56 | 0.45 | 0.45 |
| 1.50 | 1.00 | 0.55 | 0.41 | 0.41 | 1.50 | 1.00 | 0.55 | 0.41 | 0.41 | 1.50 | 1.00 | 0.55 | 0.41 | 0.41 |
| 2.00 | 1.00 | 0.54 | 0.32 | 0.32 | 2.00 | 1.00 | 0.54 | 0.32 | 0.32 | 2.00 | 1.00 | 0.54 | 0.32 | 0.32 |
| 5.00 | 1.00 | 0.53 | 0.25 | 0.19 | 5.00 | 1.00 | 0.53 | 0.25 | 0.19 | 5.00 | 1.00 | 0.52 | 0.24 | 0.19 |

to-hold control

| | | | | | | | | | | | | | | |
|---|---|---|---|---|---|---|---|---|---|---|---|---|---|---|
| | | | | | | $\nu$ | | | | | | | | |
| | 0.00 | | | | | 0.50 | | | | | 0.99 | | | |
| | $\theta$ | | | | | $\theta$ | | | | | $\theta$ | | | |
| $\gamma$ | 1 | 2 | 5 | 10 | $\gamma$ | 1 | 2 | 5 | 10 | $\gamma$ | 1 | 2 | 5 | 10 |
| 1.00 | 1.00 | 0.60 | 0.60 | 0.60 | 1.00 | 1.00 | 0.60 | 0.60 | 0.60 | 1.00 | 1.00 | 0.60 | 0.60 | 0.60 |
| 1.25 | 1.00 | 0.57 | 0.47 | 0.47 | 1.25 | 1.00 | 0.57 | 0.47 | 0.47 | 1.25 | 1.00 | 0.57 | 0.47 | 0.47 |
| 1.50 | 1.00 | 0.56 | 0.43 | 0.43 | 1.50 | 1.00 | 0.56 | 0.43 | 0.43 | 1.50 | 1.00 | 0.56 | 0.43 | 0.43 |
| 2.00 | 1.00 | 0.55 | 0.34 | 0.34 | 2.00 | 1.00 | 0.55 | 0.34 | 0.34 | 2.00 | 1.00 | 0.55 | 0.34 | 0.34 |
| 5.00 | 1.00 | 0.53 | 0.25 | 0.18 | 5.00 | 1.00 | 0.53 | 0.25 | 0.18 | 5.00 | 1.00 | 0.52 | 0.24 | 0.18 |

model-based control

| | | | | | | | | | | | | | | |
|---|---|---|---|---|---|---|---|---|---|---|---|---|---|---|
| | | | | | | $\nu$ | | | | | | | | |
| | 0.00 | | | | | 0.50 | | | | | 0.99 | | | |
| | $\theta$ | | | | | $\theta$ | | | | | $\theta$ | | | |
| $\gamma$ | 1 | 2 | 5 | 10 | $\gamma$ | 1 | 2 | 5 | 10 | $\gamma$ | 1 | 2 | 5 | 10 |
| 1.00 | 1.00 | 0.57 | 0.57 | 0.57 | 1.00 | 1.00 | 0.57 | 0.57 | 0.57 | 1.00 | 1.00 | 0.57 | 0.57 | 0.57 |
| 1.25 | 1.00 | 0.56 | 0.43 | 0.43 | 1.25 | 1.00 | 0.56 | 0.43 | 0.43 | 1.25 | 1.00 | 0.56 | 0.43 | 0.43 |
| 1.50 | 1.00 | 0.54 | 0.41 | 0.41 | 1.50 | 1.00 | 0.54 | 0.41 | 0.41 | 1.50 | 1.00 | 0.54 | 0.41 | 0.41 |
| 2.00 | 1.00 | 0.53 | 0.32 | 0.32 | 2.00 | 1.00 | 0.53 | 0.32 | 0.32 | 2.00 | 1.00 | 0.53 | 0.32 | 0.32 |
| 5.00 | 1.00 | 0.52 | 0.23 | 0.16 | 5.00 | 1.00 | 0.52 | 0.23 | 0.16 | 5.00 | 1.00 | 0.52 | 0.22 | 0.16 |

Table C.7.: Purely set-based event-triggered control, transient behavior.

| to-zero control | | | | | to-hold control | | | | | model-based control | | | | |
|---|---|---|---|---|---|---|---|---|---|---|---|---|---|---|---|
| | $\theta$ | | | | | | $\theta$ | | | | | | $\theta$ | | | |
| $\gamma$ | 1 | 2 | 5 | 10 | | $\gamma$ | 1 | 2 | 5 | 10 | | $\gamma$ | 1 | 2 | 5 | 10 |
| 1.00 | 1.00 | 0.66 | 0.47 | 0.40 | | 1.00 | 0.99 | 0.55 | 0.33 | 0.30 | | 1.00 | 1.00 | 0.46 | 0.18 | 0.14 |
| 1.25 | 0.73 | 0.52 | 0.39 | 0.35 | | 1.25 | 0.62 | 0.37 | 0.29 | 0.27 | | 1.25 | 0.50 | 0.19 | 0.10 | 0.09 |
| 1.50 | 0.59 | 0.44 | 0.34 | 0.31 | | 1.50 | 0.42 | 0.31 | 0.25 | 0.25 | | 1.50 | 0.24 | 0.14 | 0.08 | 0.07 |
| 2.00 | 0.45 | 0.36 | 0.28 | 0.25 | | 2.00 | 0.32 | 0.25 | 0.22 | 0.22 | | 2.00 | 0.12 | 0.08 | 0.07 | 0.06 |
| 5.00 | 0.22 | 0.19 | 0.16 | 0.14 | | 5.00 | 0.20 | 0.21 | 0.21 | 0.21 | | 5.00 | 0.07 | 0.06 | 0.04 | 0.04 |

Table C.8.: Purely set-based self-triggered control with predefined inputs, transient behavior.

| to-zero control | | | | | to-hold control | | | | | model-based control | | | | |
|---|---|---|---|---|---|---|---|---|---|---|---|---|---|---|---|
| | $\theta$ | | | | | | $\theta$ | | | | | | $\theta$ | | | |
| $\gamma$ | 1 | 2 | 5 | 10 | | $\gamma$ | 1 | 2 | 5 | 10 | | $\gamma$ | 1 | 2 | 5 | 10 |
| 1.00 | 1.00 | 1.00 | 0.88 | 0.73 | | 1.00 | 1.00 | 1.00 | 0.76 | 0.62 | | 1.00 | 1.00 | 1.00 | 0.76 | 0.58 |
| 1.25 | 1.00 | 1.00 | 0.71 | 0.63 | | 1.25 | 1.00 | 1.00 | 0.59 | 0.55 | | 1.25 | 1.00 | 1.00 | 0.54 | 0.50 |
| 1.50 | 1.00 | 0.82 | 0.62 | 0.54 | | 1.50 | 1.00 | 0.70 | 0.53 | 0.44 | | 1.50 | 1.00 | 0.50 | 0.43 | 0.37 |
| 2.00 | 1.00 | 0.65 | 0.51 | 0.45 | | 2.00 | 1.00 | 0.57 | 0.43 | 0.37 | | 2.00 | 1.00 | 0.50 | 0.33 | 0.27 |
| 5.00 | 1.00 | 0.54 | 0.29 | 0.25 | | 5.00 | 1.00 | 0.50 | 0.25 | 0.24 | | 5.00 | 1.00 | 0.50 | 0.20 | 0.13 |

Table C.9.: Purely set-based self-triggered control with optimized inputs, transient behavior.

| to-zero control | | | | | to-hold control | | | | | model-based control | | | | |
|---|---|---|---|---|---|---|---|---|---|---|---|---|---|---|---|
| | $\theta$ | | | | | | $\theta$ | | | | | | $\theta$ | | | |
| $\gamma$ | 1 | 2 | 5 | 10 | | $\gamma$ | 1 | 2 | 5 | 10 | | $\gamma$ | 1 | 2 | 5 | 10 |
| 1.00 | 1.00 | 0.57 | 0.57 | 0.57 | | 1.00 | 1.00 | 0.60 | 0.60 | 0.60 | | 1.00 | 1.00 | 0.57 | 0.57 | 0.57 |
| 1.25 | 1.00 | 0.56 | 0.45 | 0.45 | | 1.25 | 1.00 | 0.57 | 0.47 | 0.47 | | 1.25 | 1.00 | 0.56 | 0.43 | 0.43 |
| 1.50 | 1.00 | 0.55 | 0.41 | 0.41 | | 1.50 | 1.00 | 0.56 | 0.43 | 0.43 | | 1.50 | 1.00 | 0.54 | 0.41 | 0.41 |
| 2.00 | 1.00 | 0.53 | 0.32 | 0.32 | | 2.00 | 1.00 | 0.54 | 0.34 | 0.34 | | 2.00 | 1.00 | 0.53 | 0.32 | 0.32 |
| 5.00 | 1.00 | 0.52 | 0.24 | 0.18 | | 5.00 | 1.00 | 0.52 | 0.24 | 0.17 | | 5.00 | 1.00 | 0.52 | 0.22 | 0.15 |

Table C.10.: Lyapunov-based event-triggered control, asymptotic behavior.

to-zero control

| | ν | | | | | | | | | | | |
|---|---|---|---|---|---|---|---|---|---|---|---|---|
| | 0.00 | | | | | 0.50 | | | | | 0.99 | | | |
| | θ | | | | | θ | | | | | θ | | | |
| $\gamma$ | 1 | 2 | 5 | 10 | $\gamma$ | 1 | 2 | 5 | 10 | $\gamma$ | 1 | 2 | 5 | 10 |
| 1.00 | 0.11 | 0.11 | 0.11 | 0.11 | 1.00 | 0.11 | 0.11 | 0.11 | 0.11 | 1.00 | 0.11 | 0.11 | 0.11 | 0.11 |
| 1.25 | 0.08 | 0.08 | 0.08 | 0.08 | 1.25 | 0.08 | 0.08 | 0.08 | 0.08 | 1.25 | 0.08 | 0.08 | 0.08 | 0.08 |
| 1.50 | 0.06 | 0.06 | 0.06 | 0.06 | 1.50 | 0.06 | 0.06 | 0.06 | 0.06 | 1.50 | 0.06 | 0.06 | 0.06 | 0.06 |
| 2.00 | 0.04 | 0.04 | 0.04 | 0.04 | 2.00 | 0.04 | 0.04 | 0.04 | 0.04 | 2.00 | 0.04 | 0.04 | 0.04 | 0.04 |
| 5.00 | 0.03 | 0.03 | 0.03 | 0.03 | 5.00 | 0.03 | 0.03 | 0.03 | 0.03 | 5.00 | 0.03 | 0.03 | 0.03 | 0.03 |

to-hold control

| | ν | | | | | | | | | | | |
|---|---|---|---|---|---|---|---|---|---|---|---|---|---|---|
| | 0.00 | | | | | 0.50 | | | | | 0.99 | | | |
| | θ | | | | | θ | | | | | θ | | | |
| $\gamma$ | 1 | 2 | 5 | 10 | $\gamma$ | 1 | 2 | 5 | 10 | $\gamma$ | 1 | 2 | 5 | 10 |
| 1.00 | 0.19 | 0.19 | 0.19 | 0.19 | 1.00 | 0.19 | 0.19 | 0.19 | 0.19 | 1.00 | 0.19 | 0.19 | 0.19 | 0.19 |
| 1.25 | 0.18 | 0.18 | 0.18 | 0.18 | 1.25 | 0.18 | 0.18 | 0.18 | 0.18 | 1.25 | 0.18 | 0.18 | 0.18 | 0.18 |
| 1.50 | 0.17 | 0.17 | 0.17 | 0.17 | 1.50 | 0.17 | 0.17 | 0.17 | 0.17 | 1.50 | 0.17 | 0.17 | 0.17 | 0.17 |
| 2.00 | 0.17 | 0.17 | 0.17 | 0.17 | 2.00 | 0.17 | 0.17 | 0.17 | 0.17 | 2.00 | 0.17 | 0.17 | 0.17 | 0.17 |
| 5.00 | 0.12 | 0.12 | 0.12 | 0.12 | 5.00 | 0.12 | 0.12 | 0.12 | 0.12 | 5.00 | 0.12 | 0.12 | 0.12 | 0.12 |

model-based control

| | ν | | | | | | | | | | | |
|---|---|---|---|---|---|---|---|---|---|---|---|---|---|---|
| | 0.00 | | | | | 0.50 | | | | | 0.99 | | | |
| | θ | | | | | θ | | | | | θ | | | |
| $\gamma$ | 1 | 2 | 5 | 10 | $\gamma$ | 1 | 2 | 5 | 10 | $\gamma$ | 1 | 2 | 5 | 10 |
| 1.00 | 0.07 | 0.07 | 0.07 | 0.07 | 1.00 | 0.07 | 0.07 | 0.07 | 0.07 | 1.00 | 0.07 | 0.07 | 0.07 | 0.07 |
| 1.25 | 0.06 | 0.06 | 0.06 | 0.06 | 1.25 | 0.06 | 0.06 | 0.06 | 0.06 | 1.25 | 0.06 | 0.06 | 0.06 | 0.06 |
| 1.50 | 0.05 | 0.05 | 0.05 | 0.05 | 1.50 | 0.05 | 0.05 | 0.05 | 0.05 | 1.50 | 0.05 | 0.05 | 0.05 | 0.05 |
| 2.00 | 0.03 | 0.03 | 0.03 | 0.03 | 2.00 | 0.03 | 0.03 | 0.03 | 0.03 | 2.00 | 0.03 | 0.03 | 0.03 | 0.03 |
| 5.00 | 0.02 | 0.02 | 0.02 | 0.02 | 5.00 | 0.02 | 0.02 | 0.02 | 0.02 | 5.00 | 0.02 | 0.02 | 0.02 | 0.02 |

Table C.11.: Lyapunov-based self-triggered control with predefined inputs, asymptotic behavior.

to-zero control

| | | | | | | | | | | | | |
|---|---|---|---|---|---|---|---|---|---|---|---|---|
| | | | | | | | $\nu$ | | | | | |
| | 0.00 | | | | 0.50 | | | | 0.99 | | | |
| | $\theta$ | | | | $\theta$ | | | | $\theta$ | | | |
| $\gamma$ | 1 | 2 | 5 | 10 | $\gamma$ | 1 | 2 | 5 | 10 | $\gamma$ | 1 | 2 | 5 | 10 |
| 1.00 | 1.00 | 0.52 | 0.52 | 0.52 | 1.00 | 1.00 | 0.52 | 0.52 | 0.52 | 1.00 | 1.00 | 0.52 | 0.52 | 0.52 |
| 1.25 | 1.00 | 0.50 | 0.40 | 0.40 | 1.25 | 1.00 | 0.50 | 0.40 | 0.40 | 1.25 | 1.00 | 0.50 | 0.40 | 0.40 |
| 1.50 | 1.00 | 0.50 | 0.35 | 0.35 | 1.50 | 1.00 | 0.50 | 0.35 | 0.35 | 1.50 | 1.00 | 0.50 | 0.35 | 0.35 |
| 2.00 | 1.00 | 0.50 | 0.26 | 0.26 | 2.00 | 1.00 | 0.50 | 0.26 | 0.26 | 2.00 | 1.00 | 0.50 | 0.26 | 0.26 |
| 5.00 | 1.00 | 0.50 | 0.20 | 0.12 | 5.00 | 1.00 | 0.50 | 0.20 | 0.12 | 5.00 | 1.00 | 0.50 | 0.20 | 0.12 |

to-hold control

| | | | | | | | | | | | | |
|---|---|---|---|---|---|---|---|---|---|---|---|---|
| | | | | | | | $\nu$ | | | | | |
| | 0.00 | | | | 0.50 | | | | 0.99 | | | |
| | $\theta$ | | | | $\theta$ | | | | $\theta$ | | | |
| $\gamma$ | 1 | 2 | 5 | 10 | $\gamma$ | 1 | 2 | 5 | 10 | $\gamma$ | 1 | 2 | 5 | 10 |
| 1.00 | 1.00 | 0.50 | 0.50 | 0.50 | 1.00 | 1.00 | 0.50 | 0.50 | 0.50 | 1.00 | 1.00 | 0.50 | 0.50 | 0.50 |
| 1.25 | 1.00 | 0.50 | 0.34 | 0.34 | 1.25 | 1.00 | 0.50 | 0.34 | 0.34 | 1.25 | 1.00 | 0.50 | 0.34 | 0.34 |
| 1.50 | 1.00 | 0.50 | 0.34 | 0.34 | 1.50 | 1.00 | 0.50 | 0.34 | 0.34 | 1.50 | 1.00 | 0.50 | 0.34 | 0.34 |
| 2.00 | 1.00 | 0.50 | 0.26 | 0.26 | 2.00 | 1.00 | 0.50 | 0.26 | 0.26 | 2.00 | 1.00 | 0.50 | 0.26 | 0.26 |
| 5.00 | 1.00 | 0.50 | 0.21 | 0.20 | 5.00 | 1.00 | 0.50 | 0.21 | 0.20 | 5.00 | 1.00 | 0.50 | 0.21 | 0.20 |

model-based control

| | | | | | | | | | | | | |
|---|---|---|---|---|---|---|---|---|---|---|---|---|
| | | | | | | | $\nu$ | | | | | |
| | 0.00 | | | | 0.50 | | | | 0.99 | | | |
| | $\theta$ | | | | $\theta$ | | | | $\theta$ | | | |
| $\gamma$ | 1 | 2 | 5 | 10 | $\gamma$ | 1 | 2 | 5 | 10 | $\gamma$ | 1 | 2 | 5 | 10 |
| 1.00 | 1.00 | 0.50 | 0.50 | 0.50 | 1.00 | 1.00 | 0.50 | 0.50 | 0.50 | 1.00 | 1.00 | 0.50 | 0.50 | 0.50 |
| 1.25 | 1.00 | 0.50 | 0.34 | 0.34 | 1.25 | 1.00 | 0.50 | 0.34 | 0.34 | 1.25 | 1.00 | 0.50 | 0.34 | 0.34 |
| 1.50 | 1.00 | 0.50 | 0.34 | 0.34 | 1.50 | 1.00 | 0.50 | 0.34 | 0.34 | 1.50 | 1.00 | 0.50 | 0.34 | 0.34 |
| 2.00 | 1.00 | 0.50 | 0.24 | 0.24 | 2.00 | 1.00 | 0.50 | 0.24 | 0.24 | 2.00 | 1.00 | 0.50 | 0.24 | 0.24 |
| 5.00 | 1.00 | 0.50 | 0.20 | 0.11 | 5.00 | 1.00 | 0.50 | 0.20 | 0.11 | 5.00 | 1.00 | 0.50 | 0.20 | 0.11 |

Table C.12.: Lyapunov-based self-triggered control with optimized inputs, asymptotic behavior.

to-zero control

| | | | | | | | | | | | | | | |
|---|---|---|---|---|---|---|---|---|---|---|---|---|---|---|
| | | | | | | $\nu$ | | | | | | | | |
| | 0.00 | | | | | 0.50 | | | | | 0.99 | | | |
| | | $\theta$ | | | | | $\theta$ | | | | | $\theta$ | | |
| $\gamma$ | 1 | 2 | 5 | 10 | $\gamma$ | 1 | 2 | 5 | 10 | $\gamma$ | 1 | 2 | 5 | 10 |
| 1.00 | 1.00 | 0.50 | 0.50 | 0.50 | 1.00 | 1.00 | 0.50 | 0.50 | 0.50 | 1.00 | 1.00 | 0.50 | 0.50 | 0.50 |
| 1.25 | 1.00 | 0.50 | 0.34 | 0.34 | 1.25 | 1.00 | 0.50 | 0.34 | 0.34 | 1.25 | 1.00 | 0.50 | 0.34 | 0.34 |
| 1.50 | 1.00 | 0.50 | 0.34 | 0.34 | 1.50 | 1.00 | 0.50 | 0.34 | 0.34 | 1.50 | 1.00 | 0.50 | 0.34 | 0.34 |
| 2.00 | 1.00 | 0.50 | 0.21 | 0.21 | 2.00 | 1.00 | 0.50 | 0.21 | 0.21 | 2.00 | 1.00 | 0.50 | 0.21 | 0.21 |
| 5.00 | 1.00 | 0.50 | 0.20 | 0.10 | 5.00 | 1.00 | 0.50 | 0.20 | 0.10 | 5.00 | 1.00 | 0.50 | 0.20 | 0.10 |

to-hold control

| | | | | | | | | | | | | | | |
|---|---|---|---|---|---|---|---|---|---|---|---|---|---|---|
| | | | | | | $\nu$ | | | | | | | | |
| | 0.00 | | | | | 0.50 | | | | | 0.99 | | | |
| | | $\theta$ | | | | | $\theta$ | | | | | $\theta$ | | |
| $\gamma$ | 1 | 2 | 5 | 10 | $\gamma$ | 1 | 2 | 5 | 10 | $\gamma$ | 1 | 2 | 5 | 10 |
| 1.00 | 1.00 | 0.50 | 0.50 | 0.50 | 1.00 | 1.00 | 0.50 | 0.50 | 0.50 | 1.00 | 1.00 | 0.50 | 0.50 | 0.50 |
| 1.25 | 1.00 | 0.50 | 0.34 | 0.34 | 1.25 | 1.00 | 0.50 | 0.34 | 0.34 | 1.25 | 1.00 | 0.50 | 0.34 | 0.34 |
| 1.50 | 1.00 | 0.50 | 0.34 | 0.34 | 1.50 | 1.00 | 0.50 | 0.34 | 0.34 | 1.50 | 1.00 | 0.50 | 0.34 | 0.34 |
| 2.00 | 1.00 | 0.50 | 0.21 | 0.21 | 2.00 | 1.00 | 0.50 | 0.21 | 0.21 | 2.00 | 1.00 | 0.50 | 0.21 | 0.21 |
| 5.00 | 1.00 | 0.50 | 0.20 | 0.10 | 5.00 | 1.00 | 0.50 | 0.20 | 0.10 | 5.00 | 1.00 | 0.50 | 0.20 | 0.10 |

model-based control

| | | | | | | | | | | | | | | |
|---|---|---|---|---|---|---|---|---|---|---|---|---|---|---|
| | | | | | | $\nu$ | | | | | | | | |
| | 0.00 | | | | | 0.50 | | | | | 0.99 | | | |
| | | $\theta$ | | | | | $\theta$ | | | | | $\theta$ | | |
| $\gamma$ | 1 | 2 | 5 | 10 | $\gamma$ | 1 | 2 | 5 | 10 | $\gamma$ | 1 | 2 | 5 | 10 |
| 1.00 | 1.00 | 0.50 | 0.50 | 0.50 | 1.00 | 1.00 | 0.50 | 0.50 | 0.50 | 1.00 | 1.00 | 0.50 | 0.50 | 0.50 |
| 1.25 | 1.00 | 0.50 | 0.34 | 0.34 | 1.25 | 1.00 | 0.50 | 0.34 | 0.34 | 1.25 | 1.00 | 0.50 | 0.34 | 0.34 |
| 1.50 | 1.00 | 0.50 | 0.34 | 0.34 | 1.50 | 1.00 | 0.50 | 0.34 | 0.34 | 1.50 | 1.00 | 0.50 | 0.34 | 0.34 |
| 2.00 | 1.00 | 0.50 | 0.20 | 0.20 | 2.00 | 1.00 | 0.50 | 0.20 | 0.20 | 2.00 | 1.00 | 0.50 | 0.20 | 0.20 |
| 5.00 | 1.00 | 0.50 | 0.20 | 0.10 | 5.00 | 1.00 | 0.50 | 0.20 | 0.10 | 5.00 | 1.00 | 0.50 | 0.20 | 0.10 |

Table C.13.: Mixed set–Lyapunov-based event-triggered control, asymptotic behavior.

to-zero control

| | | | | | | | | | | | | | |
|------|------|------|------|------|------|------|------|------|------|------|------|------|------|
| | | | | | | $\nu$ | | | | | | | |
| | 0.00 | | | | | 0.50 | | | | | 0.99 | | |
| | $\theta$ | | | | | $\theta$ | | | | | $\theta$ | | |
| $\gamma$ | 1 | 2 | 5 | 10 | $\gamma$ | 1 | 2 | 5 | 10 | $\gamma$ | 1 | 2 | 5 | 10 |
| 1.00 | 1.00 | 0.44 | 0.18 | 0.13 | 1.00 | 1.00 | 0.44 | 0.18 | 0.13 | 1.00 | 1.00 | 0.44 | 0.18 | 0.13 |
| 1.25 | 0.51 | 0.24 | 0.12 | 0.09 | 1.25 | 0.51 | 0.24 | 0.12 | 0.09 | 1.25 | 0.51 | 0.24 | 0.12 | 0.09 |
| 1.50 | 0.30 | 0.15 | 0.09 | 0.06 | 1.50 | 0.30 | 0.15 | 0.09 | 0.06 | 1.50 | 0.30 | 0.15 | 0.09 | 0.06 |
| 2.00 | 0.13 | 0.09 | 0.06 | 0.06 | 2.00 | 0.13 | 0.09 | 0.06 | 0.06 | 2.00 | 0.13 | 0.09 | 0.06 | 0.06 |
| 5.00 | 0.03 | 0.03 | 0.03 | 0.03 | 5.00 | 0.03 | 0.03 | 0.03 | 0.03 | 5.00 | 0.03 | 0.03 | 0.03 | 0.03 |

to-hold control

| | | | | | | | | | | | | | |
|------|------|------|------|------|------|------|------|------|------|------|------|------|------|
| | | | | | | $\nu$ | | | | | | | |
| | 0.00 | | | | | 0.50 | | | | | 0.99 | | |
| | $\theta$ | | | | | $\theta$ | | | | | $\theta$ | | |
| $\gamma$ | 1 | 2 | 5 | 10 | $\gamma$ | 1 | 2 | 5 | 10 | $\gamma$ | 1 | 2 | 5 | 10 |
| 1.00 | 1.00 | 0.46 | 0.21 | 0.20 | 1.00 | 1.00 | 0.46 | 0.21 | 0.20 | 1.00 | 1.00 | 0.46 | 0.21 | 0.20 |
| 1.25 | 0.49 | 0.24 | 0.18 | 0.18 | 1.25 | 0.49 | 0.24 | 0.18 | 0.18 | 1.25 | 0.49 | 0.24 | 0.18 | 0.18 |
| 1.50 | 0.28 | 0.19 | 0.19 | 0.18 | 1.50 | 0.28 | 0.19 | 0.19 | 0.18 | 1.50 | 0.28 | 0.19 | 0.19 | 0.18 |
| 2.00 | 0.17 | 0.17 | 0.17 | 0.17 | 2.00 | 0.17 | 0.17 | 0.17 | 0.17 | 2.00 | 0.17 | 0.17 | 0.17 | 0.17 |
| 5.00 | 0.14 | 0.14 | 0.13 | 0.12 | 5.00 | 0.14 | 0.14 | 0.13 | 0.12 | 5.00 | 0.14 | 0.14 | 0.13 | 0.12 |

model-based control

| | | | | | | | | | | | | | |
|------|------|------|------|------|------|------|------|------|------|------|------|------|------|
| | | | | | | $\nu$ | | | | | | | |
| | 0.00 | | | | | 0.50 | | | | | 0.99 | | |
| | $\theta$ | | | | | $\theta$ | | | | | $\theta$ | | |
| $\gamma$ | 1 | 2 | 5 | 10 | $\gamma$ | 1 | 2 | 5 | 10 | $\gamma$ | 1 | 2 | 5 | 10 |
| 1.00 | 1.00 | 0.43 | 0.14 | 0.09 | 1.00 | 1.00 | 0.43 | 0.14 | 0.09 | 1.00 | 1.00 | 0.43 | 0.14 | 0.09 |
| 1.25 | 0.46 | 0.17 | 0.09 | 0.06 | 1.25 | 0.46 | 0.17 | 0.09 | 0.06 | 1.25 | 0.46 | 0.17 | 0.09 | 0.06 |
| 1.50 | 0.20 | 0.12 | 0.06 | 0.05 | 1.50 | 0.20 | 0.12 | 0.06 | 0.05 | 1.50 | 0.20 | 0.12 | 0.06 | 0.05 |
| 2.00 | 0.09 | 0.06 | 0.04 | 0.04 | 2.00 | 0.09 | 0.06 | 0.04 | 0.04 | 2.00 | 0.09 | 0.06 | 0.04 | 0.04 |
| 5.00 | 0.03 | 0.03 | 0.02 | 0.02 | 5.00 | 0.03 | 0.03 | 0.02 | 0.02 | 5.00 | 0.03 | 0.03 | 0.02 | 0.02 |

Table C.14.: Mixed set–Lyapunov-based self-triggered control with predefined inputs, asymptotic behavior.

to-zero control

| | $\nu$ | | | | | | | | | | | | | | |
| | 0.00 | | | | | 0.50 | | | | | 0.99 | | | |
| | $\theta$ | | | | | $\theta$ | | | | | $\theta$ | | | |
| $\gamma$ | 1 | 2 | 5 | 10 | $\gamma$ | 1 | 2 | 5 | 10 | $\gamma$ | 1 | 2 | 5 | 10 |
|---|---|---|---|---|---|---|---|---|---|---|---|---|---|---|
| 1.00 | 1.00 | 1.00 | 0.80 | 0.59 | 1.00 | 1.00 | 1.00 | 0.80 | 0.59 | 1.00 | 1.00 | 1.00 | 0.80 | 0.59 |
| 1.25 | 1.00 | 1.00 | 0.57 | 0.50 | 1.25 | 1.00 | 1.00 | 0.57 | 0.50 | 1.25 | 1.00 | 1.00 | 0.57 | 0.50 |
| 1.50 | 1.00 | 0.65 | 0.48 | 0.39 | 1.50 | 1.00 | 0.65 | 0.48 | 0.39 | 1.50 | 1.00 | 0.65 | 0.48 | 0.39 |
| 2.00 | 1.00 | 0.50 | 0.37 | 0.29 | 2.00 | 1.00 | 0.50 | 0.37 | 0.29 | 2.00 | 1.00 | 0.50 | 0.37 | 0.29 |
| 5.00 | 1.00 | 0.50 | 0.20 | 0.13 | 5.00 | 1.00 | 0.50 | 0.20 | 0.13 | 5.00 | 1.00 | 0.50 | 0.20 | 0.13 |

to-hold control

| | $\nu$ | | | | | | | | | | | | | | |
| | 0.00 | | | | | 0.50 | | | | | 0.99 | | | |
| | $\theta$ | | | | | $\theta$ | | | | | $\theta$ | | | |
| $\gamma$ | 1 | 2 | 5 | 10 | $\gamma$ | 1 | 2 | 5 | 10 | $\gamma$ | 1 | 2 | 5 | 10 |
|---|---|---|---|---|---|---|---|---|---|---|---|---|---|---|
| 1.00 | 1.00 | 1.00 | 0.70 | 0.52 | 1.00 | 1.00 | 1.00 | 0.70 | 0.52 | 1.00 | 1.00 | 1.00 | 0.70 | 0.52 |
| 1.25 | 1.00 | 1.00 | 0.51 | 0.46 | 1.25 | 1.00 | 1.00 | 0.51 | 0.46 | 1.25 | 1.00 | 1.00 | 0.51 | 0.46 |
| 1.50 | 1.00 | 0.56 | 0.44 | 0.34 | 1.50 | 1.00 | 0.56 | 0.44 | 0.34 | 1.50 | 1.00 | 0.56 | 0.44 | 0.34 |
| 2.00 | 1.00 | 0.50 | 0.36 | 0.29 | 2.00 | 1.00 | 0.50 | 0.36 | 0.29 | 2.00 | 1.00 | 0.50 | 0.36 | 0.29 |
| 5.00 | 1.00 | 0.50 | 0.21 | 0.20 | 5.00 | 1.00 | 0.50 | 0.21 | 0.20 | 5.00 | 1.00 | 0.50 | 0.21 | 0.20 |

model-based control

| | $\nu$ | | | | | | | | | | | | | | |
| | 0.00 | | | | | 0.50 | | | | | 0.99 | | | |
| | $\theta$ | | | | | $\theta$ | | | | | $\theta$ | | | |
| $\gamma$ | 1 | 2 | 5 | 10 | $\gamma$ | 1 | 2 | 5 | 10 | $\gamma$ | 1 | 2 | 5 | 10 |
|---|---|---|---|---|---|---|---|---|---|---|---|---|---|---|
| 1.00 | 1.00 | 1.00 | 0.73 | 0.53 | 1.00 | 1.00 | 1.00 | 0.73 | 0.53 | 1.00 | 1.00 | 1.00 | 0.73 | 0.53 |
| 1.25 | 1.00 | 1.00 | 0.51 | 0.45 | 1.25 | 1.00 | 1.00 | 0.51 | 0.45 | 1.25 | 1.00 | 1.00 | 0.51 | 0.45 |
| 1.50 | 1.00 | 0.50 | 0.42 | 0.34 | 1.50 | 1.00 | 0.50 | 0.42 | 0.34 | 1.50 | 1.00 | 0.50 | 0.42 | 0.34 |
| 2.00 | 1.00 | 0.50 | 0.34 | 0.26 | 2.00 | 1.00 | 0.50 | 0.34 | 0.26 | 2.00 | 1.00 | 0.50 | 0.34 | 0.26 |
| 5.00 | 1.00 | 0.50 | 0.20 | 0.12 | 5.00 | 1.00 | 0.50 | 0.20 | 0.12 | 5.00 | 1.00 | 0.50 | 0.20 | 0.12 |

Table C.15.: Mixed set–Lyapunov-based self-triggered control with optimized inputs, asymptotic behavior.

to-zero control

| | $\nu$ | | | | | | | | | | | | | |
|---|---|---|---|---|---|---|---|---|---|---|---|---|---|---|
| | 0.00 | | | | | 0.50 | | | | | 0.99 | | | |
| | $\theta$ | | | | | $\theta$ | | | | | $\theta$ | | | |
| $\gamma$ | 1 | 2 | 5 | 10 | $\gamma$ | 1 | 2 | 5 | 10 | $\gamma$ | 1 | 2 | 5 | 10 |
| 1.00 | 1.00 | 0.51 | 0.51 | 0.51 | 1.00 | 1.00 | 0.51 | 0.51 | 0.51 | 1.00 | 1.00 | 0.51 | 0.51 | 0.51 |
| 1.25 | 1.00 | 0.51 | 0.37 | 0.37 | 1.25 | 1.00 | 0.51 | 0.37 | 0.37 | 1.25 | 1.00 | 0.51 | 0.37 | 0.37 |
| 1.50 | 1.00 | 0.50 | 0.34 | 0.34 | 1.50 | 1.00 | 0.50 | 0.34 | 0.34 | 1.50 | 1.00 | 0.50 | 0.34 | 0.34 |
| 2.00 | 1.00 | 0.50 | 0.26 | 0.26 | 2.00 | 1.00 | 0.50 | 0.26 | 0.26 | 2.00 | 1.00 | 0.50 | 0.26 | 0.26 |
| 5.00 | 1.00 | 0.50 | 0.20 | 0.12 | 5.00 | 1.00 | 0.50 | 0.20 | 0.12 | 5.00 | 1.00 | 0.50 | 0.20 | 0.12 |

to-hold control

| | $\nu$ | | | | | | | | | | | | | |
|---|---|---|---|---|---|---|---|---|---|---|---|---|---|---|
| | 0.00 | | | | | 0.50 | | | | | 0.99 | | | |
| | $\theta$ | | | | | $\theta$ | | | | | $\theta$ | | | |
| $\gamma$ | 1 | 2 | 5 | 10 | $\gamma$ | 1 | 2 | 5 | 10 | $\gamma$ | 1 | 2 | 5 | 10 |
| 1.00 | 1.00 | 0.51 | 0.51 | 0.51 | 1.00 | 1.00 | 0.51 | 0.51 | 0.51 | 1.00 | 1.00 | 0.51 | 0.51 | 0.51 |
| 1.25 | 1.00 | 0.51 | 0.37 | 0.37 | 1.25 | 1.00 | 0.51 | 0.37 | 0.37 | 1.25 | 1.00 | 0.51 | 0.37 | 0.37 |
| 1.50 | 1.00 | 0.50 | 0.34 | 0.34 | 1.50 | 1.00 | 0.50 | 0.34 | 0.34 | 1.50 | 1.00 | 0.50 | 0.34 | 0.34 |
| 2.00 | 1.00 | 0.50 | 0.26 | 0.26 | 2.00 | 1.00 | 0.50 | 0.26 | 0.26 | 2.00 | 1.00 | 0.50 | 0.26 | 0.26 |
| 5.00 | 1.00 | 0.50 | 0.20 | 0.12 | 5.00 | 1.00 | 0.50 | 0.20 | 0.12 | 5.00 | 1.00 | 0.50 | 0.20 | 0.12 |

model-based control

| | $\nu$ | | | | | | | | | | | | | |
|---|---|---|---|---|---|---|---|---|---|---|---|---|---|---|
| | 0.00 | | | | | 0.50 | | | | | 0.99 | | | |
| | $\theta$ | | | | | $\theta$ | | | | | $\theta$ | | | |
| $\gamma$ | 1 | 2 | 5 | 10 | $\gamma$ | 1 | 2 | 5 | 10 | $\gamma$ | 1 | 2 | 5 | 10 |
| 1.00 | 1.00 | 0.51 | 0.51 | 0.51 | 1.00 | 1.00 | 0.51 | 0.51 | 0.51 | 1.00 | 1.00 | 0.51 | 0.51 | 0.51 |
| 1.25 | 1.00 | 0.51 | 0.35 | 0.35 | 1.25 | 1.00 | 0.51 | 0.35 | 0.35 | 1.25 | 1.00 | 0.51 | 0.35 | 0.35 |
| 1.50 | 1.00 | 0.50 | 0.34 | 0.34 | 1.50 | 1.00 | 0.50 | 0.34 | 0.34 | 1.50 | 1.00 | 0.50 | 0.34 | 0.34 |
| 2.00 | 1.00 | 0.50 | 0.26 | 0.26 | 2.00 | 1.00 | 0.50 | 0.26 | 0.26 | 2.00 | 1.00 | 0.50 | 0.26 | 0.26 |
| 5.00 | 1.00 | 0.50 | 0.20 | 0.12 | 5.00 | 1.00 | 0.50 | 0.20 | 0.12 | 5.00 | 1.00 | 0.50 | 0.20 | 0.12 |

Table C.16.: Purely set-based event-triggered control, asymptotic behavior.

**to-zero control**

| $\gamma$ | $\theta$ = 1 | 2 | 5 | 10 |
|---|---|---|---|---|
| 1.00 | 1.00 | 0.44 | 0.18 | 0.13 |
| 1.25 | 0.51 | 0.24 | 0.12 | 0.09 |
| 1.50 | 0.30 | 0.15 | 0.09 | 0.06 |
| 2.00 | 0.13 | 0.09 | 0.06 | 0.06 |
| 5.00 | 0.03 | 0.03 | 0.03 | 0.03 |

**to-hold control**

| $\gamma$ | $\theta$ = 1 | 2 | 5 | 10 |
|---|---|---|---|---|
| 1.00 | 1.00 | 0.46 | 0.21 | 0.20 |
| 1.25 | 0.49 | 0.24 | 0.18 | 0.18 |
| 1.50 | 0.28 | 0.19 | 0.19 | 0.18 |
| 2.00 | 0.17 | 0.17 | 0.17 | 0.17 |
| 5.00 | 0.14 | 0.14 | 0.13 | 0.12 |

**model-based control**

| $\gamma$ | $\theta$ = 1 | 2 | 5 | 10 |
|---|---|---|---|---|
| 1.00 | 1.00 | 0.43 | 0.14 | 0.09 |
| 1.25 | 0.46 | 0.17 | 0.09 | 0.06 |
| 1.50 | 0.20 | 0.12 | 0.06 | 0.05 |
| 2.00 | 0.09 | 0.06 | 0.04 | 0.04 |
| 5.00 | 0.03 | 0.03 | 0.02 | 0.02 |

Table C.17.: Purely set-based self-triggered control with predefined inputs, asymptotic behavior.

**to-zero control**

| $\gamma$ | $\theta$ = 1 | 2 | 5 | 10 |
|---|---|---|---|---|
| 1.00 | 1.00 | 1.00 | 0.80 | 0.59 |
| 1.25 | 1.00 | 1.00 | 0.57 | 0.50 |
| 1.50 | 1.00 | 0.65 | 0.48 | 0.39 |
| 2.00 | 1.00 | 0.50 | 0.37 | 0.29 |
| 5.00 | 1.00 | 0.50 | 0.20 | 0.13 |

**to-hold control**

| $\gamma$ | $\theta$ = 1 | 2 | 5 | 10 |
|---|---|---|---|---|
| 1.00 | 1.00 | 1.00 | 0.70 | 0.52 |
| 1.25 | 1.00 | 1.00 | 0.51 | 0.46 |
| 1.50 | 1.00 | 0.56 | 0.44 | 0.34 |
| 2.00 | 1.00 | 0.50 | 0.36 | 0.29 |
| 5.00 | 1.00 | 0.50 | 0.21 | 0.20 |

**model-based control**

| $\gamma$ | $\theta$ = 1 | 2 | 5 | 10 |
|---|---|---|---|---|
| 1.00 | 1.00 | 1.00 | 0.73 | 0.53 |
| 1.25 | 1.00 | 1.00 | 0.51 | 0.45 |
| 1.50 | 1.00 | 0.50 | 0.42 | 0.34 |
| 2.00 | 1.00 | 0.50 | 0.34 | 0.26 |
| 5.00 | 1.00 | 0.50 | 0.20 | 0.12 |

Table C.18.: Purely set-based self-triggered control with optimized inputs, asymptotic behavior.

**to-zero control**

| $\gamma$ | $\theta$ = 1 | 2 | 5 | 10 |
|---|---|---|---|---|
| 1.00 | 1.00 | 0.51 | 0.51 | 0.51 |
| 1.25 | 1.00 | 0.51 | 0.37 | 0.37 |
| 1.50 | 1.00 | 0.50 | 0.34 | 0.34 |
| 2.00 | 1.00 | 0.50 | 0.26 | 0.26 |
| 5.00 | 1.00 | 0.50 | 0.20 | 0.12 |

**to-hold control**

| $\gamma$ | $\theta$ = 1 | 2 | 5 | 10 |
|---|---|---|---|---|
| 1.00 | 1.00 | 0.51 | 0.51 | 0.51 |
| 1.25 | 1.00 | 0.51 | 0.37 | 0.37 |
| 1.50 | 1.00 | 0.50 | 0.34 | 0.34 |
| 2.00 | 1.00 | 0.50 | 0.26 | 0.26 |
| 5.00 | 1.00 | 0.50 | 0.20 | 0.12 |

**model-based control**

| $\gamma$ | $\theta$ = 1 | 2 | 5 | 10 |
|---|---|---|---|---|
| 1.00 | 1.00 | 0.51 | 0.51 | 0.51 |
| 1.25 | 1.00 | 0.51 | 0.35 | 0.35 |
| 1.50 | 1.00 | 0.50 | 0.34 | 0.34 |
| 2.00 | 1.00 | 0.50 | 0.26 | 0.26 |
| 5.00 | 1.00 | 0.50 | 0.20 | 0.12 |

# Bibliography

[Abdelrahim et al., 2017] Abdelrahim, M., Postoyan, R., Daafouz, J., and Nešić, D. (2017). Robust event-triggered output feedback controllers for nonlinear systems. *Automatica*, 75:96–108.

[Aggoune et al., 2014] Aggoune, W., Castillo Toledo, B., and Di Gennaro, S. (2014). Self-triggered robust control of nonlinear stochastic systems. In Djemai, M. and Defoort, M., editors, *Hybrid Dynamical Systems, Lecture Notes in Control and Information Sciences, volume 457*, pages 277–292. Springer.

[Almeida et al., 2015] Almeida, J., Silvestre, C., and Pascoal, A. M. (2015). Self-triggered state-feedback control of linear plants under bounded disturbances. *International Journal of Robust and Nonlinear Control*, 25(8):1230–1246.

[Anderson, 1958] Anderson, T. W. (1958). *An introduction to multivariate statistical analysis*. Wiley.

[Anta and Tabuada, 2010] Anta, A. and Tabuada, P. (2010). To sample or not to sample: Self-triggered control for nonlinear systems. *IEEE Transactions on Automatic Control*, 55(9):2030–2042.

[Antunes and Heemels, 2014] Antunes, D. and Heemels, W. P. M. H. (2014). Rollout event-triggered control: Beyond periodic control performance. *IEEE Transactions on Automatic Control*, 59(12):3296–3311.

[Antunes et al., 2012] Antunes, D. J., Hespanha, J. P., and Silvestre, C. J. (2012). Volterra integral approach to impulsive renewal systems: Application to networked control. *IEEE Transactions on Automatic Control*, 57(3):607–619.

[Araújo et al., 2012] Araújo, J., Fawzi, H., Mazo, M., Tabuada, P., and Johansson, K. H. (2012). An improved self-triggered implementation for linear controllers. In *Proceedings of the 3rd IFAC Workshop on Distributed Estimation and Control in Networked Systems (NECSYS)*, pages 37–42, Santa Barbara, CA, USA.

[Årzén, 1999] Årzén, K. (1999). A simple event-based PID controller. In *Proceedings of the 14th IFAC World Congress*, pages 423–428, Beijing, China.

[Åström and Bernhardsson, 2002] Åström, K. J. and Bernhardsson, B. M. (2002). Riemann and Lebesgue sampling. In *Proceedings of the 41st IEEE Conference on Decision and Control (CDC)*, pages 2011–2016, Las Vegas, NV, USA.

[Aydiner, 2014] Aydiner, E. (2014). *Robust Self-Triggered Model Predictive Control for Discrete-Time Linear Systems based on Homothetic Tubes*. Diploma thesis, University of Stuttgart.

[Bacic et al., 2003] Bacic, M., Cannon, M., Lee, Y. I., and Kouvaritakis, B. (2003). General interpolation in MPC and its advantages. *IEEE Transactions on Automatic Control*, 48(6):1092–1096.

[Bartle, 1995] Bartle, R. G. (1995). *The Elements of Integration and Lebesgue Measures*. Wiley.

[Battistelli et al., 2012] Battistelli, G., Benavoli, A., and Chisci, L. (2012). Data-driven communication for state estimation with sensor networks. *Automatica*, 48(5):926–935.

[Bemporad et al., 2002] Bemporad, A., Borrelli, F., and Morari, M. (2002). Model predictive control based on linear programming—the explicit solution. *IEEE Transactions on Automatic Control*, 47(12):1974–1985.

[Bemporad and Garulli, 2000] Bemporad, A. and Garulli, A. (2000). Output-feedback predictive control of constrained linear systems via set-membership state estimation. *International Journal of Control*, 73(8):655–665.

[Bernardini and Bemporad, 2012] Bernardini, D. and Bemporad, A. (2012). Energy-aware robust model predictive control based on noisy wireless sensors. *Automatica*, 48(1):36–44.

[Berner and Mönnigmann, 2016] Berner, P. S. and Mönnigmann, M. (2016). A comparison of four variants of event-triggered networked MPC. In *IEEE Multi-Conference on Systems and Control*, pages 1519–1524, Buenos Aires, Argentina.

[Blanchini, 1990] Blanchini, F. (1990). Control synthesis for discrete time systems with control and state bounds in the presence of disturbances. *Journal of Optimization Theory and Applications*, 65(1):29–40.

[Blanchini, 1994] Blanchini, F. (1994). Ultimate boundedness control for uncertain discrete-time systems via set-induced Lyapunov functions. *IEEE Transactions on Automatic Control*, 39(2):428–433.

[Blanchini et al., 1995] Blanchini, F., Mesquine, F., and Miani, S. (1995). Constrained stabilization with an assigned initial condition set. *International Journal of Control*, 62(3):601–617.

[Blanchini and Miani, 2008] Blanchini, F. and Miani, S. (2008). *Set-Theoretic Methods in Control*. Birkhäuser.

[Blind and Allgöwer, 2011] Blind, R. and Allgöwer, F. (2011). Analysis of networked event-based control with a shared communication medium: Part I – pure ALOHA. In *Proceedings of the 18th IFAC World Congress*, pages 10092–10097, Milano, Italy.

[Blind and Allgöwer, 2012] Blind, R. and Allgöwer, F. (2012). The performance of event-based control for scalar systems with packet losses. In *Proceedings of the 51st IEEE Conference on Decision and Control (CDC)*, pages 6572–6576, Maui, HI, USA.

[Blind and Allgöwer, 2013] Blind, R. and Allgöwer, F. (2013). On time-triggered and event-based control of integrator systems over a shared communication system. *Mathematics of Control, Signals, and Systems*, 25(4):517–557.

[Boisseau et al., 2017] Boisseau, B., Martinez, J. J., Raharijaona, T., Durand, S., and Marchand, N. (2017). Event-switched control design with guaranteed performances. *International Journal of Robust and Nonlinear Control*, 27(15):2492–2509.

[Borgers and Heemels, 2014] Borgers, D. P. and Heemels, W. P. M. H. (2014). Event-separation properties of event-triggered control systems. *IEEE Transactions on Automatic Control*, 59(10):2644–2656.

[Brockett, 1997] Brockett, R. W. (1997). Minimum attention control. In *Proceedings of the 36th IEEE Conference on Decision and Control (CDC)*, pages 2628–2632, San Diego, CA, USA.

[Brunner and Allgöwer, 2014] Brunner, F. D. and Allgöwer, F. (2014). Approximate predictive control of polytopic systems. In *Proceedings of the 19th IFAC World Congress*, pages 11060–11066, Cape Town, South Africa.

[Brunner and Allgöwer, 2016] Brunner, F. D. and Allgöwer, F. (2016). A Lyapunov function approach to the event-triggered stabilization of the minimal robust positively invariant set. In *Proceedings of the 6th IFAC Workshop on Distributed Estimation and Control in Networked Systems (NECSYS)*, pages 25–30, Tokyo, Japan.

[Brunner et al., 2017a] Brunner, F. D., Antunes, D., and Allgöwer, F. (2017). Stochastic thresholds in event-triggered control: A consistent policy for quadratic control. *Automatica*. Accepted for publication in September 2017.

[Brunner et al., 2015a] Brunner, F. D., Gommans, T. M. P., Heemels, W. P. M. H., and Allgöwer, F. (2015). Resource-aware set-valued estimation for discrete-time linear systems. In *Proceedings of the 54th IEEE Conference on Decision and Control (CDC)*, pages 5480–5486, Osaka, Japan.

[Brunner et al., 2016a] Brunner, F. D., Heemels, M., and Allgöwer, F. (2016). Robust self-triggered MPC for constrained linear systems: A tube-based approach. *Automatica*, 72:73–83.

[Brunner et al., 2014] Brunner, F. D., Heemels, W. P. M. H., and Allgöwer, F. (2014). Robust self-triggered MPC for constrained linear systems. In *Proceedings of the European Control Conference (ECC)*, pages 472–477, Strasbourg, France.

[Brunner et al., 2015b] Brunner, F. D., Heemels, W. P. M. H., and Allgöwer, F. (2015). Robust event-triggered MPC for constrained linear discrete-time systems with guaranteed average sampling rate. In *Proceedings of the 5th IFAC Conference on Nonlinear Model Predictive Control (NMPC)*, pages 117–122, Seville, Spain.

[Brunner et al., 2016b] Brunner, F. D., Heemels, W. P. M. H., and Allgöwer, F. (2016). Dynamic thresholds in robust event-triggered control for discrete-time linear systems. In *Proceedings of the European Control Conference (ECC)*, pages 923–988, Aalborg, Denmark.

[Brunner et al., 2016c] Brunner, F. D., Heemels, W. P. M. H., and Allgöwer, F. (2016). γ-invasive event-triggered and self-triggered control for perturbed linear systems. In *Proceedings of the 55th IEEE Conference on Decision and Control (CDC)*, pages 1346–1351, Las Vegas, NV, USA.

[Brunner et al., 2016d] Brunner, F. D., Heemels, W. P. M. H., and Allgöwer, F. (2016). Numerical evaluation of a robust self-triggered MPC algorithm. In *Proceedings of the 6th IFAC Workshop on Distributed Estimation and Control in Networked Systems (NECSYS)*, pages 151–156, Tokyo, Japan.

[Brunner et al., 2017b] Brunner, F. D., Heemels, W. P. M. H., and Allgöwer, F. (2017). Event-triggered and self-triggered control for linear systems based on reachable sets. Submitted to Automatica in February 2017.

[Brunner et al., 2017c] Brunner, F. D., Heemels, W. P. M. H., and Allgöwer, F. (2017). Robust event-triggered MPC with guaranteed asymptotic bound and average sampling rate. *IEEE Transactions on Automatic Control*, 62(5694–5709):11.

[Brunner et al., 2016e] Brunner, F. D., Müller, M. A., and Allgöwer, F. (2016). Enhancing output feedback MPC for linear discrete-time systems with set-valued moving horizon estimation. In *Proceedings of the 55th IEEE Conference on Decision and Control (CDC)*, pages 2733–2738, Las Vegas, NV, USA.

[Brunner et al., 2017d] Brunner, F. D., Müller, M. A., and Allgöwer, F. (2017). Enhancing output-feedback MPC with set-valued moving horizon estimation. *IEEE Transactions on Automatic Control*. Accepted for publication in October 2017.

[Cassandras, 2014] Cassandras, C. G. (2014). The event-driven paradigm for control, communication and optimization. *Journal of Control and Decision*, 1(1):3–17.

[Cerf et al., 2016] Cerf, S., Berekmeri, M., Robu, B., Marchand, N., and Bouchenak, S. (2016). Cost function based event triggered model predictive controllers application to big data cloud services. In *Proceedings of the 55th IEEE Conference on Decision and Control (CDC)*, pages 1657–1662, Las Vegas, NV, USA.

[Cervin and Henningsson, 2008] Cervin, A. and Henningsson, T. (2008). Scheduling of event-triggered controllers on a shared network. In *Proceedings of the 47th IEEE Conference on Decision and Control (CDC)*, pages 3601–3606, Cancun, Mexico.

[Chen et al., 2015] Chen, W., Wang, J., Shi, L., and Shi, D. (2015). State estimation of finite-state hidden Markov models subject to stochastically event-triggered measurements. In *Proceedings of the 54th IEEE Conference on Decision and Control (CDC)*, pages 3712–3717, Osaka, Japan.

[Chisci et al., 2001] Chisci, L., Rossiter, J. A., and Zappa, G. (2001). Systems with persistent disturbances: predictive control with restricted constraints. *Automatica*, 37(7):1019–1028.

[Chisci and Zappa, 2002] Chisci, L. and Zappa, G. (2002). Feasibility in predictive control of constrained linear systems: The output feedback case. *International Journal of Robust and Nonlinear Control*, 12(5):465–487.

[Cogill, 2009] Cogill, R. (2009). Event-based control using quadratic approximate value functions. In *Proceedings of the 48th IEEE Conference Decision and Control (CDC), 28th Chinese Control Conference(CCC)*, pages 5883–5888, Shanghai, China.

[Copp and Hespanha, 2014] Copp, D. A. and Hespanha, P. (2014). Nonlinear output-feedback model predictive control with moving horizon estimation. In *Proceedings of the 53rd IEEE Conference on Decision and Control (CDC)*, pages 3511–3517, Los Angeles, CA, USA.

[De Nicolao, 1992] De Nicolao, G. (1992). On the time-varying Riccati difference equation of optimal filtering. *SIAM Journal on Control and Optimization*, 30(6):1251–1269.

[Delfour and Mitter, 1969] Delfour, M. C. and Mitter, S. K. (1969). Reachability of perturbed systems and min sup problems. *SIAM Journal on Control*, 7(4):521–533.

[Demirel et al., 2017a] Demirel, B., Ghadimi, E., Quevedo, D. E., and Johansson, M. (2017). Optimal control of linear systems with limited control actions: threshold-based event-triggered control. *IEEE Transactions on Control of Network Systems*. Accepted for publication, DOI: 10.1109/TCNS.2017.2701003.

[Demirel et al., 2017b] Demirel, B., Leong, A. S., and Quevedo, D. (2017). Performance analysis of event-triggered control systems with a probabilistic triggering mechanism: The scalar case. In *Proceedings of the 20th IFAC World Congress*, Toulouse, France. To appear.

[Di Benedetto et al., 2013] Di Benedetto, M., Di Gennaro, S., and D'Innocenzo, A. (2013). Digital self-triggered robust control of nonlinear systems. *International Journal of Control*, 86(9):1664–1672.

[Dimarogonas et al., 2012] Dimarogonas, D. V., Frazzoli, E., and Johansson, K. H. (2012). Distributed event-triggered control for multi-agent systems. *IEEE Transactions on Automatic Control*, 57(5):1291–1297.

[Dolk et al., 2017] Dolk, V. S., Borgers, D. P., and Heemels, W. P. M. H. (2017). Output-based and decentralized dynamic event-triggered control with guaranteed $\mathscr{L}_p$-gain performance and Zeno-freeness. *IEEE Transactions on Automatic Control*, 62(1):34–49.

[Donkers et al., 2012] Donkers, M., Heemels, W., Bernardini, D., Bemporad, A., and Shneer, V. (2012). Stability analysis of stochastic networked control systems. *Automatica*, 48(5):3684–3689.

[Donkers and Heemels, 2012] Donkers, M. C. F. and Heemels, W. P. M. H. (2012). Output-based event-triggered control with guaranteed $\mathscr{L}_\infty$-gain and improved and decentralized event-triggering. *IEEE Transactions on Automatic Control*, 57(6):1362–1376.

[Donkers et al., 2014] Donkers, M. C. F., Tabuada, P., and Heemels, W. P. M. H. (2014). Minimum attention control for linear systems: A linear programming approach. *Discrete Event Dynamic Systems*, 24(2):199–218.

[Doob, 1953] Doob, J. L. (1953). *Stochastic Processes*. Wiley.

[Ebner and Trimpe, 2016] Ebner, S. and Trimpe, S. (2016). Communication rate analysis for event-based state estimation. In *Proceedings of the 13th International Workshop on Discrete Event Systems*, pages 189–196, Xi'an, China.

[Eqtami et al., 2011a] Eqtami, A., Dimarogonas, D., and Kyriakopoulos, K. (2011). Novel event-triggered strategies for model predictive controllers. In *Proceedings of the 50th IEEE Conference on Decision and Control (CDC), European Control Conference (ECC)*, pages 3392–3397, Orlando, FL, USA.

[Eqtami et al., 2010] Eqtami, A., Dimarogonas, D., and Kyriakopoulos, K. J. (2010). Event-triggered control for discrete-time systems. In *Proceedings of the American Control Conference (ACC)*, pages 4719–4724, Baltimore, MD, USA.

[Eqtami et al., 2011b] Eqtami, A., Dimarogonas, D. V., and Kyriakopoulos, K. J. (2011). Event-triggered strategies for decentralized model predictive controllers. In *Proceedings of the 18th IFAC World Congress*, pages 10068–10073, Milano, Italy.

[Eqtami et al., 2013] Eqtami, A., Heshmati-alamdari, S., Dimarogonas, D., and Kyriakopoulos, K. (2013). Self-triggered model predictive control for nonholonomic systems. In *Proceedings of the European Control Conference (ECC)*, pages 638–643, Zürich, Switzerland.

[Eqtami, 2013] Eqtami, A. M. (2013). *Event-Based Model Predictive Controllers.* Doctoral thesis, National Technical University of Athens.

[Farina et al., 2015] Farina, M., Giulioni, L., Magni, L., and Scattolini, R. (2015). An approach to output-feedback MPC of stochastic linear discrete-time systems. *Automatica*, 55:140–149.

[Feeney and Nilsson, 2001] Feeney, L. M. and Nilsson, M. (2001). Investigating the energy consumption of a wireless network interface in an ad hoc networking environment. In *Proceedings of the Twentieth Annual Joint Conference of the IEEE Computer and Communications Societies (INFOCOM)*, pages 1548–1557, Anchorage, AK, USA.

[Findeisen and Allgöwer, 2004] Findeisen, R. and Allgöwer, F. (2004). Min-max output feedback predictive control with guaranteed stability. In *Proceedings of the 16th International Symposium on Mathematical Theory of Networks and Systems (MTNS)*, Leuven, Belgium. On CD–ROM.

[Fiter et al., 2015] Fiter, C., Hetel, L., Perruquetti, W., and Richard, J. P. (2015). A robust stability framework for LTI systems with time-varying sampling. *Automatica*, 54:56–64.

[Forni et al., 2014] Forni, F., Galeani, S., Nešić, D., and Zaccarian, L. (2014). Event-triggered transmission for linear control over communication channels. *Automatica*, 50(2):490–498.

[Gallieri, 2016] Gallieri, M. (2016). $\ell_{asso}$-*MPC—Predictive Control with $\ell_1$-Regularised Least Squares.* Springer.

[Gallieri and Maciejowski, 2012] Gallieri, M. and Maciejowski, J. M. (2012). $\ell_{asso}$-MPC: smart regulation of over-actuated systems. In *Proceedings of the American Control Conference (ACC)*, pages 1217–1222, Montréal, Canada.

[Garcia and Antsaklis, 2013] Garcia, E. and Antsaklis, P. J. (2013). Model-based event-triggered control for systems with quantization and time-varying network delays. *IEEE Transactions on Automatic Control*, 58(2):422–434.

[Ge et al., 2016] Ge, X., Han, Q.-L., and Yang, F. (2016). Event-based set-membership leader-following consensus of networked multi-agent systems subject to limited communication resources and unknown-but-bounded noise. *IEEE Transactions on Industrial Electronics*, 64(6):5045–5054.

[Geiselhart et al., 2014] Geiselhart, R., Gielen, R. H., Lazar, M., and Wirth, F. R. (2014). An alternative converse Lyapunov theorem for discrete-time systems. *Systems & Control Letters*, 70:49–59.

[Georgiev and Tilbury, 2004] Georgiev, D. and Tilbury, D. M. (2004). Packet-based control. In *Proceedings of the American Control Conference (ACC)*, pages 329–336, Boston, MA, USA.

[Ghaemi et al., 2008] Ghaemi, R., Sun, J., and Kolmanovsky, I. (2008). Less conservative robust control of constrained linear systems with bounded disturbances. In *Proceedings of the 47th IEEE Conference on Decision and Control (CDC)*, pages 983–988, Cancun, Mexico.

[Gielen, 2013] Gielen, R. H. (2013). *Stability analysis and control of discrete-time systems with delay Stability analysis and control of discrete-time systems with delay*. Doctoral thesis, Technische Universiteit Eindhoven.

[Gilbert and Tan, 1991] Gilbert, E. G. and Tan, K. T. (1991). Linear systems with state and control constraints: The theory and application of maximal output admissible sets. *IEEE Transactions on Automatic Control*, 36(9):1008–1020.

[Girard, 2015] Girard, A. (2015). Dynamic triggering mechanisms for event-triggered control. *IEEE Transactions on Automatic Control*, 60(7):1992–1997.

[Glover and Schweppe, 1971] Glover, J. and Schweppe, F. (1971). Control of linear dynamic systems with set constrained disturbances. *IEEE Transactions on Automatic Control*, 16(5):411–423.

[Gommans et al., 2014] Gommans, T., Antunes, D., Donkers, T., Tabuada, P., and Heemels, M. (2014). Self-triggered linear quadratic control. *Automatica*, 50(4):1279–1287.

[Gommans, 2016] Gommans, T. M. P. (2016). *Resource-aware control and estimation: an optimization-based approach*. Doctoral thesis, Technische Universiteit Eindhoven.

[Gommans and Heemels, 2015] Gommans, T. M. P. and Heemels, W. P. M. H. (2015). Resource-aware MPC for constrained nonlinear systems: A self-triggered control approach. *Systems & Control Letters*, 79:59–67.

[Gommans et al., 2017] Gommans, T. M. P., Theunisse, T. A. F., Antunes, D. J., and Heemels, W. P. M. H. (2017). Resource-aware MPC for constrained linear systems: Two rollout approaches. *Journal of Process Control*, 51:68–83.

[Goulart and Kerrigan, 2007] Goulart, P. J. and Kerrigan, E. C. (2007). Output feedback receding horizon control of constrained systems. *International Journal of Control*, 80(1):8–20.

[Grüne and Müller, 2009] Grüne, L. and Müller, F. (2009). An algorithm for event-based optimal feedback control. In *Proceedings of the 48th IEEE Conference Decision and Control (CDC), 28th Chinese Control Conference(CCC)*, pages 5311–5316, Shanghai, China.

[Grüne et al., 2010] Grüne, L., Müller, F., Jerg, S., Junge, O., Post, M., Lehmann, D., and Lunze, J. (2010). Two complementary approches to event-based control. *at-Automatisierungstechnik*, 58(4):173–183.

[Gutman and Cwikel, 1986] Gutman, P.-O. and Cwikel, M. (1986). Admissible sets and feedback control for discrete-time linear dynamical systems with bounded controls and states. *IEEE Transactions on Automatic Control*, 3(4):373–376.

[Hahn, 1967] Hahn, W. (1967). *Stability of Motion*. Springer.

[Han et al., 2015] Han, D., Mo, Y., Wu, J., Weerakkody, S., Sinopoli, B., and Shi, L. (2015). Stochastic event-triggered sensor schedule for remote state estimation. *IEEE Transactions on Automatic Control*, 60(10):2661–2675.

[Hashimoto et al., 2015a] Hashimoto, K., Adachi, S., and Dimarogonas, D. V. (2015). Distributed aperiodic model predictive control for multi-agent systems. *IET Control Theory & Applications*, 9(1):10–20.

[Hashimoto et al., 2015b] Hashimoto, K., Adachi, S., and Dimarogonas, D. V. (2015). Time-constrained event-triggered model predictive control for nonlinear continuous-time systems. In *Proceedings of the 54th IEEE Conference on Decision and Control (CDC)*, pages 4326–4331, Osaka, Japan.

[Hashimoto et al., 2017] Hashimoto, K., Adachi, S., and Dimarogonas, D. V. (2017). Self-triggered control for constrained systems : a contractive set-based approach. In *Proceedings of the American Control Conference (ACC)*, pages 1011–1016, Seattle, WA, USA.

[Heemels and Donkers, 2013] Heemels, W. P. M. H. and Donkers, M. C. F. (2013). Model-based periodic event-triggered control for linear systems. *Automatica*, 49(3):698–711.

[Heemels et al., 2014] Heemels, W. P. M. H., Dullerud, G. E., and Teel, A. R. (2014). $\mathscr{L}_2$-Gain analysis for a class of hybrid systems with applications to reset and event-triggered control: A lifting approach. *IEEE Transactions on Automatic Control*, 61(10):1221–1226.

[Heemels et al., 2012] Heemels, W. P. M. H., Johansson, K., and Tabuada, P. (2012). An introduction to event-triggered and self-triggered control. In *Proceedings of the 51st IEEE Conference on Decision and Control (CDC)*, pages 3270–3285, Maui, HI, USA.

[Heemels et al., 2008] Heemels, W. P. M. H., Sandee, J. H., and Van Den Bosch, P. P. J. (2008). Analysis of event-driven controllers for linear systems. *International Journal of Control*, 81(4):571–590.

[Hennet, 1989] Hennet, J.-C. (1989). Une extension du lemme de Farkas et son application au problème de régulation linéaire sous contraintes. *Comptes Rendus de l'Académie des Sciences, Série I*, 308:415–419.

[Henningsson et al., 2008] Henningsson, T., Johannesson, E., and Cervin, A. (2008). Sporadic event-based control of first-order linear stochastic systems. *Automatica*, 44(11):2890–2895.

[Henriksson et al., 2015] Henriksson, E., Quevedo, D., Peters, E., Sandberg, H., and Johansson, K. (2015). Multiple-loop self-triggered model predictive control for network scheduling and control. *IEEE Transactions on Control Systems Technology*, 23(6):2167–2181.

[Herceg et al., 2013] Herceg, M., Kvasnica, M., Jones, C. N., and Morari, M. (2013). Multi-Parametric Toolbox 3.0. In *Proceedings of the European Control Conference (ECC)*, pages 502–510, Zürich, Switzerland.

[Hespanha et al., 2007] Hespanha, J. P., Naghshtabrizi, P., and Xu, Y. (2007). A survey of recent results in networked control systems. *Proceedings of the IEEE*, 95(1):138–162.

[Hetel et al., 2017] Hetel, L., Fiter, C., Omran, H., Seuret, A., Fridman, E., Richard, J.-P., and Niculescu, S. I. (2017). Recent developments on the stability of systems with aperiodic sampling: an overview. *Automatica*, 76:309–335.

[Hirata and Ohta, 2003] Hirata, K. and Ohta, Y. (2003). $\epsilon$-feasible approximation of the state reachable set for discrete-time systems. In *Proceedings of the 42nd IEEE Conference on Decision and Control (CDC)*, pages 5520–5525, Maui, HI, USA.

[Hu et al., 2007] Hu, W., Liu, G., and Rees, D. (2007). Event-driven networked predictive control. *IEEE Transactions on Industrial Electronics*, 54(3):1603–1613.

[Huang et al., 2016] Huang, J., Shi, D., and Chen, T. (2016). Dynamically event-triggered state estimation of hidden Markov models through a lossy communication channel. In *Proceedings of the 55th IEEE Conference on Decision and Control (CDC)*, pages 5122–5127, Las Vegas, NV, USA.

[IBM, 2014] IBM (2014). IBM ILOG CPLEX Optimization Studio 12.6. http://www-01.ibm.com/software/integration/optimization/cplex-optimization-studio/.

[Iino et al., 2009] Iino, Y., Hatanaka, T., and Fujita, M. (2009). Event-predictive control for energy saving of wireless networked control system. In *Proceedings of the American Control Conference (ACC)*, pages 2236–2242, St. Louis, MO, USA.

[Imer and Başar, 2006] Imer, O. C. and Başar, T. (2006). Optimal control with limited controls. In *Proceedings of the American Control Conference (ACC)*, pages 298–303, Minneapolis, MN, USA.

[Incremona et al., 2017] Incremona, G. P., Ferrara, A., and Magni, L. (2017). Asynchronous networked MPC with ISM for uncertain nonlinear systems. *IEEE Transactions on Automatic Control*, 62(9):4305–4317.

[Jia and Krogh, 2005] Jia, D. and Krogh, B. (2005). Min-max feedback model predictive control with state estimation. In *Proceedings of the American Control Conference (ACC)*, pages 262–267, Portland, OR, USA.

[Jost et al., 2015] Jost, M., Schulze Darup, M., and Mönnigmann, M. (2015). Optimal and suboptimal event-triggering in linear model predictive control. In *Proceedings of the European Control Conference (ECC)*, pages 1147–1152, Linz, Austria.

[Kalman, 1960] Kalman, R. E. (1960). A new approach to linear filtering and prediction problems. *Journal of Basic Engineering*, 82(1):35–45.

[Khashooei et al., 2017] Khashooei, B. A., Antunes, D. J., and Heemels, W. P. M. H. (2017). Output-based event-triggered control with performance guarantees. *IEEE Transactions on Automatic Control*, 62(7):3646–3652.

[Kiener et al., 2014] Kiener, G. A., Lehmann, D., and Johansson, K. H. (2014). Actuator saturation and anti-windup compensation in event-triggered control. *Discrete Event Dynamic Systems: Theory and Applications*, 24(2):173–197.

[Kim et al., 2006] Kim, T.-H., Park, J.-H., and Sugie, T. (2006). Output-feedback model predictive control for LPV systems with input saturation based on quasi-min-max algorithm. In *Proceedings of the 45th IEEE Conference on Decision and Control (CDC)*, pages 1454–1459, San Diego, CA, USA.

[Kishida et al., 2016] Kishida, M., Kögel, M., and Findeisen, R. (2016). Verifying robust forward admissibility for nonlinear systems using (skewed) structured singular values. In *Proceedings of the 55th IEEE Conference on Decision and Control (CDC)*, pages 4065–4071, Las Vegas, NV, USA.

[Kögel and Findeisen, 2014] Kögel, M. and Findeisen, R. (2014). On self-triggered reduced-attention control for constrained systems. In *Proceedings of the 53rd IEEE Conference on Decision and Control (CDC)*, pages 2795–2801, Los Angeles, CA, USA.

[Kögel and Findeisen, 2015a] Kögel, M. and Findeisen, R. (2015). Robust output feedback model predictive control using reduced order models. In *Proceedings of the 9th IFAC Symposium on Advanced Control of Chemical Processes (ADCHEM)*, pages 1009–1015, Whistler, Bristish Columbia, Canada.

[Kögel and Findeisen, 2015b] Kögel, M. and Findeisen, R. (2015). Robust output feedback predictive control with self-triggered measurements. In *Proceedings of the 54th IEEE Conference on Decision and Control (CDC)*, pages 5487–5493, Osaka, Japan.

[Kögel and Findeisen, 2015c] Kögel, M. and Findeisen, R. (2015). Self-triggered, prediction-based control of Lipschitz nonlinear systems. In *Proceedings of the European Control Conference (ECC)*, pages 2150–2155, Linz, Austria.

[Kögel and Findeisen, 2016] Kögel, M. and Findeisen, R. (2016). Output feedback MPC with send-on-delta measurements for uncertain systems. In *Proceedings of the 6th IFAC Workshop on Distributed Estimation and Control in Networked Systems (NECSYS)*, pages 145–150, Tokyo, Japan.

[Kolarijani et al., 2015] Kolarijani, A. S., Adzkiya, D., and Mazo, M. (2015). Symbolic abstractions for the scheduling of event-triggered control systems. In *Proceedings of the 54th IEEE Conference on Decision and Control (CDC)*, pages 6153–6158, Osaka, Japan.

[Kolmanovsky and Gilbert, 1995] Kolmanovsky, I. and Gilbert, E. G. (1995). Maximal output admissible sets for discrete-time systems with disturbance inputs. In *Proceedings of the American Control Conference (ACC)*, pages 1995–1999, Seattle, WA, USA.

[Kolmanovsky and Gilbert, 1998] Kolmanovsky, I. and Gilbert, E. G. (1998). Theory and computation of disturbance invariant sets for discrete-time linear systems. *Mathematical Problems in Engineering*, 4(4):317–367.

[Kuntsevich and Pshenichnyi, 1996] Kuntsevich, V. M. and Pshenichnyi, B. N. (1996). Minimal invariant sets of dynamic systems with bounded disturbances. *Cybernetics and Systems Analysis*, 32(1):58–64.

[Kushner and Yin, 2003] Kushner, H. J. and Yin, G. G. (2003). *Stochastic Approximation and Recursive Algorithms and Applications.* Applications of Mathematics, Stochastic Modelling and Applied Probability. Springer.

[Laila et al., 2002] Laila, D. S., Nešić, D., and Teel, A. R. (2002). Open- and closed-loop dissipation inequalities under sampling and controller emulation. *European Journal of Control*, 8(2):109–125.

[Lazar, 2010] Lazar, M. (2010). On infinity norms as Lyapunov functions: Alternative necessary and sufficient conditions. In *Proceedings of the 49th IEEE Conference on Decision and Control (CDC)*, pages 5936–5942, Atlanta, GA, USA.

[Lazar et al., 2006] Lazar, M., Heemels, W. P. M. H., Weiland, S., and Bemporad, A. (2006). Stabilizing model predictive control of hybrid systems. *IEEE Transactions on Automatic Control*, 51(11):1813–1818.

[Le et al., 2011] Le, V. T. H., Stoica, C., Dumur, D., Alamo, T., and Camacho, E. F. (2011). Robust tube-based constrained predictive control via zonotopic set-membership estimation. In *Proceedings of the 50th IEEE Conference on Decision and Control (CDC), European Control Conference (ECC)*, pages 4580–4585, Orlando, FL, USA.

[Lee and Kouvaritakis, 2002] Lee, Y. I. and Kouvaritakis, B. (2002). Superposition in efficient robust constrained predictive control. *Automatica*, 38(5):875–878.

[Lehmann et al., 2013] Lehmann, D., Henriksson, E., and Johansson, K. H. (2013). Event-triggered model predictive control of discrete-time linear systems subject to disturbances. In *Proceedings of the European Control Conference (ECC)*, pages 1156–1161, Zürich, Switzerland.

[Lehmann and Lunze, 2011] Lehmann, D. and Lunze, J. (2011). Extension and experimental evaluation of an event-based state-feedback approach. *Control Engineering Practice*, 19(2):101–112.

[Lemmon, 2010] Lemmon, M. (2010). Event-triggered feedback in control, estimation, and optimization. In Bemporad, A., Heemels, W. P. M. H., and Johansson, M., editors, *Networked Control Systems, Lecture Notes in Control and Information Sciences, volume 406*, pages 293–358. Springer.

[Li and Shi, 2014] Li, H. and Shi, Y. (2014). Event-triggered robust model predictive control of continuous-time nonlinear systems. *Automatica*, 50(5):1507–1513.

[Linsenmayer et al., 2016] Linsenmayer, S., Dimarogonas, D. V., and Allgöwer, F. (2016). A non-monotonic approach to periodic event-triggered control with packet loss. In *Proceedings of the 55th IEEE Conference on Decision and Control (CDC)*, pages 507–512, Las Vegas, NV, USA.

[Lipsa and Martins, 2013] Lipsa, G. M. and Martins, N. C. (2013). Remote state estimation with communication costs for first-order LTI systems. *IEEE Transactions on Automatic Control*, 56(9):2013–2025.

[Liu and Marquez, 2007] Liu, B. and Marquez, H. J. (2007). Razumikhin-type stability theorems for discrete delay systems. *Automatica*, 43(7):1219 – 1225.

[Liu et al., 2017] Liu, S., Song, Y., Wei, G., Ding, D., and Liu, Y. (2017). Event-triggered dynamic output feedback RMPC for polytopic systems with redundant channels: Input-to-state stability. *Journal of the Franklin Institute*, 354(7):2871–2892.

[Liu et al., 2013] Liu, S., Zhang, J., Liu, J., Feng, Y., and Rong, G. (2013). Distributed model predictive control with asynchronous controller evaluations. *Canadian Journal of Chemical Engineering*, 91(10):1609–1620.

[Liz and Ferreiro, 2002] Liz, E. and Ferreiro, J. (2002). A note on the global stability of generalized difference equations. *Applied Mathematics Letters*, 15:655–659.

[Lješnjanin et al., 2014] Lješnjanin, M., Quevedo, D. E., and Nešić, D. (2014). Packetized MPC with dynamic scheduling constraints and bounded packet dropouts. *Automatica*, 50(3):784–797.

[Löfberg, 2002] Löfberg, J. (2002). Towards joint state estimation and control in minmax MPC. In *Proceedings of the 15th IFAC World Congress*, pages 273–278, Barcelona, Spain.

[Löfberg, 2003] Löfberg, J. (2003). Approximations of closed-loop minimax MPC. In *Proceedings of the 42nd IEEE Conference on Decision and Control (CDC)*, pages 1438–1442, Maui, HI, USA.

[Löfberg, 2004] Löfberg, J. (2004). YALMIP: A toolbox for modeling and optimization in MATLAB. In *Proceedings of the IEEE International Symposium on Computer Aided Control Systems Design (CACSD)*, pages 284–289, Taipei, Taiwan.

[Lu et al., 2017] Lu, L., Zou, Y., and Niu, Y. (2017). Event-driven robust output feedback control for constrained linear systems via model predictive control method. *Circuits, Systems, and Signal Processing*, 36(2):543–558.

[Lunze and Lehmann, 2010] Lunze, J. and Lehmann, D. (2010). A state-feedback approach to event-based control. *Automatica*, 46(1):211–215.

[Mahmoud and Memon, 2015] Mahmoud, M. S. and Memon, A. M. (2015). Aperiodic triggering mechanisms for networked control systems. *Information Sciences*, 296(1):282–306.

[Mamduhi et al., 2013] Mamduhi, M. H., Molin, A., and Hirche, S. (2013). On the stability of prioritized error-based scheduling for resource-constrained networked control systems. In *Proceedings of the 4th IFAC Workshop on Distributed Estimation and Control in Networked Systems (NECSYS)*, pages 356–362, Koblenz, Germany.

[Maniatopoulos et al., 2012] Maniatopoulos, S., Dimarogonas, D. V., and Kyriakopoulos, K. J. (2012). A decentralized event-based predictive navigation scheme for air-traffic control. In *Proceedings of the American Control Conference (ACC)*, pages 2503–2508, Montréal, Canada.

[Mayne, 2014] Mayne, D. Q. (2014). Model predictive control: Recent developments and future promise. *Automatica*, 50(12):2967–2986.

[Mayne et al., 2006] Mayne, D. Q., Raković, S. V., Findeisen, R., and Allgöwer, F. (2006). Robust output feedback model predictive control of constrained linear systems. *Automatica*, 42(7):1217–1222.

[Mayne et al., 2009] Mayne, D. Q., Raković, S. V., Findeisen, R., and Allgöwer, F. (2009). Robust output feedback model predictive control of constrained linear systems: Time varying case. *Automatica*, 45(9):2082–2087.

[Mayne et al., 2000] Mayne, D. Q., Rawlings, J. B., Rao, C. V., and Scokaert, P. O. M. (2000). Constrained model predictive control: Stability and optimality. *Automatica*, 36(6):789–814.

[Mayne et al., 2005] Mayne, D. Q., Seron, M. M., and Raković, S. V. (2005). Robust model predictive control of constrained linear systems with bounded disturbances. *Automatica*, 41(2):219–224.

[Mazo et al., 2010] Mazo, M., Anta, A., and Tabuada, P. (2010). An ISS self-triggered implementation of linear controllers. *Automatica*, 46(8):1310–1314.

[Meng and Chen, 2012] Meng, X. and Chen, T. (2012). Optimal sampling and performance comparison of periodic and event based impulse control. *IEEE Transactions on Automatic Control*, 57(12):3252–3259.

[Mi and Li, 2016] Mi, X. and Li, S. (2016). Event-triggered MPC design for distributed systems with network communications. *IEEE/CAA Journal of Automatica Sinica*. Accepted for publication, DOI: 10.1109/JAS.2016.7510154.

[Mishra et al., 2014] Mishra, P. K., Vachani, L., and Chatterjee, D. (2014). Event triggered control for discrete time dynamical systems. In *Proceedings of the European Control Conference (ECC)*, pages 1444–1449, Linz, Austria.

[Molchanov and Pyatnitskiy, 1989] Molchanov, A. P. and Pyatnitskiy, Y. S. (1989). Criteria of asymptotic stability of differential and difference inclusions encountered in control theory. *Systems & Control Letters*, 13(1):59–64.

[Molin and Hirche, 2009] Molin, A. and Hirche, S. (2009). On LQG joint optimal scheduling and control under communication constraints. In *Proceedings of the 48th IEEE Conference Decision and Control (CDC), 28th Chinese Control Conference(CCC)*, pages 5832–5838, Shanghai, China.

[Molin and Hirche, 2013] Molin, A. and Hirche, S. (2013). On the optimality of certainty equivalence for event-triggered control systems. *IEEE Transactions on Automatic Control*, 58(2):470–474.

[Nagahara et al., 2016] Nagahara, M., Quevedo, D. E., and Nešić, D. (2016). Maximum hands-off control: A paradigm of control effort minimization. *IEEE Transactions on Automatic Control*, 61(3):735–747.

[Nghiem, 2012] Nghiem, T. X. (2012). *Green scheduling of control systems*. Doctoral thesis, University of Pennsylvania.

[Nowzari and Cortés, 2014] Nowzari, C. and Cortés, J. (2014). Team-triggered coordination for real-time control of networked cyber-physical systems. *IEEE Transactions on Automatic Control*, 61(1):34–47.

[Ong and Gilbert, 2006] Ong, C.-J. and Gilbert, E. G. (2006). The minimal disturbance invariant set: Outer approximations via its partial sums. *Automatica*, 42(9):1563–1568.

[Postoyan et al., 2015] Postoyan, R., Tabuada, P., Nešić, D., and Anta, A. (2015). A framework for the event-triggered stabilization of nonlinear systems. *IEEE Transactions on Automatic Control*, 60(4):982–996.

[Poveda and Teel, 2017] Poveda, J. I. and Teel, A. R. (2017). A robust event-triggered approach for fast sampled-data extremization and learning. *IEEE Transactions on Automatic Control*, 62(10):4949–4964.

[Rabi and Johansson, 2009] Rabi, M. and Johansson, K. H. (2009). Scheduling packets packets for for event-triggered event-triggered control scheduling control. In *Proceedings of the European Control Conference (ECC)*, pages 3779–3784, Budapest, Hungary.

[Rabi et al., 2012] Rabi, M., Moustakides, George, V., and Baras, J. S. (2012). Adaptive sampling for linear state estimation. *SIAM Journal on Control and Optimization*, 50(2):672–702.

[Raković, 2012] Raković, S. V. (2012). Invention of prediction structures and categorization of robust MPC syntheses. In *Proceedings of the 4th IFAC Conference on Nonlinear Model Predictive Control (NMPC)*, pages 245–273, Noordwijkerhout, The Netherlands.

[Raković et al., 2005] Raković, S. V., Kerrigan, E. C., Kouramas, K. I., and Mayne, D. Q. (2005). Invariant approximations of the minimal robust positively invariant set. *IEEE Transactions on Automatic Control*, 50(3):406–410.

[Raković and Kouramas, 2006a] Raković, S. V. and Kouramas, K. I. (2006). The minimal robust positively invariant set for linear discrete time systems: Approximation methods and control applications. In *Proceedings of the 45th IEEE Conference on Decision and Control (CDC)*, pages 4562–4567, San Diego, CA, USA.

[Raković and Kouramas, 2006b] Raković, S. V. and Kouramas, K. I. (2006). State decomposition principle in set invariance theory for linear discrete time systems. In *Proceedings of the 45th IEEE Conference on Decision and Control (CDC)*, pages 4544–4549, San Diego, CA, USA.

[Raković et al., 2012a] Raković, S. V., Kouvaritakis, B., Cannon, M., Panos, C., and Findeisen, R. (2012). Parameterized tube model predictive control. *IEEE Transactions on Automatic Control*, 57(11):2746–2761.

[Raković et al., 2012b] Raković, S. V., Kouvaritakis, B., Findeisen, R., and Cannon, M. (2012). Homothetic tube model predictive control. *Automatica*, 48(8):1631–1638.

[Raković and Lazar, 2014] Raković, S. V. and Lazar, M. (2014). The Minkowski-Lyapunov equation for linear dynamics: Theoretical foundations. *Automatica*, 50(8):2015–2024.

[Raković and Mayne, 2005] Raković, S. V. and Mayne, D. Q. (2005). Set robust control invariance for linear discrete time systems. In *Proceedings of the 44th IEEE Conference on Decision and Control (CDC)*, pages 975–980, Seville, Spain.

[Rao, 1973] Rao, C. R. (1973). *Linear statistical inference and its applications*. Wiley.

[Rawlings and Mayne, 2009] Rawlings, J. B. and Mayne, D. Q. (2009). *Model Predictive Control: Theory and Design*. Nob Hill Publishing, Madison, WI, USA.

[Robbins and Monro, 1951] Robbins, H. and Monro, S. (1951). A stochastic approximation method. *The Annals of Mathematical Statistics*, 22(3):400–407.

[Rockafellar and Wets, 2009] Rockafellar, R. T. and Wets, R. J.-B. (2009). *Variational Analysis*. Springer, 3rd edition.

[Schenato, 2009] Schenato, L. (2009). To zero or to hold control inputs with lossy links? *IEEE Transactions on Automatic Control*, 54(5):1093–1099.

[Schneider, 1993] Schneider, R. (1993). *Convex bodies: the Brunn-Minkowski theory*. Cambridge University Press.

[Schneider, 2013] Schneider, R. (2013). *Convex bodies: the Brunn-Minkowski theory*. Cambridge University Press, 2nd edition.

[Schulze Darup et al., 2017] Schulze Darup, M., Schaich, R. M., and Cannon, M. (2017). How scaling of the disturbance set affects robust positively invariant sets for linear systems. *International Journal of Robust and Nonlinear Control*, 27(16):3236–3258.

[Schweppe, 1968] Schweppe, F. (1968). Recursive state estimation: Unknown but bounded errors and system inputs. *IEEE Transactions on Automatic Control*, 13(1):22–28.

[Scokaert and Mayne, 1998] Scokaert, P. O. M. and Mayne, D. Q. (1998). Min-max feedback model predictive control for constrained linear systems. *IEEE Transactions on Automatic Control*, 43(8):1136–1142.

[Selivanov and Fridman, 2016] Selivanov, A. and Fridman, E. (2016). Event-triggered $H_\infty$ control: A switching approach. *IEEE Transactions on Automatic Control*, 61(10):3221–3226.

[Senejohnny et al., 2016] Senejohnny, D., Tesi, P., and De Persis, C. (2016). Resilient self-triggered network synchronization. In *Proceedings of the 55th IEEE Conference on Decision and Control (CDC)*, pages 489–494, Las Vegas, NV, USA.

[Seuret et al., 2014] Seuret, A., Prieur, C., and Marchand, N. (2014). Stability of non-linear systems by means of event-triggered sampling algorithms. *IMA Journal of Mathematical Control and Information*, 31(3):415–433.

[Seyboth et al., 2013] Seyboth, G. S., Dimarogonas, D. V., and Johansson, K. H. (2013). Event-based broadcasting for multi-agent average consensus. *Automatica*, 49(1):245–252.

[Shi et al., 2016] Shi, D., Chen, T., and Darouach, M. (2016). Event-based state estimation of linear dynamic systems with unknown exogenous inputs. *Automatica*, 69:275–288.

[Sijs et al., 2010] Sijs, J., Lazar, M., and Heemels, W. P. M. H. (2010). On integration of event-based estimation and robust MPC in a feedback loop. In *Proceedings of the 13th ACM International Conference on Hybrid Systems: Computation and Control (HSCC)*, pages 31–40, Stockholm, Sweden.

[Silvestre et al., 2015] Silvestre, D., Rosa, P., Hespanha, J. P., and Silvestre, C. (2015). Self-triggered set-valued observers. In *Proceedings of the European Control Conference (ECC)*, pages 3647–3652, Linz, Austria.

[Sinopoli et al., 2004] Sinopoli, B., Member, S., Schenato, L., Franceschetti, M., Poolla, K., Jordan, M. I., Member, S., and Sastry, S. S. (2004). Kalman filtering with intermittent observations. *IEEE Transactions on Automatic Control*, 49(9):1453–1464.

[Stengel, 1994] Stengel, R. F. (1994). *Optimal Control and Estimation*. Dover Publications.

[Stöcker and Lunze, 2013] Stöcker, C. and Lunze, J. (2013). Input-to-state stability of event-based state-feedback control. In *Proceedings of the European Control Conference (ECC)*, pages 1145–1150, Zürich, Switzerland.

[Subramanian et al., 2016] Subramanian, S., Lucia, S., and Engell, S. (2016). A non-conservative robust output feedback MPC for constrained linear systems. In *Proceedings of the 55th IEEE Conference on Decision and Control (CDC)*, pages 2333–2338, Las Vegas, NV, USA.

[Sui et al., 2008] Sui, D., Feng, L., and Hovd, M. (2008). Robust output feedback model predictive control for linear systems via moving horizon estimation. In *Proceedings of the American Control Conference (ACC)*, pages 453–458, Seattle, WA, USA.

[Syrmos et al., 1997] Syrmos, V. L., Abdallah, C. T., Doratos, P., and Grigoriadis, K. (1997). Static output feedback—a survey. *Automatica*, 33(2):125–137.

[Tabbara and Nešić, 2008] Tabbara, M. and Nešić, D. (2008). Input–output stability of networked control systems with stochastic protocols and channels. *IEEE Transactions on Automatic Control*, 53(5):1160–1175.

[Tabuada, 2007] Tabuada, P. (2007). Event-triggered real-time scheduling of stabilizing control tasks. *IEEE Transactions on Automatic Control*, 52(9):1680–1685.

[Tallapragada and Chopra, 2014] Tallapragada, P. and Chopra, N. (2014). Decentralized event-triggering for control of nonlinear systems. *IEEE Transactions on Automatic Control*, 59(12):3312–3324.

[Tarn and Rasis, 1976] Tarn, T.-J. and Rasis, Y. (1976). Observers for nonlinear stochastic systems. *IEEE Transactions on Automatic Control*, 21(4):441–448.

[Teixeira et al., 2010] Teixeira, P. V., Dimarogonas, D. V., Johansson, K. H., and Sousa, J. (2010). Multi-agent coordination with event-based communication. In *Proceedings of the American Control Conference (ACC)*, pages 824–829, Baltimore, MD, USA.

[Tiberi and Johansson, 2013] Tiberi, U. and Johansson, K. H. (2013). A simple self-triggered sampler for perturbed nonlinear systems. *Nonlinear Analysis: Hybrid Systems*, 10:126–140.

[Trimpe and Andrea, 2014] Trimpe, S. and Andrea, R. D. (2014). Event-based state estimation with variance-based triggering. *IEEE Transactions on Automatic Control*, 59(12):3266–3281.

[Varutti et al., 2010] Varutti, P., Faulwasser, T., Kern, B., Kögel, M., and Findeisen, R. (2010). Event-based reduced-attention predictive control for nonlinear uncertain systems. In *IEEE International Symposium on Computer-Aided Control System Design (CACSD), part of Multi-Conference on Systems and Control (MSC)*, pages 1085–1090, Yokohama, Japan.

[Varutti et al., 2009] Varutti, P., Kern, B., Faulwasser, T., and Findeisen, R. (2009). Event-based model predictive control for networked control systems. In *Proceedings of the 48th IEEE Conference Decision and Control (CDC), 28th Chinese Control Conference(CCC)*, pages 567–572, Shanghai, China.

[Velasco et al., 2009] Velasco, M., Martí, P., and Bini, E. (2009). On Lyapunov sampling for event-driven controllers. In *Proceedings of the 48th IEEE Conference Decision and Control (CDC), 28th Chinese Control Conference(CCC)*, pages 6238–6243, Shanghai, China.

[Wang and Lemmon, 2011] Wang, X. and Lemmon, M. (2011). On event design in event-triggered feedback systems. *Automatica*, 47(10):2319–2322.

[Wang and Lemmon, 2009] Wang, X. and Lemmon, M. D. (2009). Self-triggered seedback control systems with finite-gain $\mathscr{L}_2$ stability. *IEEE Transactions on Automatic Control*, 54(3):452–467.

[Witsenhausen, 1968] Witsenhausen, H. S. (1968). A minimax control problem for sampled linear systems. *IEEE Transactions on Automatic Control*, AC-13(1):5–21.

[Wu et al., 2013] Wu, J., Jia, Q. S., Johansson, K. H., and Shi, L. (2013). Event-based sensor data scheduling: Trade-off between communication rate and estimation quality. *IEEE Transactions on Automatic Control*, 58(4):1041–1046.

[Wu et al., 2016a] Wu, J., Ren, X., Han, D., Shi, D., and Shi, L. (2016). Finite-horizon Gaussianity-preserving event-based sensor scheduling in Kalman filter applications. *Automatica*, 72:100–107.

[Wu et al., 2016b] Wu, W., Reimann, S., Görges, D., and Liu, S. (2016). Event-triggered control for discrete-time linear systems subject to bounded disturbance. *International Journal of Robust and Nonlinear Control*, 26(9):1902–1918.

[Xu and Hespanha, 2004a] Xu, Y. and Hespanha, J. P. (2004). Communication logics for networked control systems. In *Proceedings of the American Control Conference (ACC)*, pages 572–577, Boston, MA, USA.

[Xu and Hespanha, 2004b] Xu, Y. and Hespanha, J. P. (2004). Optimal communication logics in networked control systems. In *Proceedings of the 43rd IEEE Conference on Decision and Control (CDC)*, pages 3527–3532, Paradise Island, Bahamas.

[Yajie and Wei, 2015] Yajie, L. and Wei, L. (2015). The co-design method for robust satisfactory $H_\infty/H_2$ fault-tolerant event-triggered control of NCS with $\alpha$ - safety degree. *The Open Automation and Control Systems Journal*, 7:1061–1070.

[Zavala and Biegler, 2009] Zavala, V. M. and Biegler, L. T. (2009). The advanced-step NMPC controller: Optimality, stability and robustness. *Automatica*, 45(1):86–93.

[Zhang et al., 2007] Zhang, G., Chen, X., and Chen, T. (2007). A model predictive control approach to networked systems. In *Proceedings of the 46th IEEE Conference on Decision and Control (CDC)*, pages 3339–3344, New Orleans, LA, USA.

[Zhang et al., 2014] Zhang, J., Liu, S., and Liu, J. (2014). Economic model predictive control with triggered evaluations: State and output feedback. *Journal of Process Control*, 24(8):1197–1206.

[Zhe et al., 2015] Zhe, L. I., Shaoyuan, L. I., and Jing, W. U. (2015). Event-triggered model predictive control for constrained invariant set trajectory. In *Proc. 34th Chinese Control Conf. (CCC)*, pages 4185–4190, Hangzhou, China.

[Zou et al., 2016] Zou, Y., Wang, Q., Jia, T., and Niu, Y. (2016). Multirate event-triggered MPC for NCSs with transmission delays. *Circuits, Systems, and Signal Processing*, 35(12):4249–4270.